BASIC Microcomputer Models in Biology

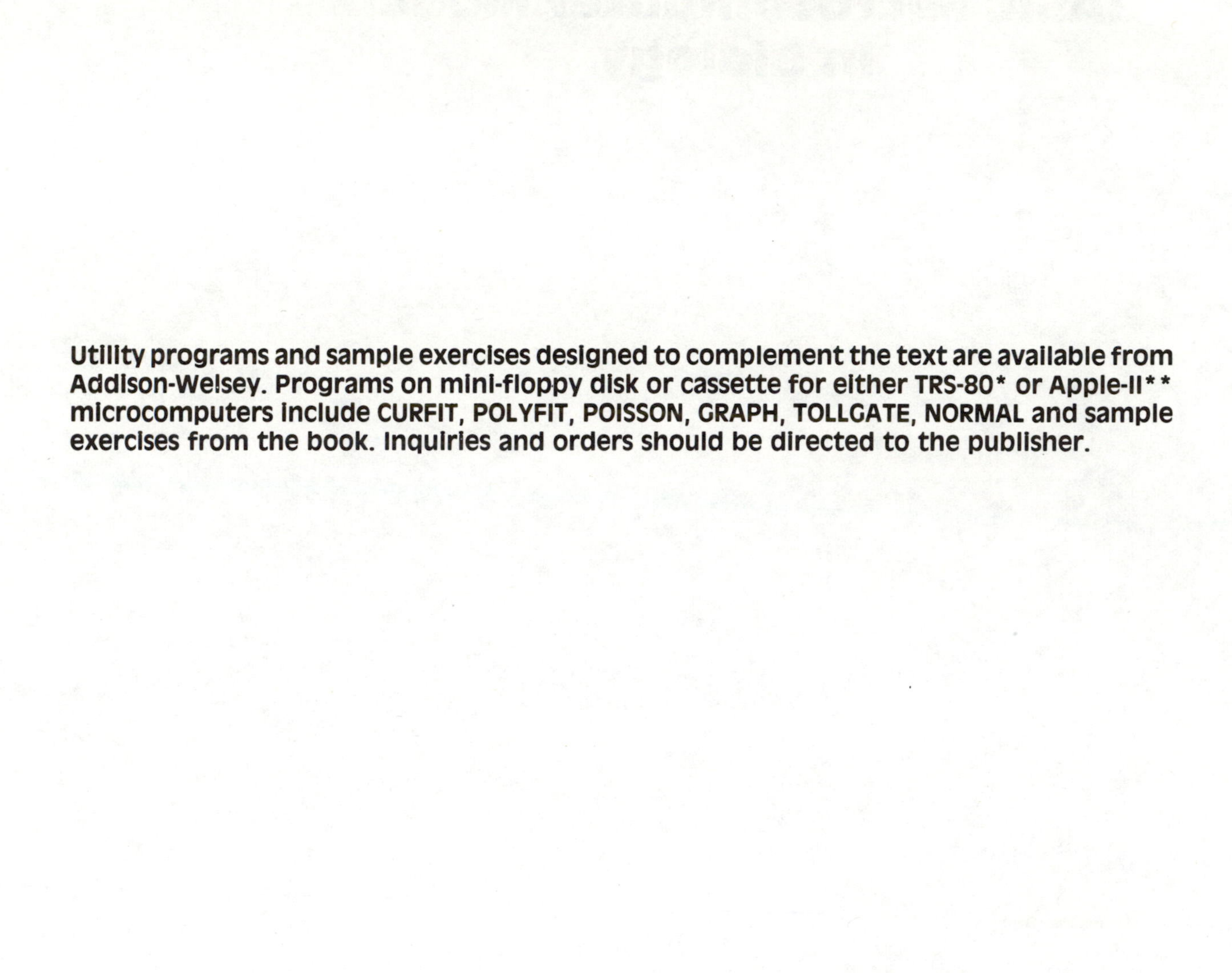

Utility programs and sample exercises designed to complement the text are available from Addison-Welsey. Programs on mini-floppy disk or cassette for either TRS-80* or Apple-II** microcomputers include CURFIT, POLYFIT, POISSON, GRAPH, TOLLGATE, NORMAL and sample exercises from the book. Inquiries and orders should be directed to the publisher.

BASIC Microcomputer Models in Biology

James D. Spain
Michigan Technological University

1982

Addison-Wesley Publishing Company
Advanced Book Program/World Science Division
Reading, Massachusetts

London · Amsterdam · Don Mills, Ontario · Sydney · Tokyo

This book was prepared in camera-ready form by Phyllis and Brian Winkel (Way With Words), Terre Haute, Indiana, on Ohio Scientific Instruments Challenger III microcomputer system using WordStar software and NEC Spinwriter printer using Math/Times Roman font.

Library of Congress Cataloging in Publication Data

Spain, James D.
BASIC microcomputer models in biology.

Bibliography: p.
Includes index.
1. Biology—Mathematical models. 2. Biology—Data processing. 3. Basic (Computer program language)
I. Title.
QH323.5.S644 574'.0724 81-19132
ISBN 0-201-10678-7 AACR2

ABCDEFGHIJ-AL-8987654321

Manufactured in the United States of America

CONTENTS

PREFACE....xi

INTRODUCTION
THE ROLE OF COMPUTER MODELING AND SIMULATION IN BIOLOGY....1
Conceptual Models....2
The Mathematical Model....3
Computer Simulation....4
The Relationship of Modeling and Simulation to the Research Process....5
The Development of Multicomponent Models....8
Computer Modeling with the Microcomputer....10
Goals and Objectives....11

PART ONE
SIMPLE MODEL EQUATIONS....12
CHAPTER 1
ANALYTICAL MODELS BASED ON DIFFERENTIAL EQUATIONS....16
1.1 The Exponential Growth Model....16
1.2 Exponential Decay....20
1.3 Newton's Law of Cooling....20
1.4 Dilution Model....22
1.5 Diffusion Across a Boundary of Unit Area....22
1.6 Von Bertalanffy Fish Growth Model....23
1.7 Stream Drift Model....23
1.8 Inhibited or Logistic Growth Model....24
1.9 Kinetics of Bimolecular Reactions....24
Conclusion....26
CHAPTER 2
ANALYTICAL MODELS BASED ON EQUILIBRIUM OR STEADY-STATE ASSUMPTIONS....27
2.1 Michaelis-Menten Model of Enzyme Saturation....28
2.2 Langmuir Absorption Model....31
2.3 The Saturation or Satiation of Predators....33
2.4 The Henderson-Hasselbalch Model of Weak Acids....35
2.5 The Effect of pH on Enzyme Activity....35
2.6 A Concerted Model for Allosteric Enzymes....39
2.7 Selection Within the Hardy-Weinberg Genetic Equilibrium....41
Conclusion....42
CHAPTER 3
FITTING MODEL EQUATIONS TO EXPERIMENTAL DATA....43
3.1 The Linear Transformation Approach to Curve Fitting....44

3.2 Using CURFIT to Evaluate Parameters for Theoretical Model Equations....................46
3.3 Standard Theoretical Equations....................50
3.4 Some Pitfalls of the Transformation Process....................50
3.5 Curve Types Resulting from Compound Equations....................55
3.6 Fitting Polynomials to Experimental Data....................56
Conclusion....................61
CHAPTER 4
FLOWCHARTING....................63
4.1 General Flowcharting Rules....................64
4.2 Templates....................64
4.3 Flowchart Symbols and Their Use....................65
4.4 Examples of Simple Flowcharts....................66
Conclusion....................69
CHAPTER 5
NUMERICAL SOLUTION OF RATE EQUATIONS....................70
5.1 The Finite Difference Approach....................71
5.2 The Euler Technique....................71
5.3 The Time Increment as a Factor in Numerical Integration....................72
5.4 The Improved Euler or Second Order Runge-Kutta Method....................73
5.5 An Iterated Second Order Runge-Kutta Method....................75
Conclusion....................75

PART II
MULTICOMPONENT SYSTEMS MODELS....................77
CHAPTER 6
BIOCHEMICAL REACTION KINETICS....................80
6.1 Kinetics of Bimolecular Reaction....................81
6.2 Chemical Equilibrium Model....................83
6.3 Kinetics of a Sequential Reaction....................85
6.4 The Chance-Cleland Model for Enzyme-Substrate interaction....................86
6.5 Autocatalytic Reactions....................87
Conclusion....................90
CHAPTER 7
DYNAMICS OF HOMOGENEOUS POPULATIONS OF ORGANISMS....................91
7.1 The Verhulst-Pearl Equation....................91
7.2 Oscillation and Overshoot in Population Density....................93
7.3 Variable Carrying Capacity....................96
7.4 Gause's Model for Competition Between Two Species....................97
7.5 The Lotka-Volterra Predation Model....................102
7.6 Modifications of the Predation Rate Term....................103
7.7 Modification of the Growth Term....................105
Conclusion....................106
CHAPTER 8
SIMULATION MODELS BASED ON AGE-CLASSES AND LIFE TABLES....................107
8.1 Age Specific Natality Rates....................107
8.2 Age Specific Survival Rates....................108
8.3 Life Tables....................110
8.4 Use of Subscripted Variables in Age-Class Models....................113
8.5 Age-Class Model with Sex Differentiated Survival and Reproduction.114
8.6 Fisheries Model Based on Age-Classes....................116
8.7 Epidemic Simulation....................118

8.8 Intraspecific Action: Cannibalism....120
Conclusion....123
CHAPTER 9
POPULATION GENETICS SIMULATIONS....124
9.1 The Effect of Selection on Random Populations....124
9.2 Balanced Genetic Load and Balanced Polymorphism....127
9.3 Balanced Mutational Load....129
9.4 Selection Against Sex-Linked Recessive Genes....130
9.5 Selection Against Recessive Genes Involving Two Loci....131
9.6 The Inbreeding Model of Wright....132
Conclusion....133
CHAPTER 10
LIGHT INTENSITY AND PHOTOSYNTHESIS....135
10.1 Variations of Light Intensity on an Annual Cycle....135
10.2 Modeling Diurnal Variation in Light Intensity....137
10.3 Effects of Cloud Cover and Backscatter....137
10.4 Reflection of Light from Water Surface....138
10.5 Absorption of Light by the Water Column....140
10.6 Effect of Light on Photosynthesis....141
10.7 Phytoplankton Productivity Curves....143
Conclusion....145
CHAPTER 11
TEMPERATURE AND BIOLOGICAL ACTIVITY....146
11.1 Temperature Variation in Terrestrial Systems....146
11.2 Temperature Variation in Aquatic Systems....148
11.3 Effect of Temperature on Chemical Reaction Rates....150
11.4 Simulation of Temperature Effect on Enzyme Activity....152
11.5 O'Neill Model for the Effect of Temperature on Biological Activity....153
11.6 Exponential Temperature-Activity Model....156
11.7 Use of the Polynomial to Model Temperature Effects....157
Conclusion....159
CHAPTER 12
MATERIAL AND ENERGY FLOW IN ECOSYSTEMS....160
12.1 The Energy Flow Concept....160
12.2 Block Diagrams and Compartment Models....162
12.3 The Silver Springs Model....164
12.4 The English Channel Marine Community....169
Conclusion....171
CHAPTER 13
SIMPLE MODELS OF MICROBIAL GROWTH....173
13.1 The Monod Model....173
13.2 The Continuous Culture or Chemostat Simulation....176
13.3 Multiple Limiting Nutrients....178
13.4 Competition of Microbes for a Single Limiting Nutrient....180
13.5 Competition for Multiple Limiting Nutrients....183
13.6 Growth Inhibition by Sub-Lethal Concentration of Toxicants....184
Conclusion....185
CHAPTER 14
COMPARTMENT MODELS IN PHYSIOLOGY....187
14.1 Transport by Simple Diffusion....187
14.2 Osmotic Pressure....189

14.3 Model of Countercurrent Diffusion....................191
14.4 A Model of Active Transport....................193
14.5 Simple Approach to Active Transport....................195
14.6 Extracellular Space Determination....................196
14.7 Modeling of Fluid-Flow Processes....................197
14.8 Mechanical Operation of the Aortic Segment of the Heart....................199
14.9 The Iodine Compartment Model....................200
Conclusion....................202
CHAPTER 15
PHYSIOLOGICAL CONTROL SYSTEMS....................203
15.1The Generalized Feedback Control System....................204
15.2 Regulation of Tyroxine Secretion by the Pituitary....................207
15.3 Temperature Control Through Sweating....................209
15.4 A Model of Temperature Control Below Thermal Neutrality....................211
15.5 The Control of Protein Synthesis by Gene Repression....................214
Conclusion....................216
CHAPTER 16
THE APPLICATION OF MATRIX METHODS TO MODELING....................217
16.1 The Matrix Approach to Linear Transport Phenomena....................217
16.2 Solution of Non-Linear Equations....................221
16.3 Multicomponent Interspecific Action....................222
16.4 The Leslie Age-Class Matrix....................224
16.5 Modifications of the Leslie Matrix Model....................225
Conclusion....................226
CHAPTER 17
DYNAMICS OF MULTIENZYME SYSTEMS....................227
17.1 Mass Action Model of Glycolysis....................228
17.2 General Mass Action Approach for Single Substrate Reactions....................232
17.3 The Rate Law Approach for Single Substrate Reaction Sequences....................234
17.4 The Rate Law Approach for Bisubstrate Reactions....................236
17.5 The Yates-Pardee Model of Feedback Control....................237
17.6 Allosteric Control of Phosphofructokinase....................238
Conclusion....................240
CHAPTER 18
MULTISTAGE NUTRIENT LIMITATIONS MODELS....................241
18.1 The Two Stage Nutrient Uptake Model....................242
18.2 The Droop Nutrient Limitation Model....................244
18.3 Control of Nutrient Uptake by the Internal Nutrient Pool....................246
18.4 Multiple Nutrient Limitation....................246
Conclusion....................248

PART III
PROBABILISTIC MODELING....................250
CHAPTER 19
MONTE CARLO MODELING OF SIMPLE STOCHASTIC PROCESSES....................252
19.1 Random Number Generators....................252
19.2 Tests of Randomness....................252
19.3 The Basic Monte Carlo Simulation....................255
19.4 Stochastic Simulation of Radioactive Decay....................258
19.5 The Monte Carlo Model of a Monohybrid Cross....................259
19.6 Monte Carlo Model of the Dihybrid Cross....................260
Conclusion....................262

CHAPTER 20
MODELING OF SAMPLING PROCESSES....263
20.1 Sampling a Community of Organisms....264
20.2 The Effect of Sample Size on Diversity Index....265
20.3 Mark and Recapture Simulation....266
20.4 Spatial Distribution of Organisms....269
20.5 Sampling from the Poisson Distribution....270
20.6 Generating Normally Distributed Random Numbers Using the Table-Lookup Method....270
20.7 Empirical Method for Generating Normally Distributed Random Numbers....272
20.8 Sampling from the Exponential Distribution....274
Conclusion....274
CHAPTER 21
RANDOM WALKS AND MARKOV CHAINS....275
21.1 The Classical Random Walk Model....275
21.2 Population Growth, the Direct Approach....277
21.3 An Unlimited Growth Model Based on Variance Estimation....278
21.4 Stochastic Simulation of an Epidemic....280
21.5 Modeling Markov Processes with a Transition Matrix....281
21.6 Genetic Drift....283
Conclusion....285
CHAPTER 22
QUEUEING SIMULATION IN BIOLOGY....287
22.1 The Toll Gate Design Problem....288
22.2 Queueing Simulation of Wolf-Moose Interaction....289
22.3 Cattle Reproduction Simulation....292
22.4 Enzyme-Substrate Interaction....293
22.5 A Queueing Simulation of the Operon Model....295
Conclusion....298

BIBLIOGRAPHY....299

APPENDIX 1
INTRODUCTION TO BASIC PROGRAMMING....309
APPENDIX 2
CURFIT: A PROGRAM FOR FITTING THEORETICAL EQUATIONS TO DATA....323
APPENDIX 3
POLYFIT: A PROGRAM FOR FITTING POLYNOMIAL EQUATIONS TO DATA....333
APPENDIX 4
KEY TO CURVE TYPES....341
APPENDIX 5
POISSON DISTRIBUTION PROGRAM....343
APPENDIX 6
GRAPH: A SUBROUTINE FOR PLOTTING DATA....345
APPENDIX 7
TABLE OF CHI SQUARE....347

INDEX....349

PREFACE

As far as I am aware, this is the first general introductory text on the subject of computer modeling for life scientists. On numerous occasions I have tried to justify why I should consider myself qualified to write a book of this nature. Such a book is clearly needed to introduce the student to the exciting and rapidly developing field of computer modeling, and yet, no general text seemed to be forthcoming. Eventually I decided that even though my credentials as a mathematician are essentially nonexistent, I could draw enough from my experiences teaching quantitative courses in biology and chemistry to write an introductory text on modeling. The exercises in the text are drawn from ten years experience developing instructional simulation models and teaching introductory courses on biological simulation techniques. The mathematics employed is generally quite simple, and hence understandable to most students of life sciences. The emphasis is on applications rather than sophisticated mathematical methodology. Few people realize the exciting things that can be done on the digital computer with relatively simple numerical methods. Even unsophisticated models can produce realistic and instructive simulation output when implemented in this way.

Neither understanding of nor appreciation for computer modeling can occur without actually programming the models and interacting with the computer. For this reason, this book consists of a series of exercises which are integrated with the descriptive material. Thus, it may seem more like a laboratory manual than a text. The computer is, in fact, a powerful laboratory tool ideally suited for developing and testing biological concepts. As students complete the exercises, they begin to see each system from a new perspective. Real understanding of a system often comes more from the process of model development than from examination of the simulation data the model produces. Students are always more enthusiastic about developing their own models than they are about working with 'canned' programs which someone else has prepared. The modeling process serves to reinforce the basic understanding of the quantitative principles of biology. Thus, the text might be described as an introduction to the numerical approach to biomathematics.

Biology is a highly integrated science and mathematics is one of the chief integrating forces. As an example, the mathematical concepts of chemical kinetics employed in biochemistry are often used in modeling population dynamics. Compartment modeling and transport processes find applications at all levels of biological organization from the sub-cellular level to the ecosystem level. For this reason, the material has been organized cumulatively. Each chapter builds in one way or another on preceding chapters. My students are encouraged to complete at least one or two representative exercises from each chapter. A diversity of material is provided so that most students can satisfy their particular interests. Experience with simple models provides them with the background to appreciate the potential role of computer modeling in understanding complex biological systems such as large ecosystem models, world models, and complex models dealing with human physiology.

The text is divided into three parts. The first deals with simple equations that model single biological systems or system components. Individual chapters illustrate the various ways in which simple model equations are derived, and used in the computer to generate simulation data. This approach gives the students time to learn computer programming techniques while doing simple examples. Those sections or exercises essential in understanding the other parts have been marked with a double asterisk (**).

The second part is concerned with deterministic models of multicomponent systems. These models generally employ numerical techniques to solve multiple equations with interdependent variables. This part is organized according to specific areas of application and/or modeling approaches within a specific area of application. To a limited extent, one may pick and choose according to his or her field of interest. However, there are certain techniques and concepts with broad applicability which should be understood irrespective of field of interest. As before, these essential concepts have been marked with the double asterisk.

The third part deals with the effects of random processes on biological systems. Its chapters are devoted to topics such as sampling processes, random walks, and queueing. The effect of discrete events occurring randomly in time is emphasized. The examples and exercises employ a direct approach called the Monte Carlo technique.

The book was originally designed to employ BASIC programming language in conjunction with time-sharing terminals. This approach was not totally successful for several reasons. Students without previous computer course work often dropped the course as a result of the initial trauma of interacting with a major computer system. For example, they experienced difficulties with account numbers, run identification numbers, passwords, lost programs and long turnaround time. The problems were often compounded by a communication breakdown between the computer center, the instructor, and the student. Other problems included crowding of the terminal room, and the difficulty of contacting the instructor when questions arose about programming exercises.

When microcomputers became available, it was evident that many of these problems could be solved by setting up a small independent computer lab in the department. Subsequent experience has proven this approach to be correct. Many more students now complete the course, and the quality of work has been enhanced because of the improved graphics capability and quick turnaround time provided by the microcomputer. Thus, the present text has been modified extensively to take advantage of the exceptional capabilities of the microcomputer. However, the general aspects of the original book have been retained so that it may be employed with time-sharing systems if these are preferred. For more detailed discription of the development of the simulation course at Michigan Tech, see Spain (1981).

References have been provided throughout the text to give access to the more important literature on the subject of modeling. However, the reader is reminded that this is an introductory text and it is not possible to include a rigorous literature review on each subject. Models developed in conjunction with my own course work have been included without any citation or source indicated. These deal with common biological systems, so it is likely that similar models are described elsewhere in the literature. I apologize in advance to those who see their pet ideas published here without proper citation. When such oversights occur, please inform me so they can be properly referenced in future editions.

I am heavily indebted to Dr. Edwin T. Williams of the Department of Chemistry and Chemical Engineeering for thoughtfully and critically reviewing early drafts of the text and for making many suggestions which have been incorporated. I would also like to thank my colleagues in the Biological Sciences Department for their comments and criticisms. Special thanks go to Dr. Janice Glime, Dr. Kenneth Kramm, and Dr. Robert Keen, especially for their help on the sections dealing with ecology, and to Dr. Martin Auer for his review of the sections on limiting factors for growth. Thanks also go to Dr. James Horton at the Biology Department, California State College in Bakersfield, and to Mrs. Lois Young for their editorial assistance.

Special gratitude goes to Dr. Brian Winkel, my close friend and colleague at the Division of Mathematics, Rose-Hulman Institute of Technology, for his comments, criticisms, and encouragement during the final stages of manuscript development, and to Phyllis Winkel, for the hours spent patiently transcribing various versions into the word-processor. Finally, thanks go to the many Michigan Tech students who contributed comments and suggestions during the development of this course material.

I believe that the objectives outlined in this preface are very desirable for an introductory course in modeling for biologists. I have tried to make this text meet those objectives and fill what many believe is a need in the curriculum of life science majors.

The reader should recognize that this is not intended to be the 'last word' in computer modeling. Rather it represents a first attempt to organize the general principles of computer modeling into a single integrated body of information. I hope that others will be stimulated to build on my beginning and ultimately produce a text which is fully representative of this exciting new field.

James D. Spain

INTRODUCTION
THE ROLE OF COMPUTER MODELING AND SIMULATION IN BIOLOGY

Future life scientists may be characterized more by their ability to use the computer for data analysis and simulation than by their ability to use the traditional microscope. Biology is potentially the most mathematical of the sciences. This results from the fact that living systems involve a complex interaction of chemical and physical processes all of which are capable of being described in mathematical terms. These systems which are much more complex than any devised by the mind of man have for the most part resisted mathematical analysis by the classical methods so successfully employed by physicists and chemists. Only recently, with the advent of the digital computer, have biologists been able to use numerical methods to deal with these multi-component systems. The resulting new interest in biomathematics has been further stimulated by the availability of the personal computer. Suddenly biologists find that they can easily bring into their office or lab a tool capable of a wide range of data analysis techniques. Statistical methods that previously required hours to accomplish using the calculator can now be completed in minutes. Thus, many techniques, such as cluster analysis and multiple regression, should soon become routinely employed by most knowledgable biologists and a necessary component of undergraduate biology instruction.

Computer modeling and simulation is another area of biomathematics which is now accessible to all biologists as a result of the microcomputer revolution. Simulations previously employed by the few who happened to have the necessary equipment and expertise are now available to anyone willing to take the time to learn the few simple programming tricks that are needed to use BASIC as a simulation language. The objective of this book is to provide an organized discussion of biological simulation techniques which may be implemented on the microcomputer.

James D. Spain, BASIC Microcomputer Models in Biology

ISBN 0-201-10678-7

Conceptual Models

Models and other analogies have always played an important role in the scientific thought process. The use of models is so ingrained in our thinking that we are often unconscious of the important distinction between models and real systems. A model may be defined as any representation of a real system. It may deal with either the structure or function of the system. It may use words, diagrams, mathematical notation, or physical structures to represent the system. It is often synonymous with concept, hypothesis, or analogy. As no model can totally represent the real system in every detail, it must always involve varying degrees of simplification. Using this broad definition, it becomes evident that essentially all science deals with the formation, examination, and improvement of conceptual models about our universe. At this point, it would be valuable to review some of the conceptual models that have contributed to our present ideas about living systems.

The atom as a unit of elemental structure and the molecule as a unit of chemical reactivity were for a very long time simply conceptual models. Only in the last few decades have we been able to observe directly a few very large molecules using electron microscopy. X-ray diffraction allows one to construct physical models of molecules based upon the diffraction patterns they cast. It is still an indirect technique for observing that which we may never be able to observe directly. When Linus Pauling (1977) worked out the α-helix structure now found to be present to some degree in most protein polypeptide chains, he experimented with various molecular configurations using paper models until he arrived at one which had repeat distances that were consistent with his x-ray data. Another classical example is the double helix structure of deoxyribose nucleic acid (DNA) ,first proposed by Watson and Crick, based upon a molecular model which fitted many of the known properties of DNA (Watson, 1968). Like most good models, it triggered a burst of experimental activity which has continued to the present.

What is often referred to as the typical animal cell or the typical plant cell is actually a diagrammatic model of the cell based upon a composite of many observations of many kinds of cells using a variety of observation techniques (Hardin, 1966). The three-dimensional structure of most organs within the body is based upon the serial analysis of hundreds of two-dimensional tissue sections using the technique of stereology (Elias and Pauly, 1966). The resulting three-dimensional organ models are often essential to the understanding of their function under normal and abnormal conditions.

The conceptual model of the gene has gone through a long history of evolution (Glass, 1963). The blending theory of heredity gave way to the concept of hereditary particles described as 'beads on a string.' Subsequently, the gene has been considered as that portion of chromosomal DNA which codes for a single polypeptide chain. Despite the great advances made in this field, it is still dominated by conceptual models based largely upon the observation of genetic effects. Each new model has led to a new set of questions, which in turn have led to a new and better understanding of the nature of the gene.

In ecology, the food chain and the food pyramid are important conceptual models which have been used to explain the flow of energy and cycling of materials within the ecosystem. The theory of evolution is based on an enormous amount of evidence, but still must be considered a conceptual model.

In biochemistry, we deal with a variety of conceptual models about enzyme action, including enzyme attachment to the substrate and enzyme responses to changes in temperature and pH. The metabolic pathways that we call glycolysis, Krebs' cycle, and the carbon cycle of photosynthesis are in reality conceptual models which, like other models, may require modification to be consistent with new findings. Each of these models has led to great spurts in experimental activity because of the questions raised. We depend on these models to make decisions about medical treatment, drug action, treatment of poisons, and nutrition of both plants and animals.

It is clear that biology has profited greatly by translating real systems into models of various types. Some of the earliest models involved careful observation and drawing in great detail the morphology of systems under investigation. This technique still provides a valuable means of forcing oneself to see details about a structure which would otherwise go unnoticed. Those who have taken the time to accurately draw some detailed biological structure on paper soon realize that they had never really examined the subject before. New morphological details suddenly appear, and interesting relationships become evident. This illustrates that the greatest benefit of a model often comes as a direct result of the thought process involved in model development. Clearly, modeling has played an important role in the understanding of most biological systems.

Life science is involved in more than just description and understanding of natural systems. Very early in history, man decided to be a manager of his environment, and one of the objectives of biology is to gain sufficient information about complex living systems to manipulate them for our benefit. Management decisions are made on the basis of the conceptual models as perceived by the manager. Whether we are talking about wildlife management or medicine, it is obvious that improvement of our conceptual models will lead to better management.

Conceptual models, by themselves, are generally lacking in rigor. They can be imprecise and interpreted differently by different people. To circumvent this disadvantage, a method has been developed for translating the conceptual model into a form which is more subject to precise description, evaluation, and validation. This form of the conceptual model is called the mathematical model.

The Mathematical Model

Mathematical models deal with the rate of change of systems of cells, organisms, populations, or molecules with time and the extent to which such systems are effected by light, pH, temperature, or other environmental factors. A mathematical model may be as simple as a single equation relating one variable to another, or it may involve the interaction of many equations having several mutually dependent variables. The latter will be referred to as a multicomponent system model. Simple mathematical models may be obtained in various ways, but the two main approaches involve either theoretical derivation, or empirical derivation through statistical analysis.

An equation or group of equations by themselves may not contribute much to the understanding of a particular phenomenon. For this reason, it is usually necessary to solve the equation for some representative values of the independent variable (for example, time, pH, or temperature) and to present the resulting information in the form of a graph. Because of the amount of data involved, this is best done by implementing the mathematical model on a computer, especially one with graphics display capability.

Computer Simulation

Simulation in its simplest form involves implementing of a mathematical model on the computer to produce simulation data. In this way, the output of the mathematical model may be readily compared with experimental data from the real system in order to evaluate the model. Because of the complex interrelationships involved, simulation is particularly essential to the understanding of the multicomponent system model.

The process of developing a simulation model forces the investigator to describe the system in simple terms. When working with the model in this way, the investigator must take into account details about the system which might otherwise go unnoticed. The objective of any kind of model is to provide a means for obtaining new insights into the operation of a system. Simulation assists in this regard by permitting experimental interaction with the model to produce verifiable responses.

There are two basic approaches to simulation using mathematical models. One involves the use of the analog computer to solve the mathematical equations by reducing them to electrical analogs such as resistors, capacitors, and amplifiers. The output of the analog computer is a time varying voltage which is recorded on either an X-Y plotter or a cathode ray tube. It is especially useful for simulation models which involve the integration of complex differential equations such as those describing growth or energy flow. However, not all systems lend themselves to the analog approach, and programming requires considerable understanding of the electronics principles involved. The other basic approach involves the simulation of a system by numerical methods employing the digital computer. This text is concerned exclusively with the latter technique.

The general approach to digital computer simulation usually involves the following steps. First, the system must be analyzed in order to determine the basic components required for the development of a conceptual model. Often this analysis results in the construction of a block diagram of the system. Next, the key variables are defined, and each is expressed in the form of a simple functional relationship with the other variables in the system. Equations are then derived establishing the actual mathematical relationship between the variables. This derivation is done either empirically, through the use of statistical methods such as curve fitting, or analytically, by deriving the equations from theoretical considerations. Next, the mathematical expressions are programmed into the computer, and various rate constants or coefficients are assigned. The simulation is then allowed to run and produce a set of simulation data. The simulation data are then compared with experimentally obtained data to validate the performance of the model. Usually, the model requires some modification at this point in order for it to simulate the real system more accurately. After sufficient verification and validation, the model may be used to perform experiments in much the same manner as one performs experiments on the real system.

In many ways, simulation may be likened to a game that one plays with the computer. The computer keeps track of all the rules of the game, follows through the play in correct sequence, and provides any random numbers that might be necessary to simulate the effects of chance. Persons working with the simulation would have various decisions to make just as they would in playing computerized football, where, depending on conditions, they would select certain strategies to optimize the chances of scoring. In the same way, an individual working with a population growth simulation would have an opportunity to decide initial population levels, growth rates, whether there is predation or not, and how predation relates to changes in population density. The simulation would be run for a while so that its behavior might be observed. Subsequently, one may wish to alter conditions in an attempt to stabilize the system or to observe the effects of different growth rates. The outcome of such a "game" will depend on the assumptions made in designing the simulation and on the initial parameters selected by the simulator.

Just as computerized football can only approach the real game in terms of complexity, the whims of chance, and multiplicity of decisions to be made, simulations can only approach the real system to varying degrees depending upon the complexity of the computer model employed. On the other hand, it is possible to simulate systems which would be almost impossible to investigate experimentally because of the magnitude of time and/or space involved. For example, investigation of a real predator-prey system could involve population estimates taken over a 10-50 year period and a 10-1000 square mile area. Even population experiments in small closed systems may require intense study over many weeks to provide meaningful results. A simulation of these same systems could be carried out in seconds on the computer.

The Relationship of Modeling and Simulation to the Research Process

Some of the concepts presented above have been summarized in a diagram relating modeling and simulation to the overall process of research and management. (See Figure 1.)

It all begins, of course, with the Real System. It must be understood that any biological system must always remain to some extent a "black box." No matter how much information we have about a particular biological entity we must always remain on the outside looking in. For this reason, the Real System is distinguished from all other components of the diagram by representing it as a circle. All of the boxes represent forms of information that in one way or another are derived from the Real System. As such, they are reflections or perceptions of the Real System. Arrows represent processes by which this information is obtained and manipulated.

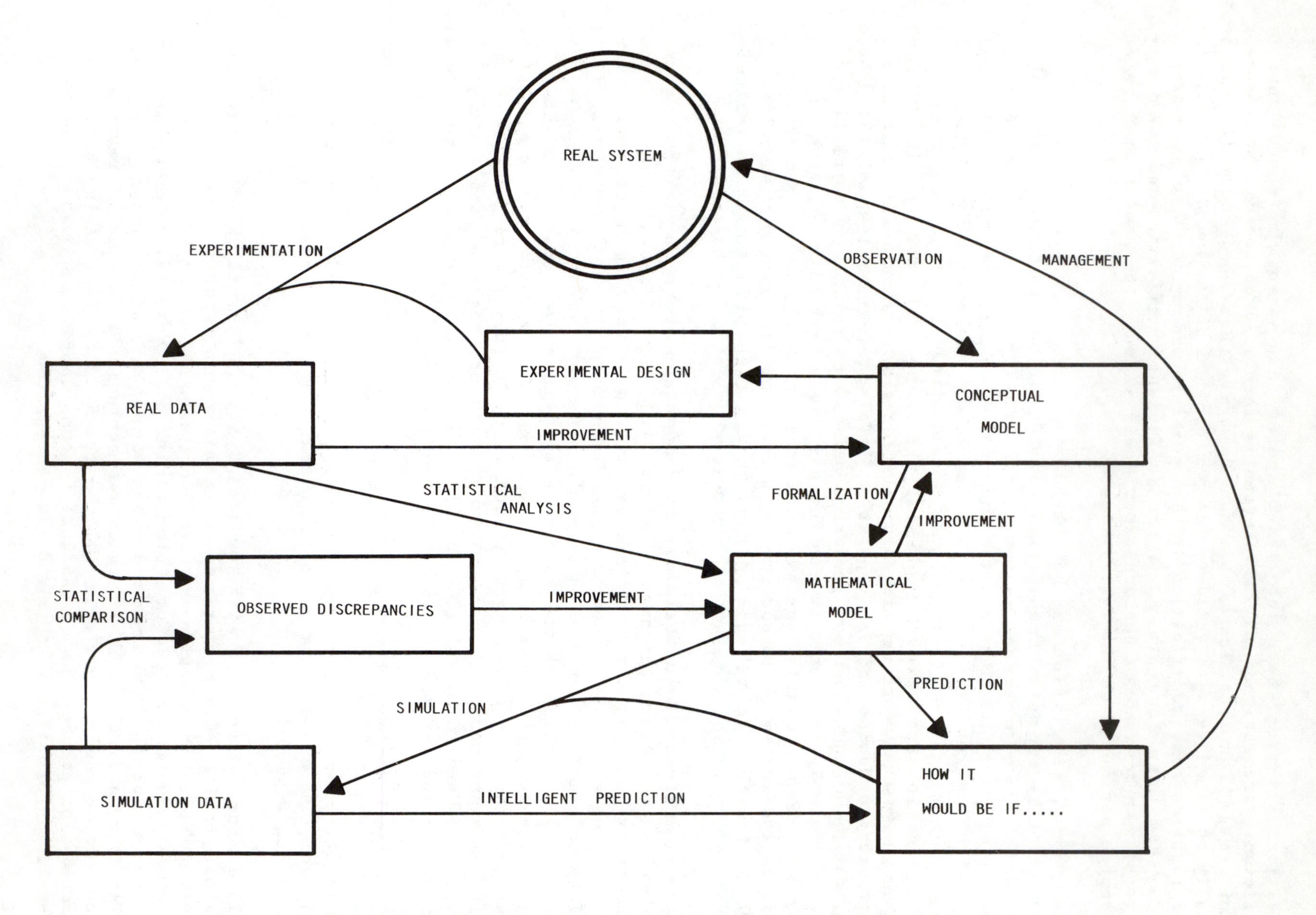

Figure 1. A diagram relating modeling and simulation to the research process. Based in part on a diagram by Gérardin (1968).

Science is primarily concerned with producing a conceptual model which most nearly reflects the Real System. The conceptual model is our picture of the Real System based on available information interpreted in the light of past experience with similar systems. Science usually becomes aware of a new biological entity as a result of investigating some other system. However, even the most casual observation generally results in the formulation of a crude conceptual model. Initially this may take the form of an analogy to some known system. The conceptual model usually suggests one or more experiments which could be performed to support or refute the model. Through experimentation one obtains data which usually throws new light on the original observation or its interpretation. In one way or another, this process generally results in the improvement of the conceptual model. This sequence describes what might be called the classical research loop.

A mathematical model results from a formalization of the conceptual model in quantitative terms, usually as an equation describing the response of a system to some variable such as time or temperature. The mathematical model may also be derived from statistical analysis of the experimental data, or by a combination of these two sources. In any case, the thinking involved in the formalization of the model generally results in an improved conceptual model. This improved conceptual model in turn will suggest further experimentation. Repeated excursions through the quantitative research loop can result in still further improvements in both the mathematical model and the conceptual model.

The process of simulation expands the power of the quantitative research loop by utilizing the mathematical model to generate simulation data which may be compared to real data in order to show discrepancies. Assuming that the model was properly programmed in the computer, these discrepancies indicate errors in the assumptions used in formulating the mathematical model, and hence flaws in the conceptual model. The errors may be so large as to require an entirely new conceptual model, or so small as to require only slight modifications of parameters in the mathematical model. In any case, new questions are raised and new experiments may be called for. The fine tuning of the model in this way results in further improvements to both the conceptual model and the mathematical model.

Thus, simulation permits one to take advantage of a more powerful research loop. Its chief advantage is that it provides a clear and unbiased test of the concepts involved in the formulation of the mathematical model and hence the conceptual model. It can therefore play a key role in the understanding of quantitative biological systems. No other technique can provide such a powerful aid to the development and testing of concepts.

If this was the only value of simulation and modeling in biology, it would be more than sufficient justification for their use. However, simulation provides another important benefit. It allows one to perform "experiments" on the system model that lie outside the range of normal experimentation. In other words, this allows us to ask the question "what would happen to the system if...?". This process is what we might call the enlightened management loop.

By simulating various management strategies on the mathematical model, we are able to make intelligent predictions about their success or failure on the Real System. There are, of course, less effective management loops. For example, the loop farthest to the right might describe the politically expedient management loop in which observation leads to conceptual model and thence directly to prediction and management.

Computer simulation is particularly valuable for systems which involve multiple, non-linear interactions. This is especially true in the fields which deal with the management or control of multicomponent systems such as ecological systems and physiological systems. Both of these examples usually contain numerous variables interrelated to each other by positive and negative feedback loops. Forrester (1969) has shown that human decisions about such systems are usually counterintuitive in that perturbations of the system often suggest corrective measures which produce effects which are exactly opposite from those desired. It is essential therefore that management decisions be based on something more than simplistic conceptual models. Watt (1970) gives this as one of the most important justifications for simulation in the area of ecosystems management. It is clear that in many instances one is unable to experiment with the real system. On the other hand, it is important that the control strategy be correct because large populations of humans or other organisms are often involved. In all such cases, computer simulation and systems analysis techniques provide a logical basis for the management decisions which must be made.

Modeling and simulation do not provide the answer to all questions because there are many systems which by their non-quantitative nature do not lend themselves to mathematical modeling. However, it is clear that for certain systems, modeling and simulation can provide an extremely powerful aid to both understanding and management.

The Development of Multicomponent Models

One might wonder how it is possible to model complex systems or how they would be verified if they could be modeled. This is not as difficult as it may seem initially. Multicomponent system models are composed of clusters of submodels or modules which have multiple interlocking dependencies. Each submodel may be developed independently and validated by comparing its simulation output with that of the comparable real system. Only then would the submodels be merged into a single entity capable of acting like the total system. This approach makes it possible to break down the most complex problem into a number of manageable components.

All systems may be viewed as consisting of a hierarchy of organization (Walters, 1971). Each component of the system embodies sub-components which in turn contain sub-sub-components in a hierarchical chain (Figure 2).

Fortunately, it is not necessary to understand how all sub-components of a system operate in order to model the system. A large number of components and sub-components are normally lumped into one or more sub-models which define the net effect of these sub-components on the system as a whole. This process of lumping reduces the resolution of the model, but has the effect of focusing the major attention on primary controlling factors.

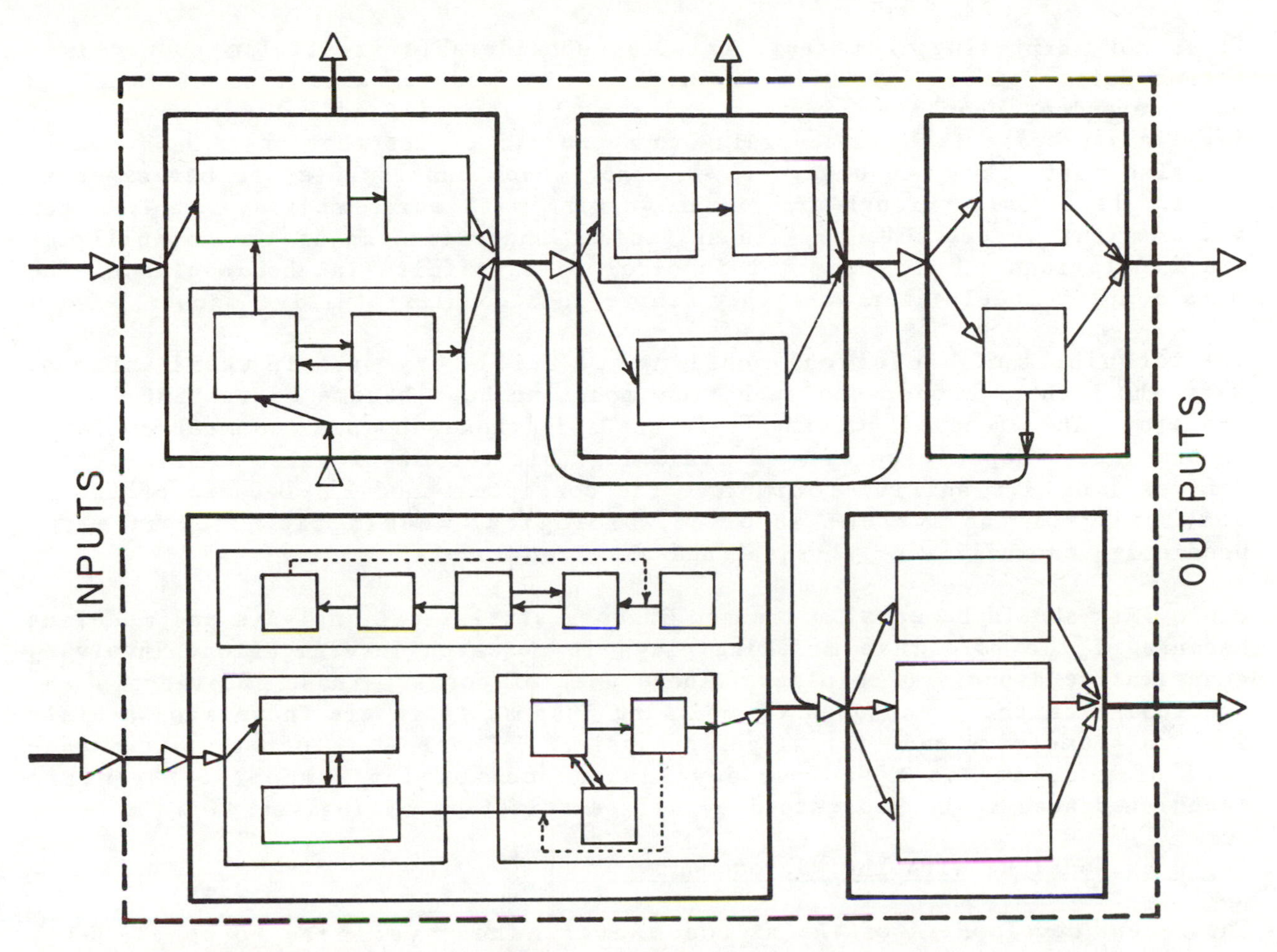

Figure 2. The hierarchy of organization within a hypothetical living system. Based on a diagram by C. J. Walters in Odum (1971).

When we are modeling an intermediate level of the hierarchical chain, higher levels are usually considered to be external to the system of interest. The behavior of an external variable is usually approximated and incorporated into the model as an environmental forcing function. If it subsequently becomes evident that such an approximation plays a major role in determining model behavior, it may become necessary to expand the model to include these other hierarchical levels.

It is a mistake to think that simulation data must duplicate exactly the data obtained by experimentation with the real system. A simulation model is, by definition, not the real system and it is unreasonable to expect simulation data to match that of the real system point by point (Meadows, et al., 1972). It is usually sufficient if the model is able to show general correspondence to the main behavioral modes exhibited by the real system. One should also be aware that it is possible to make a simulation so complicated that it contributes little or nothing to the understanding of the system being studied. A model is usually constructed in order to help identify the major controlling factors of a system and these may become hidden in an overly complicated model.

It is not surprising that there is often considerable skepticism about simulation data produced by multicomponent system models. This type of model is most dependent upon the computer and its ability to handle large amounts of data in a precise fashion according to a program of instructions. Such models are also most likely to behave in a manner which runs counter to our expectations. It is only through experience working with more familiar simple model systems that we are able to gain sufficient understanding of the capabilities and limitations of the computer in order to have faith in the results of the more complex models even when they behave in a counterintuitive manner.

The techniques of biological modeling are still very much in their infancy. Many who perhaps expect too much from modeling have become disenchanted with its slow rate of development. There are also those who put too much emphasis on predictive capability without realizing that the major value of simulation derives from its ability to improve the conceptual model. Because of the complexity of the systems involved, biological models may never have the predictive capability of physical and chemical models.

Biologists should be equally trained in both statistical analysis and modeling because of the role that modeling plays in research investigations involving quantitative aspects of biology. There are, of course, those individuals who specialize in the techniques of modeling just as there are those who specialize in biometrics and statistical analysis. However, in order to take advantage of modeling as a tool in developing and testing concepts, the basic techniques need to be understood by all quantitative biologists.

Computer Modeling with the Microcomputer

The recent development of the microcomputer makes available to practically every biologist a tool that may be used for developing and implementing a wide range of complex simulations. BASIC, employed on most microcomputers, is a simple language which is easily adapted to most modeling strategies. Deterministic techniques, often involving numerical integration of non-linear differential equations, may be conveniently programmed when one understands the few simple principles involved. The random number generator provided on most microcomputers makes it easy to program Monte Carlo simulations of random processes. Although matrix manipulation is not quite as convenient to accomplish on microcomputers, it can be very useful for certain types of models. Once one has a clear understanding of modeling principles, he or she can easily adapt to specialized simulation languages when the need arises in more advanced situations.

Sometimes people are concerned that microcomputers may not have sufficient memory to be used for simulations. This is rarely a problem as many simulations are not particularly long. The 16K bytes provided by most microcomputers are more than adequate for all of the introductory models described in this book. Most simulations are accomplished by repetition of a few operations.

Although BASIC is relatively slow compared to compiled languages such as FORTRAN and PASCAL, time is not an important factor when dealing with microcomputers. Most deterministic simulations are completed in a few minutes. In extreme cases, such as large Monte Carlo models, one could presumably leave the computer running overnight! The only cost incurred would be the cost of power.

Thus, the microcomputer provides a very convenient alternative to those of the time-sharing terminal. Although the emphasis in this text is on the use of microcomputer BASIC, with few modifications, the same techniques could be employed on most time sharing systems.

Goals and Objectives

The objective of this text is to introduce the concepts of modeling and simulation using simple mathematical techniques. It is also designed to employ simple programming techniques so that the major emphasis can be on simulation rather than programming. The most effective way to learn computer modeling is by actually developing, programming, and testing simulations. Therefore, this book will introduce the basic concepts of modeling by presenting a variety of simple biological systems capable of being modeled using a direct approach that does not require an extensive background in mathematics.

A major goal of the text is to present the concept that biology is essentially a mathematical science that ultimately will be described in mathematical terms. It is hoped that once students recognize the importance of mathematics to life sciences, they will be encouraged to expand their capabilities far beyond the scope of this book by taking advanced coursework in both mathematics and in quantitative biology.

PART ONE
SIMPLE MODEL EQUATIONS

The most basic principle of biological modeling is that an equation may serve as an analog or model of a simple biological process. This arises from the fact that almost any biological process may be described by a response curve which relates the intensity or amount of some causative agent or stimulus to the intensity or amount of the biological response which it produces. By convention, the intensity or amount of the stimulus, called the independent variable x, is expressed along the horizontal axis of a graph, and the intensity or amount of the response, called the dependent variable y, is expressed along the vertical axis. Although it is possible to invert the x and y axes on a graph we try to retain this conventional relationship as much as possible. This relationship is shown in Figure 1.

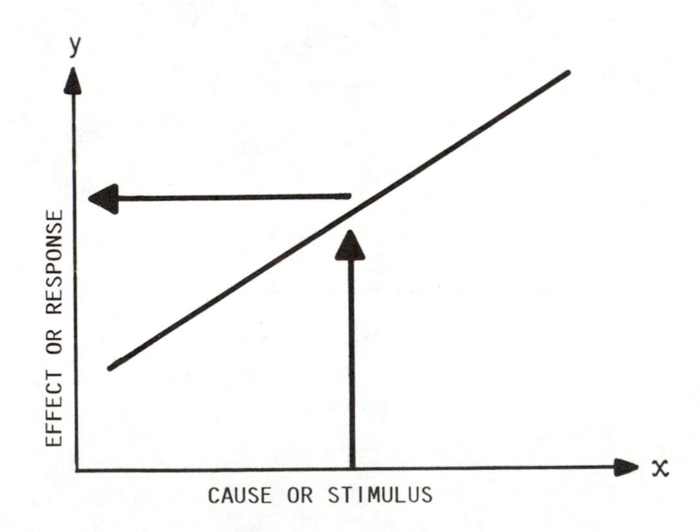

Figure 1. Graph of simple cause and effect relationship

James D. Spain, BASIC Microcomputer Models in Biology
ISBN 0-201-10678-7

Experimental data resulting from the response of a biological system under different conditions may be used to plot this cause and effect (input-output) curve. The curve expresses in a quantitative fashion how a biological system performs under a given set of conditions. It is therefore a diagrammatic model of the system. Curves are also used to describe the response of equations in relating one variable to another. Thus, it is possible, in one way or another, to find an equation which will generate output data that are mathematically indistinguishable from data produced by the biological system. The equation, therefore, becomes the ultimate model of the system or system component. The proper equation can thus stand in place of or substitute for a system component in terms of its relationship to other components of the system, and is interchangeable from the mathematical standpoint. This substitution principle is illustrated in Figure 2.

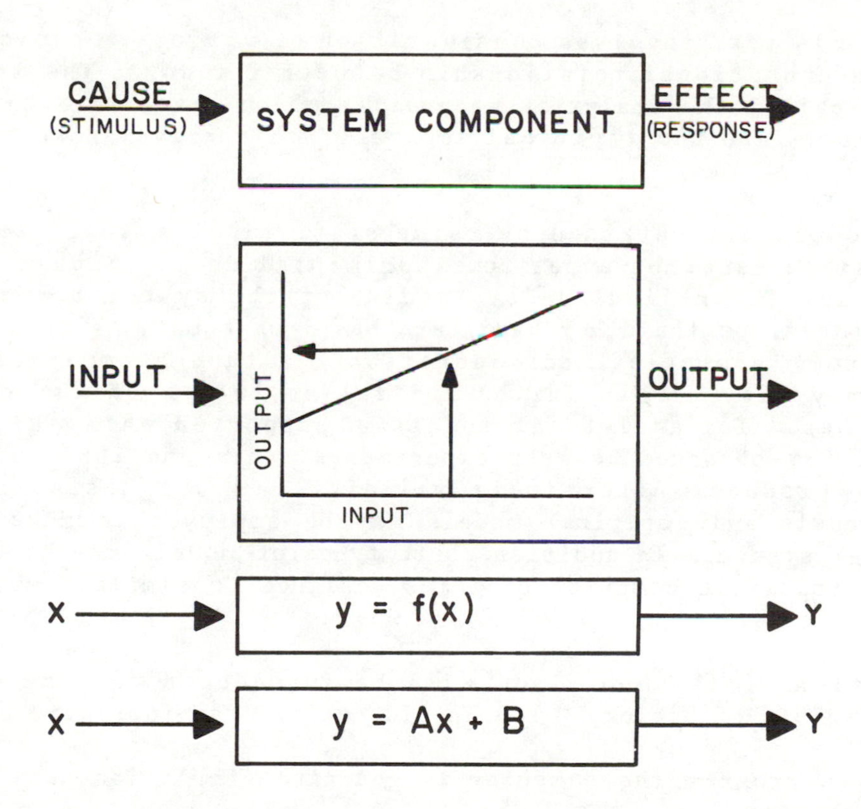

Figure 2. Stepwise replacement of a biological system component by a model equation.

The process of finding the appropriate model equation first involves identifying the functional relationship of the system. From examination of the data we are usually able to set up a simple function equation stating that the output variable, y, is a function of some input variable, x. This general function equation has the form $y = f(x)$. Sometimes the output variable, y, is a function of more than a single input variable so that the function equation has the form $y = f(x,z)$. For the most part, this book is concerned with equations involving only a single independent variable other than time.

After finding an equation which effectively models the biological system, it still is necessary to find appropriate values for the model parameters, i.e. the coefficients and exponents in the model equation. These parameters may be obtained either by trial and error, or by a more organized procedure such as curve-fitting. Sometimes the model is so complex that trial and error is the only technique available. Once values have been assigned to model parameters, it is possible to validate the model by comparing simulation data with experimental data obtained from the real system. The process of model validation is to be distinguished from program verification, which refers to the process of testing a computer program to make sure it is implementing the model in the way that is was intended.

Part One of this text involves consideration of various techniques used to determine the functional relationship between x and y. The two main approaches are called the analytical approach, which gives rise to theoretical model equations, and the empirical approach, which gives rise to empirical model equations.

Empirical models are obtained by using statistical methods to fit one of several possible equation types to experimental data. Such models do not depend upon the theoretical understanding of the system being described. Analytical models, on the other hand, are based on equations that are derived primarily from theoretical considerations, although empirical modeling techniques may play a role in the initial evolution of the theory. The theoretical basis for analytical models is supported when a high degree of correspondence is obtained between experimental data and the simulation data such models produce. After their validity has been established, both analytical models and empirical models may be employed to make predictions about the real system. In addition, both types of models may be coupled with other simple models in computer programs designed to simulate multicomponent biological systems.

In this section of the book, you will be introduced to some of the most fundamental models of biology. You should achieve the following objectives.

1. Know how to program the computer to generate simulation data from simple model equations.

2. Know how to plot data correctly, either with the computer, or by hand.

3. Understand and become familiar with about ten curve types which are fundamental to biological systems modeling.

4. Know how to use curve fitting techniques to determine the values for coefficients and exponents of both theoretical and empirical model equations.

5. Understand the analytical techniques which are employed in developing theoretical model equations.

6. Know how to read computer flowcharts and translate them into working programs.

7. Know how to program a simple Euler numerical integration on the computer.

When you have accomplished these objectives, you will be ready to proceed to Part Two in which you will learn how simple equations are combined to model multicomponent systems.

CHAPTER 1
ANALYTICAL MODELS BASED ON DIFFERENTIAL EQUATIONS

Analytical models are classified according to the type of analysis or theoretical derivation on which they are based. In this chapter, we will be discussing those which are based on the solution of differential equations. Differential equations define a rate of change of some dependent variable with respect to some independent variable, usually time or distance. To show how differential equations are utilized to develop model equations, let us examine the development of the model for unlimited population growth.

1.1 The Exponential Growth Model**

The advantage of using differential equations results from the ease with which they are usually obtained from a function equation. In this example, we start with the general observation that population growth is a function of population density. That is, the more bacteria or yeast we have, the faster they reproduce. From this generalization, we can write a simple function equation for population growth.

$$G = f(N) \quad (1.10)$$

where G is growth rate, and N is population density. If we assume that growth is a direct function of N, i.e. that growth rate is directly proportional to population, and no other factors are involved, then the growth rate equation could take the form:

$$G = k \cdot N \quad (1.11)$$

where k is a proportionality constant.

** The double asterisk marks sections fundamental to the understanding of modeling.

James D. Spain, BASIC Microcomputer Models in Biology

ISBN 0-201-10678-7

The curve for this equation would be a straight line with slope = k. That is, as N increases, G increases in direct proportion. The equation has very limited value in this form, since we are mainly interested in the population density at any time during the growth process, rather than growth rate. To convert the growth rate equation to a form which gives this kind of information, we must first write it as a differential equation in which growth is redefined as dN/dt. This expression symbolizes the derivative of N with respect to time, t, and means the instantaneous rate of change of population density with respect to time. Since this is equivalent to G in Equation 1.11, our equation now has the form:

$$\frac{dN}{dt} = kN \qquad (1.12)$$

We now have an equation in terms of the two variables we want, N and t. It is now possible to reconstruct the population growth curve equation from this by the process of integration. What we will be doing is using the equation for the slope of the growth curve (1.12) to reconstruct the equation for the growth curve. Mathematicians know how to perform the process of integration only as a result of their experience with the reverse process of differentiation. These experiences have led to a large number of integration rules. A few useful integration rules are presented in Table 1.

Table 1. Common Integration Rules

1. $\int a\, dx = \int a\, dx = ax + c$

2. $\int x^n\, dx = \frac{x^{n+1}}{n+1} + c \qquad (n \neq -1)$

3. $\int \frac{dx}{x} = \ln x + c$

4. $\int e^x\, dx = e^x + c$

5. $\int e^{ax}\, dx = \frac{e^{ax}}{a} + c$

6. $\int \frac{dx}{a + bx} = \frac{\ln(a + bx)}{b} + c$

Note that these rules are written for the indefinite integral (no limits on the integral), hence each contains an integration constant, c.

The first step in the integration process is to collect all terms dealing with the dependent variable, N, on the left side of the equal sign and the independent variable and associated constants on the right side. In this example the result is as follows:

$$\int_{N_0}^{N_t} \frac{dN}{N} = \int_0^t k\,dt \tag{1.13}$$

The integration symbol gives instructions to perform an integration according to the rules. In this case, we will be integrating N and t between the limits of t = 0 and t = t. The result of integrating without limits (according to rules 1 and 3 (Table 1)) is:

$$\ln N = k \cdot t + c \tag{1.14}$$

When integrating between limits, we take the integral at the upper limit minus the integral at the lower limit, first doing the left side of the equation, then doing the right. The result is the following:

$$\ln N_t - \ln N_0 = k \cdot t - k \cdot 0 \tag{1.15}$$

Since the difference between logs equals the log of the quotient we obtain:

$$\ln(N_t/N_0) = k \cdot t \tag{1.16}$$

By taking exponents of both sides we get the forms:

$$\frac{N_t}{N_0} = e^{kt} \tag{1.17}$$

and

$$N_t = N_0 \cdot e^{kt} \tag{1.18}$$

Finally, we have arrived at a very useful model equation. This is the exponential growth equation which describes the population density at any time based on the initial population density, N_0, and the growth rate constant, k. This curve has the classic form shown in Figure 1.1. It may be used over a limited range to describe many kinds of positive feedback or self generating systems. These include the growth of bacteria, rabbits, people, money and energy demand.

Much of what was said in this first example may have been superfluous if you have had a good course in calculus. However, it was necessary to go through this derivation step by step in order for you to understand what is involved in developing an analytical model equation from a differential equation through the process of integration.

Please do not become discouraged if this derivation was unfamiliar to you! The actual process of deriving the model equations presented in Chapters 1 and 2 is not crucial to the applications of computer modeling.

Although you should have a basic understanding of this process, we are primarily interested in using these equations as practical tools to generate simulation data. The equations in this chapter are not in themselves particularly new or exciting. The objective is to give you ample opportunity to learn computer programming using equations with which you should already be somewhat familiar.

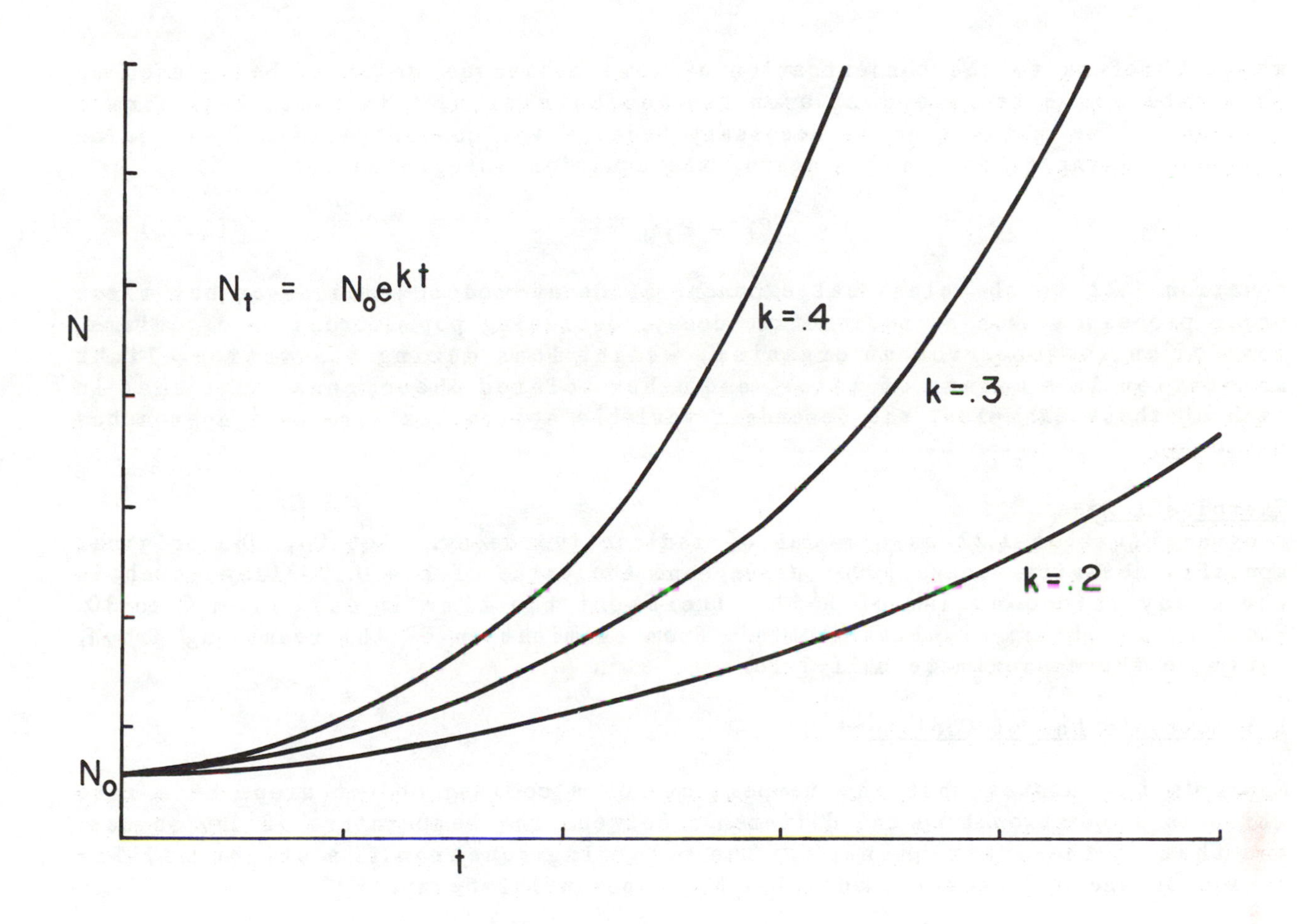

Figure 1.1. The exponential growth curve.

<u>Exercise 1.1</u>**
Write a computer program for Equation 1.18 which will permit you to simulate unlimited growth of bacteria. Increment time from 0 to 50 hours. Assume $k = .02/hr.$ and $N_0 = 2/ml$. Prepare a graph of the population density at various times up to 50 hrs.

In each of the models programmed in Chapters 1 and 2, you should attempt to utilize the special computer subroutine which is available for plotting data. When you get your simulation output, examine the curve carefully to make sure it is consistent with what you know about the system being modeled.

1.2 Exponential Decay

Many systems decrease with time at a rate which is proportional to the concentration or amount of material present at any time. This decrease is expressed by the following differential equation:

$$\frac{dC}{dt} = -kC \qquad (1.21)$$

where C refers to the concentration of some substance which is being used up at a rate which is dependent upon its concentration, t is time, k is a rate constant. The minus sign is necessary because the concentration, C, is being reduced. Using rules 1 and 3 again, the equation integrates to:

$$C_t = C_0 \cdot e^{-kt} \qquad (1.22)$$

Equation 1.22 is the classical exponential decay model which describes first order processes such as radioactive decay, declining populations of organisms, loss of an isotope from an organism, weight loss during starvation, light absorption in a column of water, and other related phenomena. Note that in each of these examples, the dependent variable approaches zero as t approaches infinity.

Exercise 1.2**
Program Equation 1.22 as a model of radioactive decay. Let C_0, the original specific activity, equal 100μ curies, and the value of $k = 0.1733$/day which is the decay rate constant of I^{124}. Increment the time in days from 0 to 30. Plot the resulting simulation data. From examination of the resulting graph, estimate the approximate half-life.

1.3 Newton's Law of Cooling**

Newton's Law states that the temperature of a cooling object drops at a rate which is proportional to the difference between the temperature of the object, and that of the environment, C. The resulting equation is a slight modification of the differential equation for exponential decay:

$$\frac{dT}{dt} = -k(T - C) \qquad (1.31)$$

Since the temperature is a function of heat content, we may also write an equation to describe the rate of heat loss from a body.

$$\frac{dH}{dt} = -k(T - C) \qquad (1.32)$$

Using rule 6, Equation 1.31 integrates to the cooling equation :

$$T_t = (T_0 - C)e^{-kt} + C \qquad (1.33)$$

where T_t is the temperature of the body at any time t, $(T_0 - C)$ is the difference between the initial body temperature and the environmental temperature, and k is the cooling rate constant.

```
50  REM  MODEL OF EXPONENTIAL DECAY
60  REM  EXERCISE 1.2
70  REM  SUBMITTED BY JOHN DOE
80  REM  10/10/82
110  REM  NO = INITIAL NUMBER OF RADIOACTIVE ATOMS
120  REM  N  = NUMBER OF ATOM REMAINING AT TIME T
130  REM  K  = DECAY CONSTANT
140  REM  F  = THE FRACTION OF ATOMS REMAINING AT TIME T
145  REM  P  = PERCENT OF ATOMS REMAINING
200  READ K,NO
210  REM   ALTERNATE GRAPHICS FOR PRINTERS UNABLE TO ACCEPT SCREEN DUMP
220  PRINT "TIME",,"PERCENT ATOMS REMAINING"
230  PRINT  TAB( 7);"0--------20--------40--------60--------80--------100"

240  FOR T = 0 TO 20
250 N = NO *  EXP ( - K * T)
260 F = N / NO
270 P = 100 * F
280 Y =  INT (P * 50 / 100)
300  HTAB 2: PRINT T;: HTAB 7: PRINT "!"; TAB( Y);"*"
310  NEXT T
320  DATA  .1733,1000
330  END

]RUN
TIME                            PERCENT ATOMS REMAINING
      0--------20--------40--------60--------80--------100
 0    !                                                 *
 1    !                                         *
 2    !                                  *
 3    !                            *
 4    !                       *
 5    !                    *
 6    !                *
 7    !             *
 8    !           *
 9    !         *
 10   !       *
 11   !      *
 12   !     *
 13   !    *
 14   !   *
 15   !  *
 16   !  *
 17   ! *
 18   ! *
 19   !*
 20   !*
```

Figure 1.2. Sample program of exponential decay model.

The form of this curve is the same as that of the exponential decay curve except that the temperature curve approaches the value of C as t approaches infinity.

Exercise 1.3**
Use Equation 1.33 to simulate the cooling of a human corpse assuming no heat production after death. Assume also that initial body temperature was 37° C, environmental temperature is 10° C, and k = 0.06/hr. Increment time in hours up to 20. Generate some simulation data, and plot temperature as a function of time.

1.4 Dilution Model

When a solute is washed out of a constant volume container by continuous influx of a solution containing a constant solute concentration, we obtain a curve which is almost identical to the previous one. The form of the integrated curve is:

$$C_t = (C_0 - C_I)e^{-(Rt/V)} + C_I \tag{1.41}$$

where C_t is the concentration of solute in the outflow at any time t, C_0 is the initial concentration in the container with volume, V, C_I is the input concentration, and R is the rate of flow. Again, the concentration is seen to decline exponentially to a value of C_I as t approaches infinity.

Exercise 1.4
The differential equation which was used in deriving the dilution model is as follows:

$$\frac{dC}{dt} = \frac{-R}{V}(C - C_I)$$

Show how Equation 1.41 results from the integration of this differential equation. Use the limits 0 to t and C_0 to C_t.

1.5 Diffusion Across a Boundary of Unit Area**

Another equation of this same form may be used to describe the process of diffusion. The rate of change in solute concentration in a cell as a result of passive diffusion into an environment having constant solute concentration is:

$$\frac{dC}{dt} = -k(C - C_0) \tag{1.51}$$

where C represents concentration of solute inside a cell of unit volume and unit surface area, C_0 represents the constant environmental concentration, and k is the diffusion rate constant. The integrated form of the equation is:

$$C_t = (C_I - C_0)e^{-kt} + C_0 \tag{1.52}$$

where C_t is the concentration in the cell at time, t, and C_I represents the initial concentration of solute in the cell.

Exercise 1.5**
Program Equation 1.52 to model the process of diffusion from a cell of unit volume, having an initial concentration of 100 units/unit vol., placed in an environment with a solute concentration of 50 units/unit vol. Assume a rate constant k = 0.3/min. Generate simulation data, and present this data in the form of a time plot.

1.6 Von Bertalanffy Fish Growth Model

This classical model which relates fish length to age is based on the assumption that fish growth is proportional to the difference between actual length and a theoretical maximum length. That is, fish grow fast initially but slow up as they approach their maximum length. The differential equation describing this process is

$$\frac{dL}{dt} = -k(L - L_m) = k(L_m - L) \tag{1.61}$$

where L is length at any time t, and L_m is the theoretical length that would be attained as t approaches infinity. The equation integrates to:

$$L = (L_I - L_m)e^{-kt} + L_m \tag{1.62}$$

where L is the length at any time, t, expressed in years, L_I is the initial length at t = 0, and k is the growth rate constant.

Exercise 1.6
Write a program to simulate the growth of striped bass over the period between 1 and 30 years. Assume L_m for striped bass is 60 in., k is 0.096/yr and L_I is 1 in. Plot the resulting lengths as a function of time.

1.7 Stream Drift Model

McLay (1970) developed the following model to describe how benthic invertebrates enter the drift of a stream.

$$N_x = N_0 \cdot e^{-Rx} \tag{1.71}$$

where N_0 is the population density of animals at the source, N_x is the density at distance x (in meters) downstream and R is a constant unique to the organism and the stream.

Exercise 1.7
Program this model for chironomids in a stream where R = 0.13/m. Generate some simulation data for various distances from the source which had a density of 1000 organisms/m^2. What is the differential equation on which this model is based?

1.8 Inhibited or Logistic Growth Model**

The model of inhibited growth takes into account the fact that a population does not normally grow beyond some natural limit density, L. The result is the following equation:

$$\frac{dN}{dt} = kN(L - N) \qquad (1.81)$$

Note that as N approaches L, the term inside the parentheses approaches zero, and so does the growth rate, dN/dt. This equation can be integrated to the following:

$$N_t = \frac{N_0 L}{N_0 + (L - N_0)e^{-Lkt}} \qquad (1.82)$$

where N_t is the population density at time, t, and N_0 is the original population density. This equation describes the sigmoid shaped curve for population growth in a limited environment. This equation may also be used to model the spread of disease, growth obtained from a given amount of nutrient, and other resource-limited processes. Equation 1.81 and related equations will be discussed in detail in Chapter 7.

Exercise 1.8**

Program the inhibited population growth curve based on Equation 1.82 to obtain some simulation data. Assume the following parameters:

N_0 = 10 organisms/unit area
L = 500 organisms/unit area
k = .0006/wk·organisms/unit area

Increment t from 0 to 40 weeks. Plot the population density as a function of time.

1.9 Kinetics of Bimolecular Reactions**

Assume two chemical reactants, A and B, interact to form product, P, as described in the following reaction:

$$A + B \xrightarrow{k} P$$

The rate at which reactant B is used up is a function of the concentrations of both A and B as follows:

$$\frac{d[B]}{dt} = -k[A][B] \qquad (1.91)$$

where [B] and [A] refer to the concentrations of B and A respectively, and k is the reaction rate constant.

If [A] and [B] are assumed to be equal, then this equation becomes:

$$\frac{d[B]}{dt} = -k[B]\cdot[B] = -k[B]^2 \qquad (1.92)$$

which we arrange to a form that may be integrated:

$$\int_{B_0}^{B_t} \frac{d[B]}{[B]^2} = \int_0^t -kdt \tag{1.93}$$

This integrates to:

$$\frac{1}{[B]_t} = kt + \frac{1}{[B]_0} \tag{1.94}$$

and is rearranged to solve for $[B]_t$:

$$[B]_t = \frac{[B]_0}{[B]_0 \cdot kt + 1} \tag{1.95}$$

Equations 1.94 and 1.95 describe the characteristics of second order reaction kinetics, where the rate is proportional to the product of the concentration of two reactants or the square of one reactant. When data from such reactions are plotted $1/[B]$ as a function of t, the result is a straight line with slope equal to k and y intercept equal to $1/[B]_0$ as predicted by Equation 1.94.

In bimolecular reactions, if the concentration of one reactant, A, greatly exceeds the other, then its concentration is not significantly changed by the reaction. It therefore becomes a constant which may be combined with the rate constant to produce a new constant, k', as follows:

$$\frac{d[B]}{dt} = -k[A][B] = -k'[B] \tag{1.96}$$

This integrates to:

$$\ln[B]_t = -k't + \ln[B]_0 \tag{1.97}$$

and

$$[B]_t = [B]_0 \cdot e^{-k't} \tag{1.98}$$

We recognize this equation to be that of the exponential decay curve (1.22). It is also called the first order kinetics equation. It is characteristic of reactions where the rate is dependent upon the concentration of a single chemical species. Reactions are diagnosed as being first order if a straight line is obtained when $\ln[B]$ is plotted as a function of t, as predicted by Equation 1.97. Note that the slope of this straight line plot provides a means of obtaining the values of k' and k.

The terms zero order, first order, and second order were first employed to describe chemical reaction kinetics. However, they are now generally employed to describe any rate process which is constant (zero order), is dependent on the concentration of a single variable (first order), or is dependent on the product of two variables or the square of one variable (second order).

Equations 1.96 and 1.98 are especially important because they illustrate the fact that the reaction order is affected when one of the reactants is held constant, either because of an initial high concentration relative to the second reactant, or because it is maintained at a steady-state concentration by other reactions. This principle will have a bearing on several of the multicomponent models developed in later chapters.

Exercise 1.9**

Write a program employing Equation 1.95 to simulate the change in concentration of B with time. Assume $[B]_0 = 100$ M, k = 0.1/sec. M. Increment time in seconds between zero and 30. Make a graph of [B] as a function of t and 1/[B] as a function of t.

Conclusion

This brief introduction to some of the basic analytical models developed from differential equations is intended to show the importance of this technique for describing biological phenomena. In this chapter analytical methods were used to convert simple differential equations into mathematical models. In Chapter 5 you will be introduced to the simpler and more powerful technique of numerical integration which may be applied with equal ease to differential equations of almost any degree of complexity.

For additional information on analytical methods of integration, refer to Bittinger (1976), DeSapio (1976), Milhorn (1966), or any good introductory calculus text.

CHAPTER 2

ANALYTICAL MODELS
BASED ON EQUILIBRIUM OR STEADY-STATE ASSUMPTIONS

In the last chapter we demonstrated that rate equations may be converted to simple biological models through the process of integration. In this chapter we will continue to make use of differential equations, except that we will use equilbrium or steady-state assumptions to eliminate the differential by setting one rate process equal to another. Although such models are only accurate under the assumed conditions, they provide a reasonable approximation of system behavior. Before deriving any models using this approach, let us first examine the nature of the equilibrium condition and compare it with the steady-state condition.

Equilibrium models are based upon the assumption that two or more processes are competing with one another in such a way that the net change in the variables approaches zero. True equilibrium involves an isolated or closed system. As an example assume two substances, A and B, which are interconvertible and isolated from the environment. At equilibrium, the rate of formation of A (flux, f_2) and the rate of breakdown of A (flux, f_1) will be such that the net flux equals zero.

$$A \underset{f_2}{\overset{f_1}{\rightleftharpoons}} B$$

The fluxes may be described as

$$f_1 = k_1[A] \qquad (2.01)$$

and

$$f_2 = k_2[B] \qquad (2.02)$$

where k_1 and k_2 are rate constants and [A] and [B] are concentrations.

James D. Spain, BASIC Microcomputer Models in Biology

ISBN 0-201-10678-7

At equilibrium, the net flux equals

$$(f)_{net} = 0 = k_2[B] - k_1[A] \tag{2.03}$$

and

$$f_1 = f_2 \tag{2.04}$$

By rearranging Equation 2.03, we obtain the equilibrium constant expression:

$$k_2[B] = k_1[A] \tag{2.05}$$

$$\frac{k_1}{k_2} = \frac{[B]}{[A]} = K \tag{2.06}$$

The advantage of the equilibrium assumption is that it allows us to use a constant to relate one concentration to another. In other words, if we know K and [A], we may calculate [B].

In contrast, steady-state conditions are associated with open systems. Let us again consider substances A and B, now arranged in a steady-state relationship.

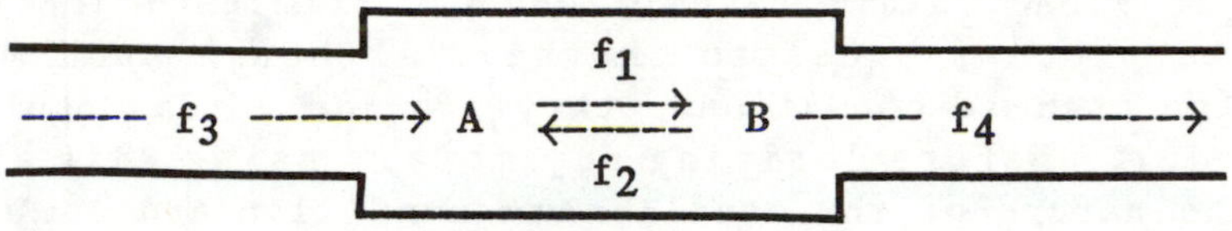

$$\xrightarrow{f_3} A \underset{f_2}{\overset{f_1}{\rightleftarrows}} B \xrightarrow{f_4}$$

Note that two additional fluxes are added, f_3, which describes the rate at which A enters the system, and f_4, which describes the rate at which B leaves the system. At steady-state, the net change in [A] or [B] will be zero, but f_1 will no longer equal f_2. Instead, the following 3 equations would be true:

$$f_1 = f_2 + f_3 \tag{2.07}$$

$$f_1 = f_2 + f_4 \tag{2.08}$$

$$f_3 = f_4 \tag{2.09}$$

By making either equilibrium or steady-state assumptions, we are thus able to write equations which assist in relating one variable to another. Even though strict steady-state or equilibrium conditions are rarely achieved in biological systems, these assumptions provide an important means of deriving certain types of models as the following examples will show.

2.1 Michaelis-Menten Model of Enzyme Saturation**

As you will see in subsequent chapters, the phenomenon of saturation is very fundamental to many biological systems. Models based on this approach are as important to the understanding of living systems as are the exponential growth and decay models of Chapter 1.

The general theory of enzyme action was developed by Michaelis and Menten in 1913. The model which follows was derived by Briggs and Haldane and further developed by Lehninger (1975) based on the following assumptions:

The enzyme is assumed to react with the substrate to form an enzyme-substrate compound which breaks down to form product and free enzyme according to the reaction

$$E + S \underset{k_2}{\overset{k_1}{\rightleftarrows}} ES \underset{k_4}{\overset{k_3}{\rightleftarrows}} E + P$$

where E and S refer to enzyme and substrate, ES is the enzyme-substrate compound and P is product. The four rate constants involved are k_1, k_2, k_3 and k_4. The initial reaction velocity is assumed to depend on a rate-limiting step involving breakdown of ES.

$$v = k_3[ES] \qquad (2.10)$$

The rate of formation of ES is given by

$$\frac{d[ES]}{dt} = k_1[E]\cdot[S] + k_4[E]\cdot[P] \qquad (2.11)$$

Under initial conditions, when [P] = 0, then

$$\frac{d[ES]}{dt} = k_1[E]\cdot[S] = k_1([E_t]-[ES])\cdot[S] \qquad (2.12)$$

where total enzyme = $[E_t] = [E] + [ES]$.

The rate of breakdown of ES is

$$-\frac{d[ES]}{dt} = k_2[ES] + k_3[ES] = (k_2 + k_3)\cdot[ES] \qquad (2.13)$$

At steady-state, rate of formation equals rate of breakdown:

$$k_1([E_t]-[ES])\cdot[S] = (k_2 + k_3)\cdot[ES] \qquad (2.14)$$

Rearranging

$$\frac{([E_t]-[ES])\cdot[S]}{[ES]} = \frac{k_2 + k_3}{k_1} = K_m \qquad (2.15)$$

where K_m = the Michaelis-Menten constant.

Solving for [ES] we get

$$[ES] = \frac{[E_t]\cdot[S]}{K_m + [S]} \qquad (2.16)$$

Since v is proportional to [ES] (Equation 2.10), v approaches a maximum velocity, V_{max}, as the enzyme becomes totally saturated as ES.

Hence

$$\frac{v}{V_{max}} = \frac{[ES]}{[E_t]} = \frac{[S]}{K_m + [S]}$$

so that
$$v = \frac{V_{max} \cdot [S]}{K_m + [S]} \qquad (2.17)$$

Equation 2.17 is the Michaelis-Menten equation which describes the hyperbolic curve seen in Figure 2.1. Substrate concentration is normally expressed in moles per liter, and reaction velocity is usually expressed as moles of product per minute, assuming constant conditions of pH, temperature, and enzyme concentration. If enzyme concentration is increased, V_{max} will be found to increase in direct proportion. K_m is defined to be equal to the substrate concentration when $v = V_{max}/2$, and would therefore have units of moles/liter. Since $V_{max}/2$ is reached at the same substrate concentration independent of the size of V_{max}, K_m is independent of enzyme concentration. If v is expressd in units of moles of product/mole of enzyme·min., then V_{max} is equal to the turnover number. The units of substrate concentration and K_m are unaffected.

Equation 2.17 may be transformed to an equation of a straight line by taking the reciprocal of both sides and rearranging as follows:

$$\frac{1}{v} = \frac{K_m + [S]}{V_{max} \cdot [S]} \qquad (2.18)$$

$$\frac{1}{v} = \frac{K_m}{V_{max}} \cdot \frac{1}{[S]} + \frac{1}{V_{max}} \qquad (2.19)$$

From this equation it is seen that when the reciprocal of enzyme reaction velocity is plotted against the reciprocal of substrate concentration (1/v as a function of 1/[S]), the data should fall on a straight line with slope equal to K_m/V_{max}, and intercept on 1/v axis equal to $1/V_{max}$. It is through this method that these constants are usually obtained. For further information about the derivation of the Michaelis-Menten equation or its applications consult Lehninger (1975) or any other biochemistry text.

The general system encountered in enzyme catalysis is not unique. Basically it involves a processing machine (the enzyme molecule) which converts a resource material (the substrate) into another form (the product). The enzyme is effective as a processor only during that portion of time that it is in contact with the substrate. As the concentration of resource material or substrate is increased, the processor or enzyme becomes more effective until its maximum effectiveness is reached when it is saturated with the resource essentially 100 % of the time. A good analogy to this would be a machine tool in a factory. The tool can turn out product only as fast as it is provided with raw stock. However as the supply of raw stock is increased to the point where the machine is working essentially 100 % of the time, it has achieved its maximum efficiency. There is good reason to believe that the mathematical description of this process would be similar to the Michaelis-Menten equation.

<u>Exercise 2.1</u>**
Write a program which utilizes the Michaelis-Menten model (Equation 2.17) to simulate the initial reaction velocity of an enzyme at different substrate concentrations. Vary substrate from 0 to 40 mM. Use a value of 10 mM for K_m and 100 μmoles of product/min for V_{max}. Graph the resulting simulation data.

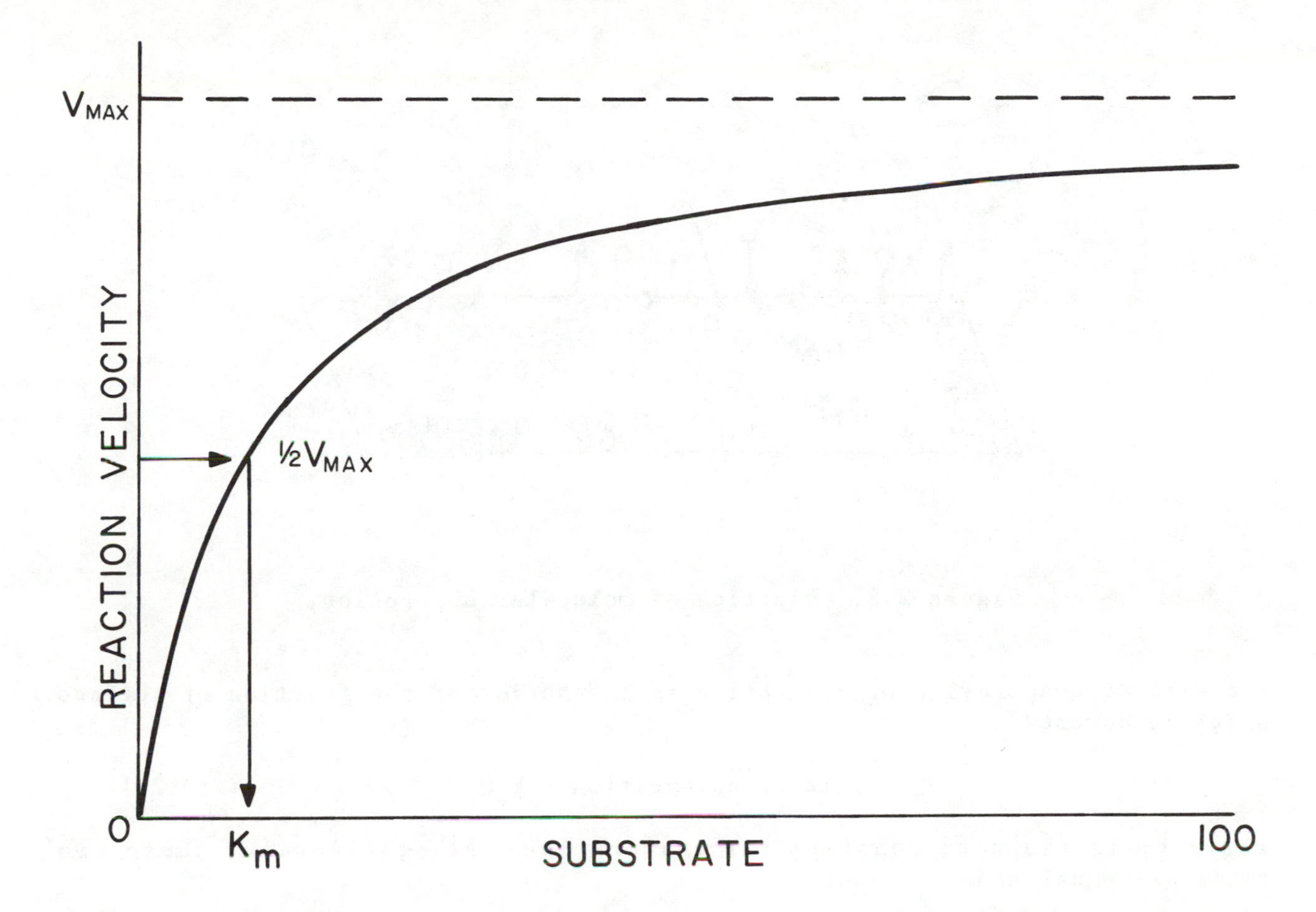

Figure 2.1. Typical curve of enzyme reaction rate as a function of substrate concentration.

2.2 Langmuir Absorption Model

Irving Langmuir (1916) used a sightly different approach to describe the general saturation process. He was concerned with the absorption of gas molecules on an absorbing surface. He postulated that in such a process, molecules would be absorbed in a layer only a single molecule thick, and thus there would be only a limited number of sites for absorption. This concept is diagrammed in Figure 2.2.

Langmuir also assumed an equilibrium condition in which the rate of absorption or condensation equaled the rate of desorption or evaporation. He further assumed that the rate of absorption or condensation was a function of both the gas pressure (or concentration) and the fraction of free surface available. The following equation was assumed to describe the condensation process:

$$\text{rate of condensation} = k_1(1 - \emptyset)P \qquad (2.20)$$

where k_1 is a rate constant, $\emptyset$ is the fraction of surface covered by absorbing molecules, so $1 - \emptyset$ is the fraction of free surface, and P is the gas pressure.

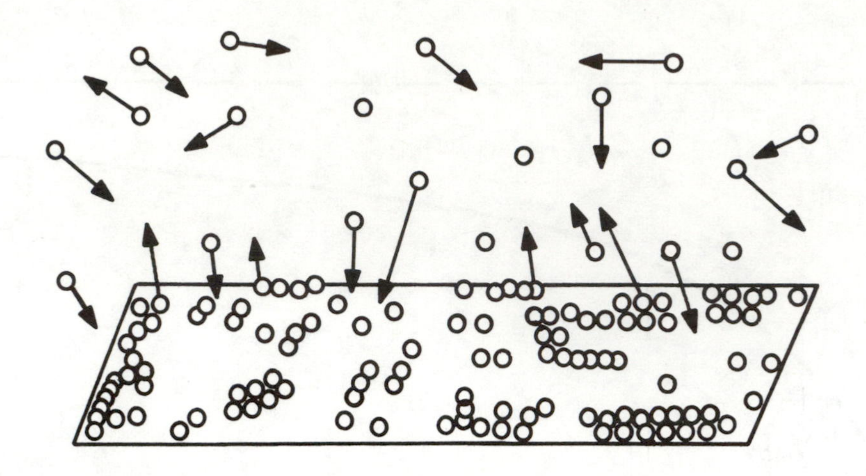

Figure 2.2. Kinetics of molecular absorption.

The rate of evaporation or desorption is a function of the fraction of surface which is covered

$$\text{rate of evaporation} = k_2\emptyset \tag{2.21}$$

where k_2 is the rate constant for evaporation. At equilibrium, these two rates are equal and

$$k_1(1 - \emptyset)P = k_2\emptyset \tag{2.22}$$

$$k_1P - k_1\emptyset P = k_2\emptyset \tag{2.23}$$

$$k_1P = \emptyset(k_2 + k_1P) \tag{2.24}$$

$$\emptyset = \frac{k_1P}{k_2 + k_1P} \tag{2.25}$$

$$\emptyset = \frac{P}{\frac{k_2}{k_1} + P} \tag{2.26}$$

Since $\emptyset = a/A_T$, where a is area covered, A_T is total area available, and $k_2/k_1 = K$, the equation may also take the form

$$a = \frac{A_T \cdot P}{K + P} \tag{2.27}$$

Note the striking similarity between this equation and that of the Michaelis-Menten model. The concept of saturating limited sites is fundamental to both the Langmuir model and the Michaelis-Menten model.

2.3 The Saturation or Satiation of Predators**

Let us now apply the same kind of assumptions to a system in which a limited number of predator units (sites) are being saturated with units of prey. The two processes involved are the capture and eating of prey by predators, and the assimilation of prey by predators. The overall process would be diagrammed as follows:

$$N + P \underset{a}{\overset{c}{\rightleftarrows}} P'$$

where N is prey density, P is the density of hungry predators, and P' is density of full or satiated predators. Assuming that the rate of capture is a function of both the density of prey and the density of unsatiated predators, we obtain the equation

$$\text{rate of capture} = N \cdot P \cdot c \qquad (2.30)$$

where c is a second order rate constant for the capture process. The rate of assimilation and return of predators to the hunting mode would be given by the equation

$$\text{rate of assimilation} = P' \cdot a \qquad (2.31)$$

where a is a first order rate constant for assimilation of prey by the satiated predator. It is assumed that this system reaches a steady-state condition in which the rate of assimilation equals the rate of capture.

Hence

$$P' \cdot a = N \cdot P \cdot c \qquad (2.32)$$

If we assume the total predator population, P_T, equals P plus P', then

$$P' \cdot a = N \cdot (P_T - P') \cdot c \qquad (2.33)$$

and collecting terms gives

$$P' \cdot a = N \cdot P_T \cdot c - N \cdot P' \cdot c$$

$$P' \cdot (a + Nc) = N \cdot P_T \cdot c$$

Solving for P'

$$P' = \frac{N \cdot P_T}{\frac{a}{c} + N} \qquad (2.34)$$

By combining the capture and assimilation constants into a single constant, k, we obtain

$$P' = \frac{P_T \cdot N}{k + N} \qquad (2.35)$$

This familiar hyperbolic equation shows that predators and prey follow a saturating relationship much like that which exists between enzymes and substrate molecules.

Now let us utilize this relationship to evaluate the rate of predation or capture under steady-state conditions. The change in prey population density due to predation in a given time interval, is described by the following restatement of Equation 2.30:

$$\frac{\Delta N_p}{\Delta t} = N \cdot P \cdot c \qquad (2.36)$$

where ΔN_p is the change in prey density due to predation.

Since P is equal to $P_T - P'$, and P' is described by the hyperbolic expression, we can write the following:

$$\frac{\Delta N_p}{\Delta t} = N \cdot (P_T - \frac{P_T \cdot N}{k + N}) \cdot c$$

$$= cN \cdot P_T(1 - \frac{N}{k + N})$$

$$\frac{\Delta N_p}{\Delta t} = c \cdot P_T \cdot (\frac{k \cdot N}{k + N}) = P_T \cdot (\frac{a \cdot N}{k + N}) \qquad (2.37)$$

This hyperbolic relationship was recognized empirically by Holling (1959) and was further investigated by Tanner (1975). It is a very useful model which allows us to calculate the loss due to predation as a function of total predator density and prey density as follows:

$$\Delta N_p = P_T \cdot (\frac{a \cdot N}{k + N}) \cdot \Delta t \qquad (2.38)$$

The applications of Equation 2.38 to the modeling of population dynamics are discussed further in Chapter 7.

<u>Exercise 2.2</u>**
Assume that a constant predator population of 15 wolves is preying on a moose population which fluctuates between 100 and 1000, both existing in a 250 square mile habitat. During a year when the average number of moose was 500, the pack of 15 wolves killed 125 moose. Other evidence suggested that the value for k was about 200 moose. Use this information and Equation 2.38 to plot the curve relating predation loss to moose population density. It is necessary to rearrange the equation to solve for constant a.

<u>Exercise 2.3</u>
Perform a dimension analysis on Equation 2.38 to determine the units of a and k. Determine the units of capture constant, c, from Equation 2.36. What is the limiting value for the rate of predation when N is very large compared with k (i.e., let k disappear from the denominator in Equation 2.37)? What is the limiting value for the rate of predation when N is very small compared with k?

2.4 The Henderson-Hasselbalch Model of Weak Acids**

The ionization of weak acids may be described by the following chemical equilibrium:

$$HA \rightleftharpoons H^+ + A^-$$

where HA is any ionizable weak acid, H^+ refers to hydrogen ions and A^- refers to the conjugate base of the weak acid. This gives rise to the following equilibrium constant expression:

$$K = \frac{[H^+][A^-]}{[HA]} \tag{2.41}$$

where K is the equilibrium constant, $[H^+]$, $[A^-]$, $[HA]$ refer to the concentrations of H^+, A^-, and HA at equilibrium.

Solving for $[H^+]$ gives

$$[H^+] = \frac{K[HA]}{[A^-]} \tag{2.42}$$

Taking logarithms (base 10) of both sides and then changing signs gives

$$-\log[H^+] = -\log K + -\log\left(\frac{[HA]}{[A^-]}\right) \tag{2.43}$$

By definition

$$-\log[H^+] = pH = pK + \log\left(\frac{[A^-]}{[HA]}\right) \tag{2.44}$$

The general form of this equation, is called the Henderson-Hasselbalch equation:

$$pH = pK + \log\left(\frac{[Base]}{[Acid]}\right) \tag{2.45}$$

For the purpose of simulating this equation it is valuable to express it in relation to a single variable, F_B, the fraction of the acid which is in the base form.

$$pH = pK + \log\left(\frac{(F_B)}{(1-F_B)}\right) \tag{2.46}$$

Exercise 2.4**
Write a program which will permit you to use Equation 2.46 to simulate the titration of a mono-basic weak acid. Assume you are working with acetic acid which has a pK value of 4.8. Observe the effect of varying the fraction of base from 0.01 to 0.99 in increments of 0.01. Graph the resulting data as a titration curve with pH as a function of the fraction of base.

2.5 The Effect of pH on Enzyme Activity

A model of the typical optimum pH curve for enzyme action may be based upon the following assumptions. The interaction between an enzyme and its substrate requires that the groups of the enzyme active site are in a particular state of ionization. For example, if a carboxylic acid group of the enzyme were involved, only one form of the group would make an effective bond with the substrate ($R-COO^-$ but not R-COOH, or vice versa). Once the enzyme-substrate compound is formed, only one particular ionized form undergoes a reaction to produce the product. Finally, the relative rate of formation of product is dependent on the fraction of total enzyme which is present as enzyme-substrate compound in the proper ionization state.

These concepts are embodied in the following overall equilbrium reaction discussed in Mahler and Cordes (1966).

$$\begin{array}{ccccc} & & A & & \\ & & + & & \\ EH_2 & \underset{}{\overset{K_1}{\rightleftharpoons}} & EH^- + H^+ & \overset{K_2}{\rightleftharpoons} & E^= + 2H^+ \\ & & \Big\downarrow\!\Big\uparrow K_a & & \\ AEH_2 & \overset{K_3}{\rightleftharpoons} & AEH^- + H^+ & \overset{K_4}{\rightleftharpoons} & AE^= + 2H^+ \\ & & \Big\downarrow k & & \\ & & P & & \end{array}$$

In this system EH_2, EH and E are various ionized forms of the enzyme, of which only EH is effective in binding with substrate, A, to make the enzyme-substrate compound. AEH_2, AEH, and AE are various ionized forms of the enzyme-substrate compound, of which only AEH is effective in breaking down to form product, P. K_1, K_2, K_3, and, K_4 are ionization constants, K_a is the disassociation constant for the enzyme-substrate compound, and k is the rate constant for product formation.

Since the rate of formation of the product is a function of AEH concentration, dP/dt = k[AEH], the maximum rate would be obtained if the enzyme were all in the form of AEH. The fraction of maximum rate would be obtained from calculating how much enzyme is in this form relative to the total. This value is called the saturation fraction.

$$\bar{Y} = \frac{[AEH]}{[E]+[EH]+[EH_2]+[AE]+[AEH]+[AEH_2]} \qquad (2.51)$$

where $\bar{Y}$ is the fraction of enzyme saturated as AEH, [AEH] is the concentration of AEH, and [E]+[EH]+... is the total concentration of enzyme in all possible states.

Each of these concentrations is related to EH in the following manner.

$$K_a = \frac{[A][EH]}{[AEH]} \quad , \quad [AEH] = \frac{[A][EH]}{K_a} = \alpha[EH] \qquad \text{where } \alpha = \frac{[A]}{K_a}$$

$$K_1 = \frac{[EH][H]}{[EH_2]} \quad , \quad [EH_2] = \frac{[EH][H]}{K_1}$$

$$K_2 = \frac{[E][H]}{[EH]} \quad , \quad [E] = \frac{K_2[EH]}{[H]}$$

$$K_3 = \frac{[AEH][H]}{[AEH_2]} \quad , \quad [AEH_2] = \frac{[AEH][H]}{K_3} = \frac{[A][EH][H]}{K_aK_3} = \frac{\alpha[EH][H]}{K_3}$$

$$K_4 = \frac{[AE][H]}{[AEH]} \quad , \quad [AE] = \frac{K_4[AEH]}{[H]} = \frac{K_4[A][EH]}{K_a[H]} = \frac{\alpha K_4[EH]}{[H]}$$

By substitution:

$$\bar{Y} = \frac{[EH]\alpha}{[EH](1 + \frac{K_2}{[H]} + \frac{[H]}{K_1} + \frac{\alpha K_4}{[H]} + \frac{\alpha[H]}{K_3} + \alpha)} \qquad (2.52)$$

$$= \frac{\alpha}{1 + \frac{K_2}{[H]} + \frac{[H]}{K_1} + \alpha(1 + \frac{K_4}{[H]} + \frac{[H]}{K_3})} \qquad (2.53)$$

Note that the dissociation constant of the enzyme substrate compound is a function of the two rate constants. As such, it is related to the Michaelis-Menten constant discussed in Section 2.1.

$$K_a = \frac{[A][EH]}{[AEH]} = \frac{k_2}{k_1}$$

While $$K_m = \frac{k_2 + k_3}{k_1}$$ See Equation 2.15

and thus $K_m \cong K_a$ when $k_3 << k_2$.

The ratio $[A]/K_a$ or α is therefore very close to being the ratio of substrate concentration to K_m. When $[A] = K_m$, ($\alpha \cong 1$), then $v \cong V_{max}/2$.

Exercise 2.5

Program the enzyme-pH model (Equation 2.53) to determine the saturation fraction (essentially the fraction of maximum velocity) for different pH values under two different concentrations of substrate relative to the dissociation constant of the enzyme-substrate compound ($\alpha = 1$ and $\alpha = 10$).

Test your simulation with the following input parameters:

$K_1 = 0.0001$ $\qquad K_3 = 0.0005$

$K_2 = 0.000001$ $\qquad K_4 = 0.000001$

Vary the pH from .1 to 14, then convert to $[H^+]$ and calculate $\bar{Y}$. Plot $\bar{Y}$ as a function of pH to express the enzyme activity curve as in Figure 2.3.

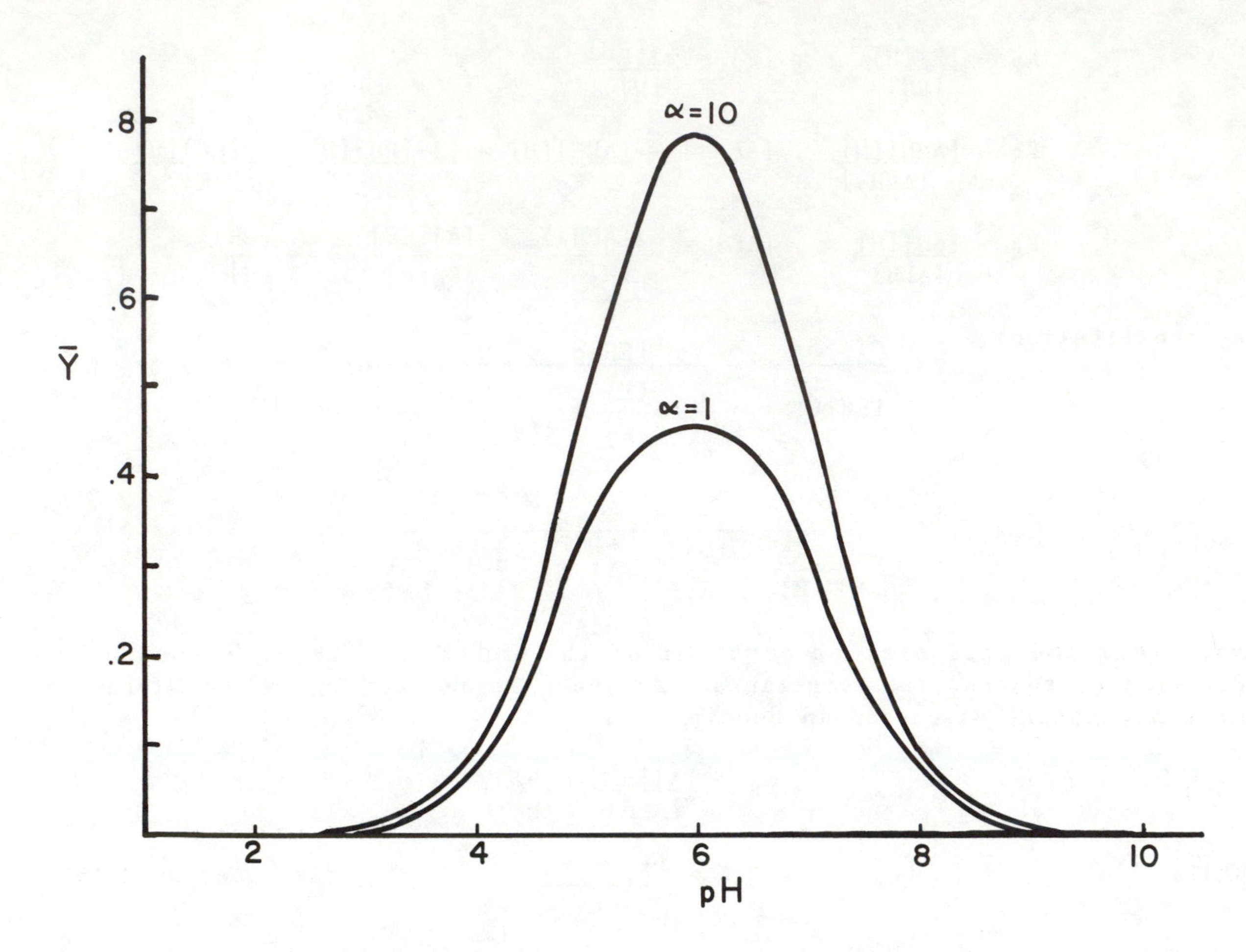

Figure 2.3. Simulated optimum pH curve based upon equations in Section 2.5. Assumed values are: $K_1 = 10^{-4}$, $K_2 = 10^{-8}$, $K_3 = 10^{-5}$, $K_4 = 10^{-7}$, $\alpha = 1$ and 10.

Exercise 2.6

The model of pH effect on enzyme activity presented above does not take into account the fact that the substrate may also be a weak acid, which must be in a particular ionized form to interact with the enzyme. Using the equations provided below, derive the equation for saturation fraction, $\bar{Y}$, in terms of α, K_1, K_2, K_3 and [H].

$$AH \underset{}{\overset{K_1}{\rightleftharpoons}} A^- + H^+$$

$$+$$

$$H_2E \overset{K_2}{\rightleftharpoons} HE^- + H^+ \overset{K_3}{\rightleftharpoons} E^= + 2H^+$$

$$HE^- \overset{K_a}{\rightleftharpoons} AEH \rightarrow P$$

$$K_1 = \frac{[A][H]}{[AH]} \qquad K_3 = \frac{[E][H]}{[HE]}$$

$$K_2 = \frac{[HE][H]}{[H_2E]} \qquad K_a = \frac{[HE][A]}{[AEH]}$$

$$\alpha = \frac{A\ \text{total}}{K_a} = \frac{[A] + [AH]}{K_a}$$

Exercise 2.7

Program the equation for $\bar{Y}$ obtained in Exercise 2.6. Use these constants.

$$K_1 = 10^{-6},\ K_2 = 10^{-5},\ K_3 = 10^{-7}, \text{ and } \alpha = 1 \text{ and } 10.$$

Vary the pH from .1 to 14 in increments of 0.1 pH units and generate a plot of $\bar{Y}$ as a function of pH.

2.6 A Concerted Model for Allosteric Enzymes**

One of the most important concerns of modern biochemistry is the control of biochemical reaction pathways. Some of this control is exerted on the synthesis of enzymes, and some of it is exerted on existing enzymes in such a way that their activity as catalysts is modulated by other metabolites. This latter form of control results when certain metabolites interact with the enzyme molecule at sites other than the catalytic site. The process is therefore called allosteric (other site) control, and the enzymes which respond to this type of control are called allosteric enzymes. A model describing allosteric enzymes was proposed by Monod, et al. (1965). This model and other models were recently reviewed by Newsholme and Start (1976) and by Savageau (1976).

Most allosteric enzymes have been found to consist of multiple subunits each having a catalytic site and one or more allosteric control sites. The typical curve relating substrate concentration to allosteric enzyme activity or reaction velocity has a sigmoid shape rather than the hyperbolic shape exhibited by other enzymes. This observation supports the assumption that there is strong cooperativity between enzyme subunits. The Monod model assumes that the subunits are identical, and held together in a symmetrical fashion. It also assumes that the enzyme exists in two conformations, one catalytically active, and the other inactive, and that an equilibrium exists between the active and inactive conformations. Because of the symmetry of the molecule, all subunits are assumed to shift from one conformation to the other in a concerted fashion. Hence this is called the concerted or symmetrical model.

Any ligands that bind with the active conformation tend to shift the equilibrium in a direction which increases the proportion of enzyme that is catalytically active. Both substrate molecules and positive modulators cause this affect. Negative modulators, on the other hand, bind with the inactive enzyme and have the effect of reducing the proportion of enzyme in the active form, thus reducing its overall catalytic activity. The following model is derived for the simple case of an allosteric enzyme composed of two subunits. Effects of modulators other than substrate are ignored.

We will define the active form as E and the inactive form as F. These exist in equilibrium as follows:

$$E \rightleftharpoons F$$

The equilibrium constant is designated as L, and is defined as follows:

$$L = \frac{[F]}{[E]} \tag{2.60}$$

The enzyme interacts with the substrate to form two species of ES, one for each of the two possible sites that may be occupied by the substrate molecule. The free enzyme may be defined by the following expressions where K is the dissociation constant for the enzyme-substrate compound, assumed to be the same for all species:

$$ES \rightleftarrows E + S \qquad K = [E][S]/[ES] \tag{2.61}$$

$$ES' \rightleftarrows E + S \qquad K = [E][S]/[ES'] \tag{2.62}$$

$$ES_2 \rightleftarrows ES + S \qquad K = [ES][S]/[ES_2] \tag{2.63}$$

To simplify, substrate concentration is expressed in terms of the dissociation constant as follows:

$$\alpha = \frac{[S]}{K} \tag{2.64}$$

The following equation defines the saturation fraction, $\bar{Y}$, in terms of the concentration of bound substrate and the total concentration of binding sites available:

$$\bar{Y} = \frac{[ES] + [ES'] + 2[ES_2]}{2([E] + [ES] + [ES'] + [ES_2] + [F])} \tag{2.65}$$

This equation is then substituted so that all terms are expressed relative to the concentration of free enzyme, [E]. The following equation results:

$$\bar{Y} = \frac{[E]\alpha + [E]\alpha + 2[E]\alpha^2}{2([E] + [E]\alpha + [E]\alpha + [E]\alpha^2 + [E]L)} = \frac{2[E](\alpha + \alpha^2)}{2[E](1 + 2\alpha + \alpha^2 + L)} \tag{2.66}$$

By cancelling terms and factoring, we obtain:

$$\bar{Y} = \frac{\alpha(1 + \alpha)}{L + (1 + \alpha)^2} \tag{2.67}$$

The general form describing this relationship for any number of subunits, n, is the following:

$$\bar{Y} = \frac{\alpha(1 + \alpha)^{n - 1}}{L + (1 + \alpha)^n} \tag{2.68}$$

Positive and negative modulators affect the equilibrium between the active and inactive forms of the enzyme and thus affect the L term of the equation as follows:

$$\bar{Y} = \frac{\alpha(1 + \alpha)^{n - 1}}{\frac{L(1 + \beta)^n}{(1 + \gamma)^n} + (1 + \alpha)^n} \tag{2.69}$$

In this equation $\beta = [I]/K_i$, expresses the concentration of the negative modulator or inhibitor, [I], in relation to the dissociation constant for the inhibitor-enzyme complex, K_i. In a similar way, $\gamma = [A]/K_a$, expresses the concentration of positive modulator or activator, [A], in relation to the dissociation constant for the activator-enzyme complex, K_a.

The inhibitor is assumed to bind to the enzyme only in the inactive form, and the activator is assumed to bind to the enzyme only in the active form. Equation 2.69 may be used to model both homotrophic and heterotrophic effects of most allosteric enzymes. It is very useful for modeling control of multi-enzyme systems.

Exercise 2.8**
Write a program using Equation 2.69 to simulate an allosteric enzyme with four subunits and four active sites. Determine the values for saturation fraction when α is varied from 0.01 to 20.01 in increments of 0.5. Plot simulation data resulting from each of the following assumed conditions:

	L	β	γ
a)	1000	0	0
b)	1000	2	0
c)	1000	0	100

Note the characteristic sigmoid shape of the curve indicating homotrophic effect in the absence of allosteric activator or inhibitor. The curves for b and c should show marked heterotrophic effects.

2.7 Selection Within the Hardy-Weinberg Genetic Equilibrium

According to the Hardy-Weinberg principle, an equilibrium is attained in a population gene pool any time random mating takes place. This allows one to calculate the rate at which genes with zero fitness would be lost from the population gene pool. Further discussion of the Hardy-Weinberg principle and various related population genetics models is reserved for Chapter 9. The following is offered as an example of an analytical model dealing with genetic equilibrium.

If the proportion of recessive genes is q, and mating is random, there will be q^2 recessive homozygotes. If these are eliminated through selection, the only source of recessive offspring would be the heterozygotes. In the next generation, the proportion of recessive genes in the population would be defined by the equation

$$q_1 = \frac{2q_o(1 - q_o)}{2q_o(1 - q_o) + (1 - q_o)^2} \cdot \frac{1}{2} = \frac{q_o}{1 + q_o} \qquad (2.71)$$

where q_o is the original proportion of recessive genes. It can be shown that for n generations, the equation becomes

$$q_n = \frac{q_o}{1 + nq_o} \qquad (2.72)$$

It is therefore possible to determine the recessive gene frequency after any number of generations of total selection against homozygous recessives.

Exercise 2.9
Assume that a deleterious recessive gene is in the population at a frequency of 0.10 (10 %). Use Equation 2.72 to simulate the process of selection through 40 generations. Graph the gene frequency as a function of time.

Conclusion

Equilibrium or steady-state models can only be used to describe phenomena which are known to exist at or near equilibrium or steady-state conditions. Because of this restriction, these models are not as broadly applicable as those involving differential equations. Typically, they would be employed where the establishment of equilibrium or steady-state is rapid relative to the other processes which make up a simulation model. In this way, a sub-model based upon the equilibrium or steady-state approach might be combined with sub-models based upon the differential equations approach to form the overall simulation model. Applications of this type will appear in subsequent chapters.

Notice that many of the derivations described in this chapter depend on the use of rate equations. In the equilibrium or steady-state approach the differentials are eliminated when the rate of formation is set equal to the rate of breakdown. The technique of formulating rate equations is thus fundamental to the development of many mathematical models. More examples will be presented in subsequent chapters.

CHAPTER 3

FITTING MODEL EQUATIONS TO EXPERIMENTAL DATA

In the previous two chapters it was shown that model equations of various forms may be derived analytically from theoretical considerations. However, it should be noted that these techniques only determine the form of the equation. Specific values for coefficients and exponents which are necessary to implement these model equations must be obtained through the process of curve fitting. This chapter will describe this technique.

Information on the general process of curve fitting is presented in many good statistics books. Hoerl (1964) describes various curves and equation types, and shows how they may be fitted graphically. The use of least squares methods for curve fitting is described in simple terms by Croxton and Cowden (1955), Spiegel (1961), and Zar (1974). Watt (1968) discusses various curve fitting methods as they apply to biological modeling. Daniel and Wood (1971) have an excellent presentation on the criterion of goodness of fit when comparing several model equations for a given set of data. The present chapter attempts to synthesize ideas and techniques from these sources into a general procedure that may be employed for most modeling situations.

Before we begin a discussion of curve fitting techniques, it is necessary to make a clearer distinction between theoretical and empirical model equations. Theoretical equations are those which have been derived from analytical considerations or at least have the potential for analytical derivation, perhaps only by analogy to known processes. Empirical equations, on the other hand, are those which have neither direct nor implied theoretical origin. This situation often occurs when there is a complex cause and effect relationship between the variables which the equation is intended to describe. Suppose for example, one needed an equation describing the density of water at different temperatures as a component of a lake stratification model. There is, of course, a theoretical basis for this temperature effect. However, it is neither pertinent nor essential to the lake stratification model. In such

James D. Spain, BASIC Microcomputer Models in Biology

ISBN 0-201-10678-7

cases, an empirical model equation obtained directly from experimental data may be used to provide the needed information. The most commonly employed empirical equations are in the polynomial class. A polynomial that will be used frequently in this text is the third degree polynomial or cubic equation.

$$y = A + Bx + Cx^2 + Dx^3$$

The curve fitting procedure would be employed to determine the best values for the coefficients, A, B, C, and D. From a mathematical standpoint theoretical equations and empirical equations are equally useful in describing the function of a system component. It is generally true, however, that a modeler would prefer to employ equations having a theoretical origin, because they usually contribute more to the understanding of the system being studied.

3.1 The Linear Transformation Approach to Curve Fitting

One general technique for fitting theoretical equations to experimental data takes advantage of the fact that many simple equations of this type may be rearranged to a linear form. The experimental data may then be transformed and fitted to the straight line by the linear least squares technique. For example, suppose we wished to model growth of biomass based on the following hypothetical data:

Time (t)	0	1	2	3	4
Biomass (y)	9.9	32	89	271	808

The plotted data appear to follow the exponential growth curve which is described by the equation

$$y = Ae^{kt}$$

In order to obtain the best values for the coefficient A, and the exponent k, we would convert the equation to the linear form by taking the natural log of both sides, which would yield

$$\ln(y) = \ln(A) + kt$$

Now, if y values from the data are transformed to ln(y) and plotted as a function of t, then one should obtain a line with slope k and intercept ln(A) (the equation for a straight line is $y = Mx + b$, where b and M are the intercept and slope). This transformation is often employed to determine growth rates of microorganisms. In the present example, the transformed data become

t	0	1	2	3	4
ln(y)	2.3	3.4	4.5	5.6	6.7

When these data are plotted we obtain the straight line shown in Figure 3.1.

We may now go back and use the derived values of A and k in the exponential form of the equation to get the values for biomass predicted by the model for each value of t.

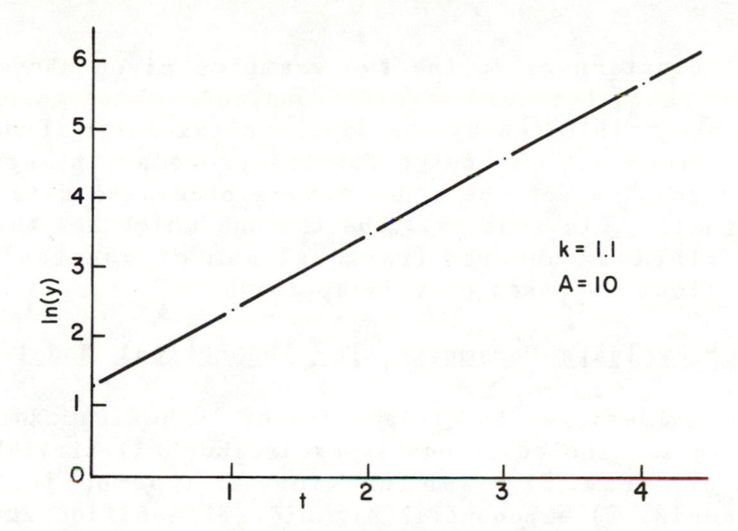

Figure 3.1. Graph of hypothetical growth data

Time	0	1	2	3	4
Actual y	10	32	89	271	808
Predicted y	10	30	90	270	810
Deviation	.1	-2	1	-1	+2

Since the deviations or residuals are small and uniformly distributed, the equation with fitted values of A and k appears to predict the values for growth within the range of data. It therefore appears to be a good model for the growth of these organisms, although there are several other statistics we should examine before accepting it as the best fit.

A similar transformation procedure is commonly employed in biochemistry when enzyme kinetics data are used to determine K_m and V_{max} in the Michaelis-Menten equation:

$$v = \frac{V_{max} \cdot [S]}{K_m + [S]}$$

One linear form of the Michaelis-Menten equation is the following:

$$\frac{1}{v} = \frac{K_m}{V_{max}} \cdot \frac{1}{[S]} + \frac{1}{V_{max}}$$

To obtain a linear plot with enzyme data, v is transformed to 1/v and [S] to 1/[S]. The data for most enzymes plot on a straight line with intercept $1/V_{max}$ and slope K_m/V_{max}.

The procedure may sound complicated, but, in fact, it is rather simple since many computer programs have been written to assist in performing the curve fitting process. A program which utilizes the linear least squares technique was developed specifically for this text and is provided in Appendix 2. This CURFIT program is designed to be run on either the Apple II Plus or Radio Shack TRS-80 microcomputer.

The technique is relatively simple when we know the system is described by a single theoretical equation as in the two examples given above. It is often the case, however, that there are several equations which result in curves of a form similar to that exhibited by the experimental data. Under these conditions, one should carry out the curve fitting procedure using each potential model equation and then select the model having predicted data which deviates least from the actual. This model will be the one which has the minimum value for the sum of deviations squared (residual sum of squares). The general procedure which follows is based on this approach.

3.2 Using CURFIT to Evaluate Parameters for Theoretical Model Equations.

The CURFIT program is designed to fit any one of 10 basic model equations to a set of x,y data points. The equations types include 1) straight line, 2) exponential growth or decay, 3) power function, 4) hyperbolic, 5) exponential saturation, 6) sigmoid, 7) exponential sigmoid, 8) modified inverse, 9) modified power function and 10) maxima function. The computer retains a set of key statistics on each fit so that the best fit may be selected by examination of the summary data. If the microcomputer is provided with a printer, you will be able to obtain hard-copy output of the summary data, the statistics on the best-fitting model, and a curve showing the location of data points in relation to the regression line (Apple only).

The following step by step procedure assumes that you are using the CURFIT program in conjunction with either a TRS-80 or Apple II microcomputer.

Data entry

CURFIT accepts up to 100 (x,y) data points. In order to make sure all points have been entered, the program asks you to count the total and enter this number first. Because of the nature of the transform process employed, neither zero nor negative numbers may be entered as data. Data containing such numbers should be normalized to positive real numbers by adding a constant to each value. If you have a zero data point that you think essential, enter 0.00001 instead. However, this procedure can cause problems (see Section 3.4). If you make an error on data entry, enter either E or Ø. This will cause the program to cycle back so that you can enter the correct value. Enter the x value first then the y value, separating them by a comma. Press RETURN (or ENTER) after typing in each entry.

Approximating curve

After all data have been entered, the CURFIT program plots the points on the video display. This data plot will be referred to as an approximating curve or scatter plot. The general shape of the approximating curve should suggest some equations which might fit the data. Match the form of the approximating curve to one or more of the standard curve types given in Section 3.3. If the curve type is not obvious, use the key to common curve types which has been provided in Appendix 5. After you have been through the key a couple of times, further use will probably be unnecessary. The key is based on the following characteristics of the approximating curve: linearity, presence of maxima or minima, direction of slope, and presence or absence of inflection points.

Selecting equations for fitting

The following equations may be fitted to the data:

Number	Descriptive Name	Equation
1.	Straight line	$y = Ax + B$
2.	Exponential growth or decay	$y = Ae^{nx}$
3.	Power function	$y = Ax^n$
4.	Hyperbolic	$y = Ax/(B + x)$
5.	Exponential saturation	$y = A(1 - e^{nx})$
6.	Sigmoid (forward or reverse)	$y = A/(1 + Bx^n)$
7.	Exponential sigmoid	$y = A/(1+ Be^{nx})$
8.	Modified inverse	$y = A/(B + x)$
9.	Modified power function or cutoff	$y = Ax^n + B$
10.	Maxima function	$y = Ax \cdot e^{nx}$

Equations are selected by entering the equation number. For some equations it will be necessary to estimate either A or B. These parameters are usually an asymptote or intercept of the approximating curve. Simply make an educated guess and enter the value. The best value of A or B may be obtained by repeating the curve-fitting process several times and accepting the constant which results in the highest value for the F statistic or lowest residual mean square.

Transformation of y and/or x

The CURFIT program converts the non-linear equation being fitted into a linear form by making the appropriate transformation of the data. This allows it to be fitted by the linear least squares process. See theory above for a more complete explanation. Linear forms of each equation and the required transforms are listed in Table 3.1.

Least squares fit of transformed data

The sums of x, y, xy, x^2 and y^2 are calculated in order to obtain the slope and intercept, which are used to define A, B, or N. The fitted values for these parameters are displayed, along with the model equation.

Calculation of model y and comparison with real data

The values for x and y are read from an array, and the value of x is used to calculate the predicted value of $y(\hat{y})$ using the coefficients and exponents derived above. The program prints out x, y, $\hat{y}$, and the deviation between y and $\hat{y}$ (residual). These values are summed and squared to provide data needed to calculate various statistics which follow.

Sum of deviations squared ($SS_{residual}$)

The residual sum of squares is defined by Zar (1974) as follows:

$$SS_{residual} = \Sigma(y_i - \hat{y}_i)^2$$

This is calculated using the direct approach as residuals are being determined.

Variance about the fitted line (residual mean square)

Variance or residual mean square is defined for the straight line by Zar (1974) as follows:

$$MS_{residual} = \frac{SS_{residual}}{DF_{residual}} = \frac{SS_{residual}}{n - 2}$$

Daniel and Wood (1971) define the residual mean square as

$$MS_{residual} = \frac{SS_{residual}}{n - K - 1}$$

where K is defined as the number of "constants fitted" or "independent variables" in the fitted equation. As such, K does not include the intercept, as it is not considered an independent variable. The computer program calculates the $MS_{residual}$ using the following equation:

$$MS_{residual} = \frac{SS_{residual}}{n - P}$$

where P is the number of parameters (coefficients and exponents) obtained by fitting. Therefore, it is consistent with both of the above sources. Actually, the least squares program can only fit two parameters, one by the slope, and one by the intercept. Additional parameters, such as A or B, when fixed by the user, are thus not included in calculating the degrees of freedom.

Standard error of the regression estimate

This is also called residual root mean square and is defined as follows:

$$SE_{regression} = RRMS = \sqrt{MS_{residual}}$$

The F-statistic

This is defined by Daniel and Wood as follows:

$$F = \frac{\dfrac{SS_{regression}}{K}}{\dfrac{SS_{residual}}{n - K - 1}} = \frac{SS_{regression}}{SS_{residual}} \cdot \frac{n - K - 1}{K}$$

It is calculated using the following equation:

$$F = \frac{SS_{regression}}{SS_{residual}} \cdot \frac{n - P}{P - 1}$$

Plot of the residuals

The user has the option of obtaining a plot of the residuals in order to evaluate the goodness of fit. See Section 3.4 for further information concerning the interpretation of the residual plot.

Plot of data compared to model

The user has the option of examining a regression plot which compares the model to the original data. This allows one to visually inspect the goodness of fit.

Examination of other models.
After all the statistics have been printed for the specific model fitted, there is an opportunity to fit other equations to the same data set. Simply respond to the prompt.

Summary data
After you have examined several model equations, a summary of the data is displayed so that the relevant statistics may be compared.

It is generally held that the best fitting equation will have the minimum residual mean square. This statistic is better than the residual sum of squares since it takes into account the different number of degrees of freedom associated with different model equations. Other statistics which may be employed to evaluate goodness of fit include the F value and the standard error of the regression estimate.

The coefficient of determination of the model equation, R^2, compares the variance of the predicted y values with the variance of the actual y values. Therefore, it is not a direct function of fit when non-linear equations are involved. For non-linear equations, higher R^2 will often result from poorer fitting equations. In certain instances, R^2 values above 1.0 are obtained when the variance of data predicted by the non-linear model actually exceeds that of original data. For this reason, the coefficient of determination for the model equation is of little value in deciding the best fitting equation. The R^2 for the straight-line transform also has questionable value, especially when x is not distributed uniformly (See Section 3.4).

Hard copy
When the you have completed the curve fitting process, the program allows you to obtain a print out of the summary data and the complete statistics on the best fitting model equation.

Retaining data for use on POLYFIT
When you have obtained the best fit to the data, CURFIT is set up to store your data on the disk so that it may be called up by a companion program called POLYFIT. Simply respond to the request at the end of the program and save the data by using a file name which would be unique to your data.

Sample data for testing CURFIT
The following sets of data have been synthesized from standard equation types. Fit at least two theoretical equations to each. Decide on the best fit based on the residual mean square or F statistic.

Exercise 3.1

x	1	2	5	10	15	20	25	30	35	40	45	50
y	5	9	20	33	43	50	56	60	64	67	69	71

Exercise 3.2

x	1	2	5	10	15	20	25	30	35	40	45	50
y	1	4	20	50	69	80	86	90	92	94	95	96

Exercise 3.3

x	0	1	2	5	10	15	20	25	30	35	40	45	50
y	100	91	83	67	50	40	33	29	25	22	20	18	17

Exercise 3.4

x	0	1	2	5	10	15	20	25	30	35	40	45	50
y	110	100	92	71	47	32	24	18	15	13	12	11	10.7

3.3 Standard Theoretical Equations

Table 3.1 describes the 10 standard types of theoretical equations that are provided in the CURFIT program. Each equation has been presented in both the non-linear standard form and the linear form which is employed in the program for the least squares fit to the straight line. Brackets in linear equations indicate transformations used for both x and y. The equation parameters are defined in terms of the slope and intercept of the fitted line. The equations are listed according to the numbers used in the program, generally from the simplest to the more complex.

Figures 3.2 and 3.3 show some of the characteristic curves resulting from the standard curve types. Parameter values have been provided to assist you in understanding the role they play in determining curve shape.

3.4 Some Pitfalls of the Transformation Process

When x is transformed to either 1/x or ln(x) there is a tendency to change the distribution of x values so that transforms of low values of x become relatively far apart, and high values become clustered. Daniel and Wood (1971) point out that all data points weigh equally in determining the mean of y, while the slope is influenced more heavily by y values associated with extreme values of x. Also, a cluster of points has little more influence than a single point. In Figure 3.4, uniformly distributed values of x have been transformed to 1/x. The result is that points are clustered at low values of 1/x (high values of x), causing high values of 1/x to have great influence in determining the slope of the line. It is virtually a certainty that the least squares line will pass through the extreme point! It is ironic that this point comes from the lowest value of x, and in the hyperbolic curve, the lowest value of y. Because this value is a small number, and because it is positioned on the steepest part of the hyperbolic curve, it would be expected to have the greatest percent error!

Table 3.1. Standard equation types

Equation				Parameters defined by slope and intercept		
No.	Name	Standard form	Linear form	A	B	n
1.	Straight line	$y = Ax + B$	$[y] = A[x] + B$	S	I	
2.	Exponential	$y = Ae^{nx}$	$[\ln y] = \ln A + n[x]$	Exp(I)		S
3.	Power Function	$y = Ax^n$	$[\ln y] = \ln A + n[\ln x]$	Exp(I)		S
4.	Hyperbola	$y = Ax/(B + x)$	$\left[\frac{x}{y}\right] = \frac{B}{A} + \frac{1}{A}[x]$	1/S	I/S	
5.	Exponential Saturation	$y = A(1 - e^{nx})$	$[\ln(A - y)] = \ln A + n[x]$	Exp(I)		S
6.	Sigmoid	$y = A/(1 + Bx^n)$	$\left[\ln\left(\frac{A}{y} - 1\right)\right] = \ln B + n[\ln x]$	estimate	Exp(I)	S
7.	Exponential sigmoid	$y = A/(1 + Be^{nx})$	$\left[\ln\left(\frac{A}{y} - 1\right)\right] = \ln B + n[x]$	estimate	Exp(I)	S
8.	Modified Inverse	$y = A/(B + x)$	$\left[\frac{1}{y}\right] = \frac{B}{A} + \frac{1}{A}[x]$	1/S	I/S	
9.	Modified Power Function	$y = Ax^n + B$	$[\ln(y - B)] = \ln A + n[\ln x]$	Exp(I)	estimate	S
10.	Maxima Function	$y = Ax \cdot e^{nx}$	$\left[\ln\left(\frac{y}{x}\right)\right] = \ln A + n[x]$	Exp(I)		S

S = slope of fitted line, I = intercept of fitted line

Transformed values of x and/or y are enclosed in brackets in the linear form of the equation.

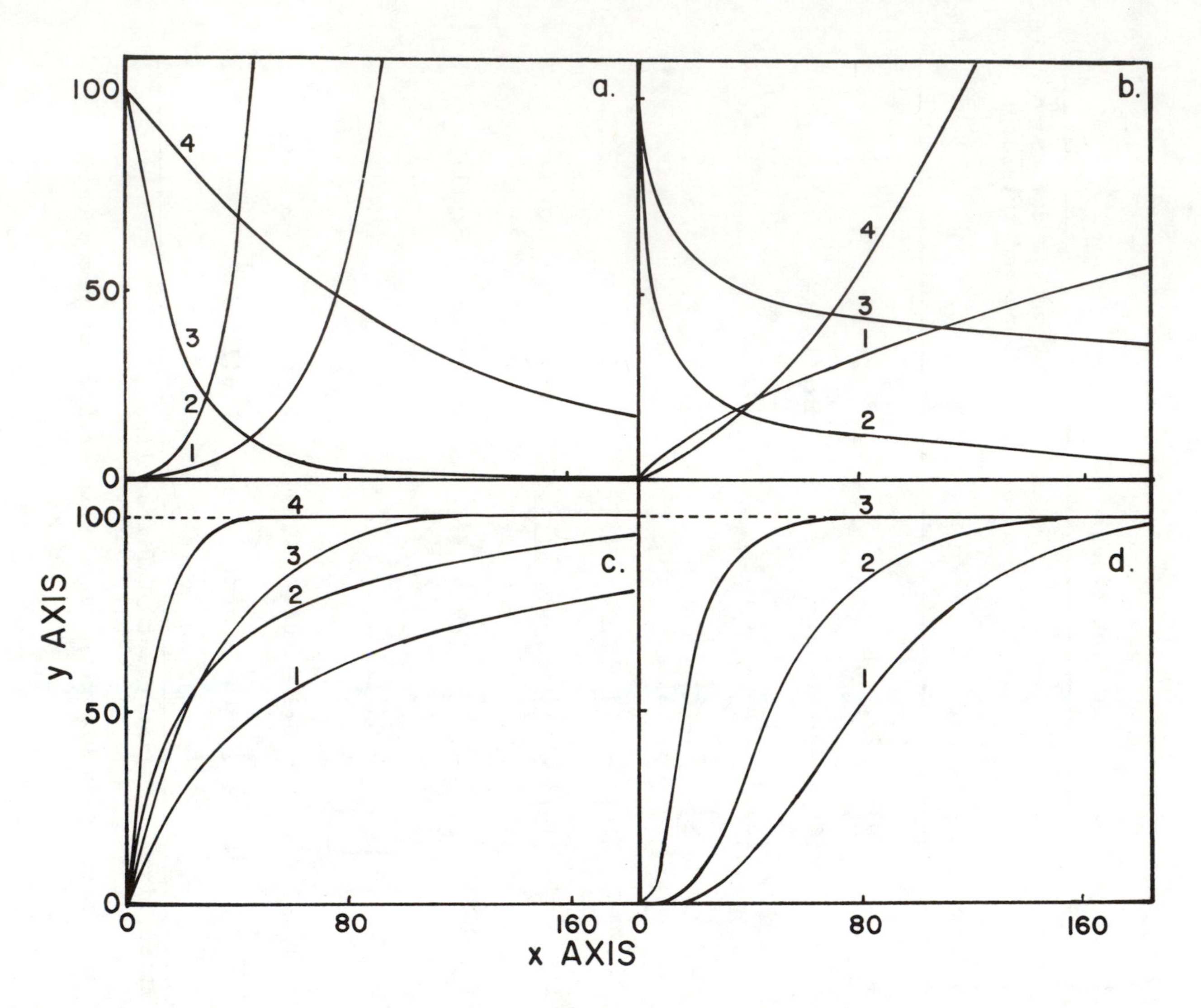

Figure 3.2. Curves resulting from Standard Equation Types, I. The equations illustrated in this figure include

a. Exponential equation:

1) A = 1, n = 0.05 2) A = 1, n = 0.1
3) A = 100, n = -0.05 4) A = 100, n = -0.01

b. Power function equation:

1) A = 1.5, n = 0.7 2) A = 100, n = -0.5
3) A = 100, n = -0.2 4) A = 0.7, n = 1.5

c. Saturation equations:

1) Hyperbolic, A = 100, B = 50 2) Hyperbolic, A = 100, B = 20
3) Exponential, A = 100, n = -0.1 4) Exponential, A = 100, n = -0.03

d. Sigmoid equation:

1) A = 100, n = -3, B = 500,000 2) A = 100, n = -3, B = 100,000
3) A = 100, n = -3, B = 5000

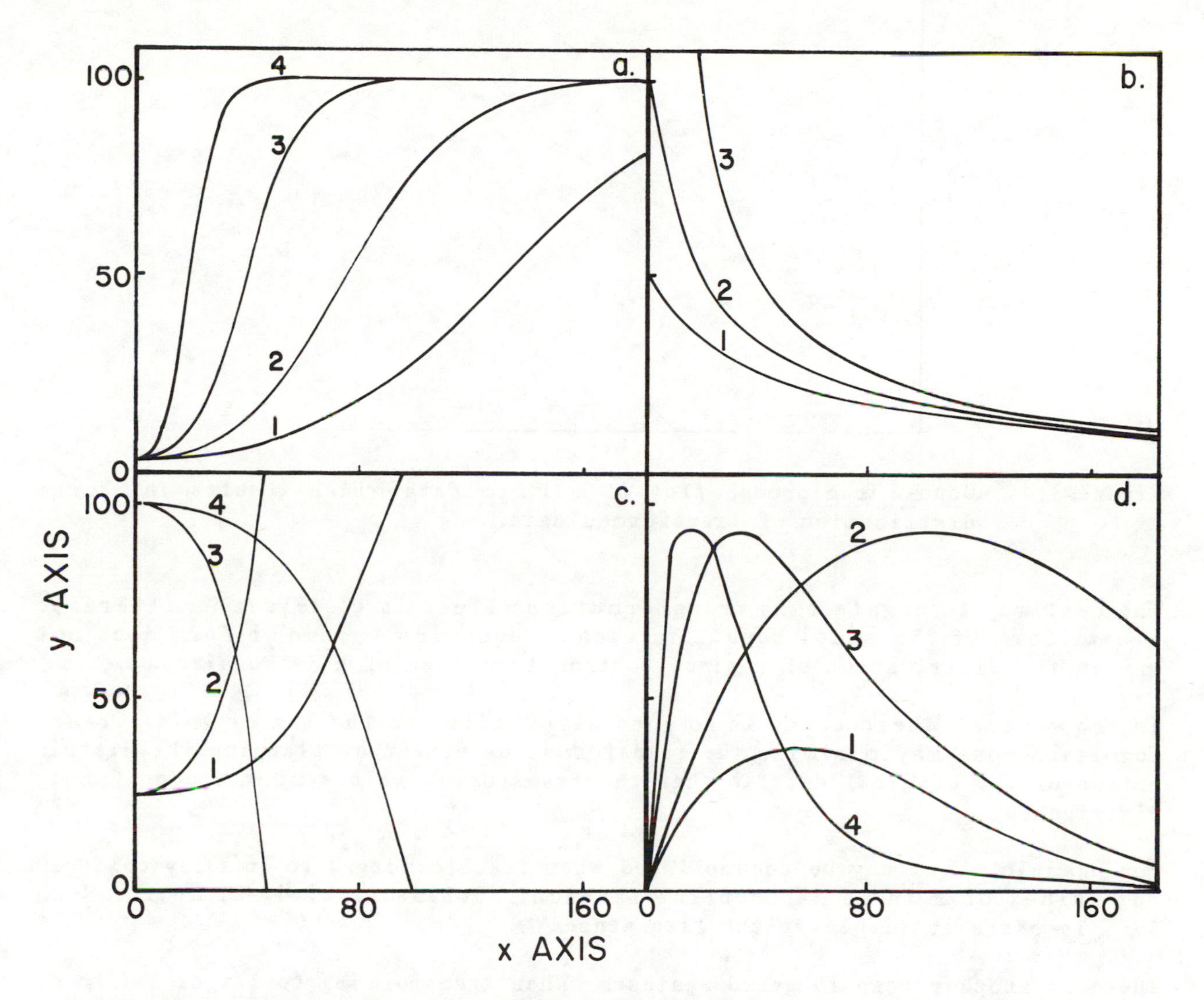

Figure 3.3. Curves resulting from Standard Equation Types, II. The equations illustrated in this figure include

a. Exponential sigmoid equation:
1) A = 100, B = 50, n = -0.03 2) A = 100, B = 50, n = -0.05
3) A = 100, B = 50, n = -0.1 4) A = 100, B = 50, n = -0.2

b. Modified inverse equation:
1) A = 2000, B = 40 2) A = 2000, B = 20
3) A = 2000, B = 0

c. Modified power function equation:
1) A = 0.00001, B = 0.25, n = 3 2) A = 0.0001, B = 0.25, n = 3
3) A = -0.0001, B = 100, n = 3 4) A = -0.00001, B = 100, n = 3

d. Maxima function equation:
1) A = 2, n = -0.02 2) A = 2.5, n = -0.01
3) A = 7.5, n = -0.03 4) A = 15, n = -0.06

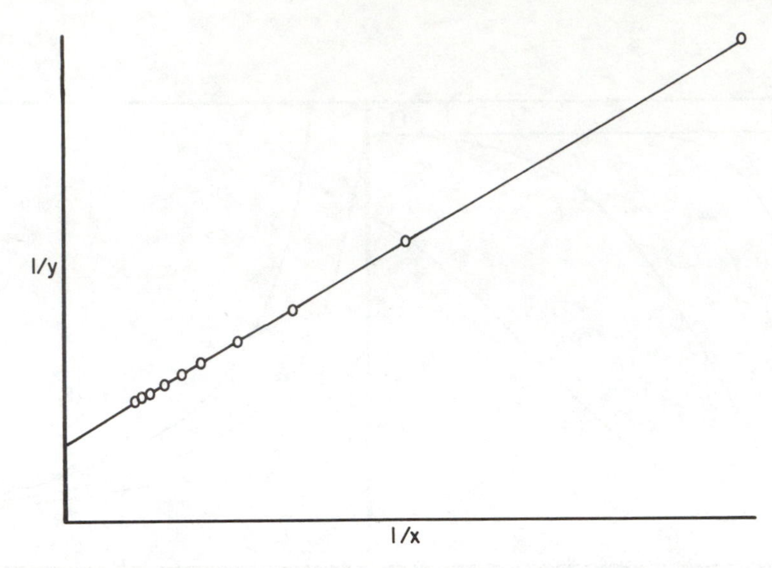

Figure 3.4. Double reciprocal plot of uniform data which results in skewed distribution of transformed data.

The best way to negate this transformation effect is to select an alternate linear form of the model equation, such as Equation 4. Such a form does not affect the distribution of x since no transformation of x is required.

In those cases where there is no good alternative to the 1/x or ln(x) transformation, one may minimize the transformation effect by planning the distribution of the original data so that the transformed values of x are uniformly distributed.

For example, this may be accomplished when transforming x to ln(x) by collecting values of x in an exponential fashion, such as 1, 2, 4, 8, 16,.... Such data is often available in the literature.

There is another trap to guard against. When transforming to 1/x or ln(x), do not attempt to use x values close to zero! Of course, x = 0 is impossible to transform either to ln(x) or 1/x. However, one might be tempted to approximate this by substituting a point close to zero such as x = .00001. This, on transformation to 1/x, becomes 100,000, and to ln(x) becomes −11.5. Either of these values would tend to become the major influence in determining the slope of the transformed line. Unless you need an accurate fit in the region close to zero, it is better to drop this point from the data.

There are two ways to recognize errors of the type described above. First, you may check the linearity of the transformed data by plotting points on linear graph paper. Some curve fitting programs provide this type of plot as part of the output. In CURFIT, you may obtain information about linearity by examining the pattern of the residuals when plotted as a function of x. If these residuals are not uniformly distributed over the range, nonlinearity is indicated. If a plot of residuals as a function of x tends to be wedge shaped, it indicates strong influence by one of the extreme points as described above. If this plot is U shaped, it provides an indication that transformed data are not linear.

3.5 Curve Types Resulting from Compound Equations

Curves which show a maximum or a minimum require special treatment. It is generally recognized that if such a case exists, there are two or more forces competing with each other. The resulting curve may be described by a compound equation resulting from multiplying or adding two simpler equations. For example, if a curve shows a maximum, with the initial portion of the curve showing a concave-up appearance, and the portion following the maximum shows a concave-down appearance, this could be the result of two processes described by Equations 6 and 9. The product of these two equations results in the following equations

$$y = \frac{A}{1 + Bx^n} \cdot (Cx^m + D) \tag{3.51}$$

This equation gives a curve of the type shown in Figure 3.5.

Equation 10 in the CURFIT program is also a compound type equation. The form was

$$y = Ax \cdot e^{nx}$$

Notice that it is the product of the straight line equation and the exponential growth or decay equation. This equation is used to model the effects of light intensity on photosynthesis in Section 10.5.

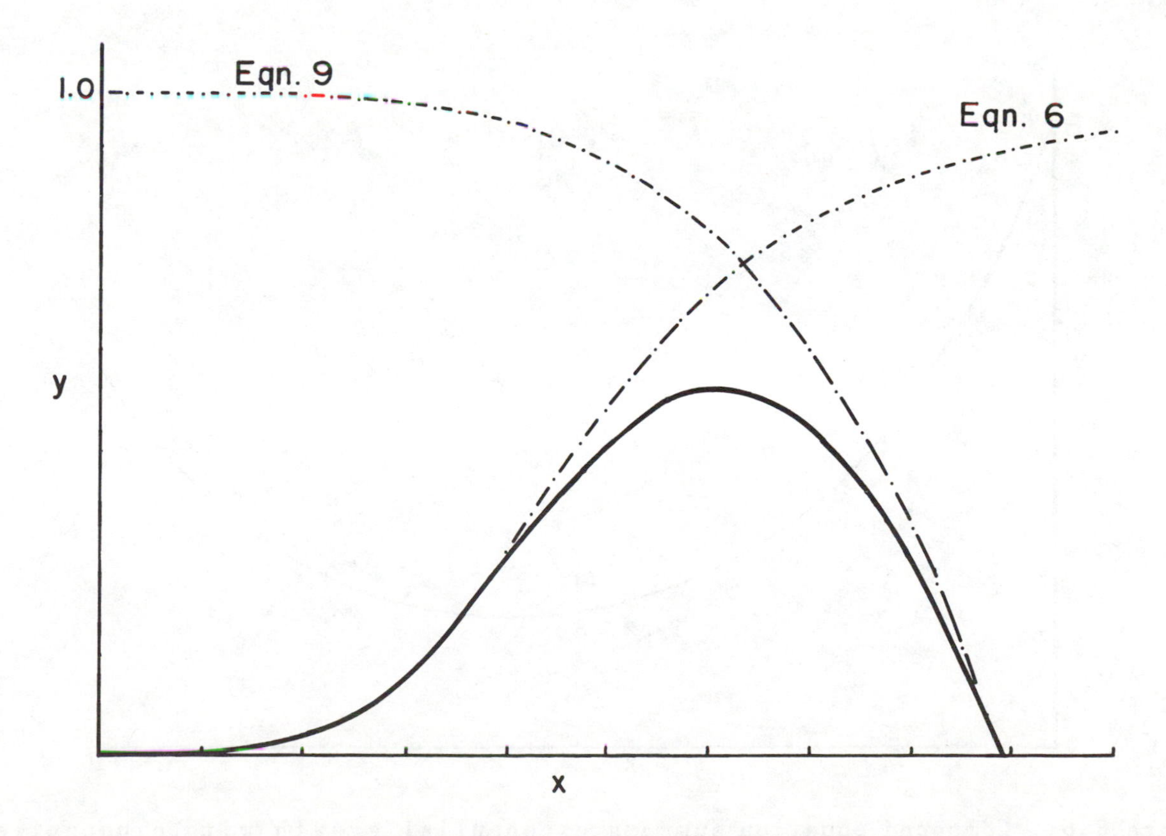

Figure 3.5. Compound curve described by Equation 3.51.

A slightly different functional relationship would result if the compound curve was the product of Equations 2 and 7 as follows:

$$y = e^{nx} \cdot (1 - Ae^{mx}) \qquad (3.52)$$

In this case, m must be greater than n, otherwise no maximum occurs. The equation above provides a reasonable model of the effect of temperature on enzyme action and other types of biological activity. See Section 11.5 for further details.

In curves which show a minimum, the compound equation results from summing two other equations, such as exponential growth and exponential decay. The resulting equation is

$$y = Be^{-nx} + Ae^{mx} \qquad (3.53)$$

This equation gives the curve shown in Figure 3.6.

Most equations of the compound type would have to be fitted in some iterative fashion. Equation 10 is an exception because of its relative simplicity.

3.6 Fitting Polynomials to Experimental Data

Fitting empirical equations to experimental data is generally easier to accomplish than fitting theoretical equations because there are fewer decisions to make.

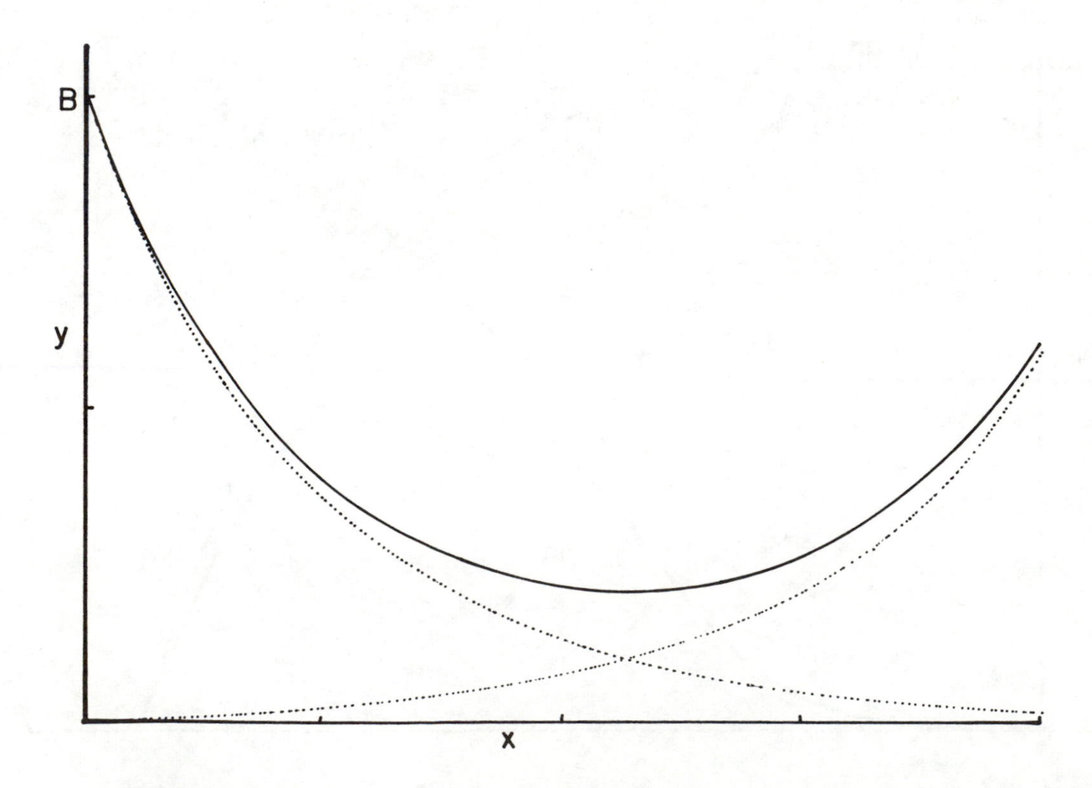

Figure 3.6. Compound equation summing exponential growth with exponential decay.

However, from a theoretical standpoint, polynomials and other empirical equations have little significance, and for this reason are to be used sparingly. If you jump immediately to an empirical equation without at least trying to employ a theoretical one, you might be passing up an opportunity to better understand a component of the system. An analytical equation type often provides strong insight into the theoretical basis for the relationship between two variables. However, assuming you have satisfied yourself that an empirical equation is called for, there is a whole family of polynomials to choose from. They include linear regression (1st order), quadratic (2nd order), cubic (3rd order), quartic (4th order), quintic (5th order), and so on. Ideally, you should fit these equations by polynomial regression analysis, starting with the lowest order and stop adding terms when no significant improvement in fit is obtained. See Zar (1974) for an example of such a fitting process.

In Appendix 3 of this text you have been provided with a curve fitting program called POLYFIT, which allows you to fit up to five different polynomial equations to a set of x,y data. These include straight line, quadratic, cubic, quartic, and quintic equations. The computer retains the key statistics on each fit so that the best fit may be selected by examining the summary data. After analysis is complete, the summary data and the full statistics on the best fitting model may be recorded on a line printer.

As with CURFIT, you may enter up to 100 data points. However, at least 7 data points must be used if the quintic is to be fitted. Because the program does not employ the transformation technique used in CURFIT, there is no restriction on entering either zero or negative numbers. However, because x values are exponentiated, there is a restriction that the absolute value of x be no greater than 100. This restriction is the result of raising x to the 10th power during the formation of the data matrix. Subsequent inversion of this matrix can result in large errors and/or overflow. Therefore, if you want to try fitting higher order polynomials, normalize the values of x to a maximum of 100.

Once data has been entered into the computer, it is simply a matter of selecting one of the polynomials to be fitted. From that point, the computer proceeds with the following steps:

Calculation of coefficients

This program is based on the method for calculating the coefficients of a cubic equation given in Croxton and Cowden (1955). The technique involves the simultaneous solution of r + 1 equations which define Σy, Σxy, Σx^2y, $\Sigma x^3y, \ldots, \Sigma x^{r+1}$, where r is the order of the polynomial being fitted. The POLYFIT computer program solves the equations using a standard matrix inversion and multiplication procedure.

The sums of n, x, x^2, x^3, x^4, x^5, $x^6, \ldots, x^{2r}$ are stored as elements of a square matrix [A]. The sums of y, xy, x^2y, $x^3y, \ldots, x^ry$ are stored as elements of a vector, [S]. The calculation proceeds with a matrix inversion and a vector multiplication in which the coefficients of the polynomial become the elements of the product vector, [C]. These are printed out on the video screen.

Correlation coefficient

These calculations are based on the equation given for the cubic by Croxton and Cowden (1955).

$$R^2 = \frac{a_0\Sigma y + a_1\Sigma xy + a_2\Sigma x^2 y + a_3\Sigma x^3 y - \bar{y}\Sigma y}{\Sigma(y)^2 - \bar{y}\Sigma y}$$

Since $$\bar{y}\Sigma y = (\Sigma y)^2/n$$

we have $$R^2 = \frac{a_0\Sigma y + a_1\Sigma xy + a_2\Sigma x^2 y + a_3\Sigma x^3 y - (\Sigma y)^2/n}{\Sigma(y)^2 - (\Sigma y)^2/n}$$

where $\bar{y}$ is the mean of all values of y.

R^2 is used to calculate the correlation coefficient, $R = \sqrt{R^2}$.

Predicted value of y, residual, and residual sum of squares

The coefficients are used, together with the original values of x to calculate the predicted value of y. This is used together with the observed value of y to calculate the residual, the square of the residual, and the sum of deviations squared, ($SS_{residual}$).

The residual mean square

This is calculated by the equation:

$$MS_{residual} = \frac{SS_{residual}}{DF} = \frac{SS_{residual}}{(n - (r + 1))}$$

The standard error of the regression estimate

This is calculated from the equation

$$S_r = \sqrt{MS_{residual}}$$

The F-statistic

The F-statistic is found by using the following equation:

$$F = \frac{MS_{regression}}{MS_{residual}} = \frac{\frac{SS_{regression}}{DF}}{\frac{SS_{residual}}{DF}} = \frac{r^2(\Sigma y^2 - (\Sigma y)^2/n)}{\frac{SS_{residual}}{(n - (r + 1))}}$$

The number of degrees of freedom, which is employed in calculating the above three statistics, is equal to the number of data points minus one greater than the order of the polynomial being fitted. For example, for the fifth order polynomial, the DF is n - 6. This is the reason for requiring at last 7 data points, as stated above.

Plot of model regression compared to data

After each equation is fitted and the statistics are displayed, you may see the model regression plot displayed along with data points. This allows you to evaluate the goodness of fit visually.

Fitting additional polynomials

After the above information is presented, you may fit additional polynomials to the same data set. When no additional equations are requested the computer will display a summary of key statistics for each of the equations already fitted.

Hard copy of the results

If a printer is available, hard copy output of each of the following will be provided:

1) a summary of key statistics on each equation fitted,
2) full statistics on the best fitting equation, and
3) regression curve with data points (available only on a computer provided with Apple Silentype printer).

Retaining data for use on CURFIT

To facilitate the analysis of the same data in both CURFIT and POLYFIT, provision has been made to store it on a disk file which may be called up by either program. At the end of each program you will be given an opportunity to store data. Be prepared to give a file name which will be unique to your data. After file name entry, the program will automatically store the data.

Exercises for Chapter 3

The following data sets were obtained from various literature sources. For each exercise, select at least two theoretical equation types which appear to fit the form of the curve. Then use the CURFIT program to fit the data. Obtain hard copy output for the best fitting equation. In situations where it is not possible to obtain a good fit using CURFIT, you may resort to using the POLYFIT program.

Exercise 3.5

Parasitism of sawfly cocoons by wasps (Holling, 1959)

Host density (cocoons/ft^2)	.5	1	2	4	8	16
Number parasitized	3	4.5	8.2	12	17	24

Exercise 3.6

Predation of cocoons by mice (Holling, 1959)

Number of cocoons/acre (x 10^3)	40	150	260	390	480	800	1100
Number opened/day/mouse	10	10	30	50	75	230	240

Exercise 3.7

Effect of thyroxine secretion upon basal metabolic rate (Guyton, 1971)

Secretion (μg/day)	0	50	100	150	200	300	400
Change in BMR (%)	-45	-20	-10	-5	0	+6	+10

Exercise 3.8

Drosophila fecundity (Pearl, 1932)

Population density (pairs/bottle)	1	2	4	8	16	32	64	128
Eggs (per 1000 female-hrs)	890	600	510	560	350	180	90	110

Exercise 3.9

Movement of bluegills from aquarium compartment A to compartment B (Ellgaard, et al., 1975)

Time (min)	0	1	2	3	4	5	6	7	8	9	10	15	20
No. in A	40	35	32.7	30.7	29.5	28.7	27.4	26.7	26.5	25.7	25.4	23.6	21.3
No. in B	0	5	7.3	9.3	10.5	11.3	12.6	13.3	13.5	14.3	14.6	16.4	18.7

Exercise 3.10

Rate of phytoplankton photosynthesis as a function of phosphate concentration (Harvey, 1963)

P added (μg/liter)	0	4	8	20	38
O_2 evolved (cm^3/liter/14 hr)	.2	1.0	1.6	2.3	2.7

Exercise 3.11

Bluegill biomass at various times (Kitchell, et al., 1974)

Time (days)	0	30	60	90	120	150	180	210	240	270	300	360
Biomass (g)	2	2	2	4	7	15	30	38	45	48	49	50

Hint: This is definitely a sigmoid curve. However, it has such a long lag period that it doesn't fit well. Try transforming x to x - 60.

The next series of exercises involve data taken from complex processes which are very difficult to fit theoretical equations to. Fit at least two empirical equations to each exercise. Determine the coefficients of the best fitting equation.

Exercise 3.12

Aortic pressure as a function of aortic volume for humans (Sias and Coleman, 1971)

Volume (ml)	10	25	50	75	100	125	150
Pressure (mm Hg)	2	6	18	38	62	98	138

Exercise 3.13

Oxygen consumption of potato beetles (Prosser and Brown, 1967)

Temp (C)	10	15	20	25	30	35	38	40	42	45
O_2 (mm^3/g/hr)	120	160	210	250	300	420	620	700	680	340

Exercise 3.14
Effect of light on photosynthesis of marine Chlorophyta (Ryther, 1956)

Light (10^3 ft. candles)	.2	.3	.5	1	2	3	4	5	7	8
Rel. Photosyn. (Pi/P_{max})	.3	.6	.9	1.0	.85	.75	.6	.35	.2	.15

Exercise 3.15
The density of water at different temperatures is crucial to certain simulations in limnology. The following data is taken from the Handbook of Chemistry and Physics (1970).

Temp., C°	Density, g/ml	Temp., C°	Density, g/ml	Temp., C°	Density, g/ml
0	0.99987	5	0.99999	10	0.99973
1	0.99993	6	0.99997	11	0.99963
2	0.99997	7	0.99993	12	0.99952
3	0.99999	8	0.99988	13	0.99940
4	1.00000	9	0.99981	14	0.99927

Exercise 3.16
What form of curve would be expected from each of the following relationships?

a) effect of ventilation rate on arterial CO_2 concentration,

b) effect of environmental temperature on thyroid hormone production,

c) effect of prey density upon predation rate,

d) effect of time on fish growth rate, and

e) relationship between trunk diameter and trunk volume for a given tree species.

Conclusions

This chapter has demonstrated how experimental data may be employed to determine the coefficients and/or exponents of the best fitting theoretical or empirical equation. In each case we have assumed that the cause and effect relationship between the dependent variable and the independent variable was not in question. The non-linear regression, curve fitting techniques are being employed to fit equations to the data, and not to establish the significance of the cause and effect relationship. The resulting equation of best fit is to be used solely as a means of predicting the value of a given dependent variable based upon a known value for the independent variable. The statistical parameters which are derived have been employed for the purpose of recognizing the condition of best fit. If you wish to make use of statistical parameters to test hypotheses about the relationship of one variable to another refer to Zar (1974) or some other reference on statistics.

When a particular equation is found to fit experimental data, the equation type often provides some insight into the theoretical basis for the relationship. For example, Equation 2 implies exponential growth as discussed in Section 1.1, Equation 4 implies a saturation process discussed in Section 2.7, and Equation 2 is the same as the form observed in conjunction with cooling, dilution and diffusion, discussed in Sections 1.3 to 1.5.

In some instances it is likely that model equations were derived before the theoretical basis for the process was fully recognized, and that the curve-fitting process may have led to the theoretical basis, which in turn led to the derivation of the analytical model.

CHAPTER 4
FLOWCHARTING

In the first three chapters, we have been concerned with models involving single equations. Computer programming these exercises in order to generate simulation data is reasonably straightforward. In subsequent chapters, simulation models involving several equations will be used. Without some general means of communicating the complex algorithms involved, programming could become very confusing. One approach would be to employ a programming language such as BASIC or FORTRAN to describe the operational sequence. However, programming languages are designed primarily for communication between man and computer, and typically contain more information than is needed to explain programming logic. What is needed is a simple technique for communication between people. Flowcharting fills this need by diagrammatically displaying the logic used on the computer. No specific programming language is employed. Instead, flowcharts employ general terminology understandable even to individuals who do not have specific familiarity with computers or computer programming languages. The flowcharting technique is easily learned, since it involves only a few simple rules. It is used to describe unambiguously how information is handled by a computer. Because it is a graphic technique, it can be scanned more quickly than any other way of representing a computer program.

Flowcharts are useful at all stages in the development of simulation models. A rudimentary flowchart is first used in the program definition stage. Typically this is done on a piece of scratch paper, where the basic input, process, and output steps of the program may be sketched. Gradually, as the flow of the program takes form, the sequence of steps is established. With a rough flowchart, one may run through the sequence to test the correctness of logic employed. When one is satisfied that the logic is basically correct, the flowchart can be translated into programming language so that it may be run on the computer. The flowchart is used again in finding and debugging mistakes

James D. Spain, BASIC Microcomputer Models in Biology

ISBN 0-201-10678-7

in programming logic. After one has obtained an operational program, the flowchart should be redrawn using a template to trace the outlines. This final corrected flowchart then becomes a part of the documentation for the simulation model. As such it may be used to explain to others the logic employed. It can also be very valuable to the modeler, should modifications of the program be required at a later date. The flowcharting techniques and rules for their use have been summarized from Chapin (1971) and Boillot, et al. (1975).

4.1 General Flowcharting Rules

1. Use simple terminology. Be terse but unambiguous.

2. Keep the flowchart as simple and uncluttered as possible.

3. Keep the general flow pattern from top to bottom, and from left to right. Reverse flows should be marked by arrow heads for clarity. Alternatively, connectors may be employed for reverse flow.

4. The start terminal should be at the top left, and the stop terminal should be at the lower right. Entry connectors should project to the left, and exit connectors should project down or to the right.

5. Use connectors to avoid crossing of flowlines.

6. Draw only one entrance flowline to any outline symbol. (Where more than a single flowline is involved, join them prior to the entry point.)

7. Only the decision block and the incrementing block should have more than one exit flowline. On the decision block these should be labeled with a 'Yes' or 'No' answer to the question posed in the decision outline.

8. Processes should be expressed with algebraic symbolism except that the 'Replaced by' ($\leftarrow$--) symbol may be employed instead of the equals symbol.

4.2 Templates

During the development stages of a simulation program, one usually draws flowchart outlines freehand without much concern for neatness. However, when neater, more easily read flowcharts are needed, a template of flowchart outlines is almost essential. A variety of templates are available at most office supply stores. Select one with outlines as close to those in the symbol guide as possible. It should be almost transparent, but not colorless, and have registration lines to assist in lining up the outlines. A good template should have several sizes of openings for the outlines, especially the process and decision block outlines.

4.3 Flowchart Symbols and Their Use

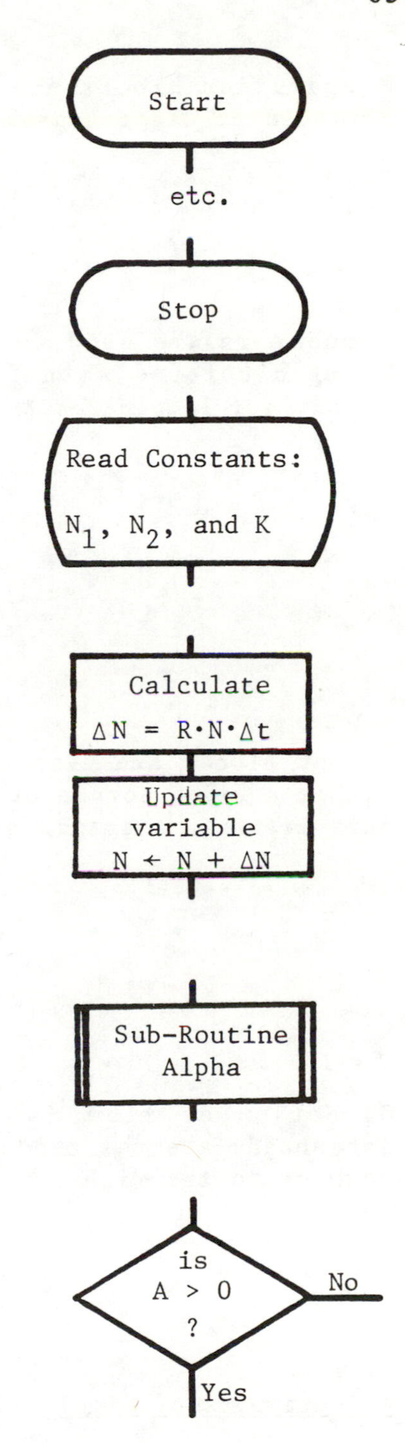

Terminal blocks are used for indicating where programs start or stop. Start is usually at top-left and stop is usually down or to the right.

Read blocks are used to indicate the assignment of numerical values to constants and initial values to variables.

Process blocks are used to indicate computational operations. They would contain the equations that normally appear in LET statements of BASIC.

Subroutine blocks are used for sections of the program which are to be used more than once.

Decision blocks are used for conditional branching of the program. These are analogous to IF...THEN statements in BASIC.

Incrementing or decrementing blocks are used for looping a specified number of times. These are analogous to FOR...NEXT statements in BASIC.

Z

Increment t
from 0 to t_{max}
stepping Δt

Stop

Preparation blocks are used for initializing a variable or other housekeeping operation.

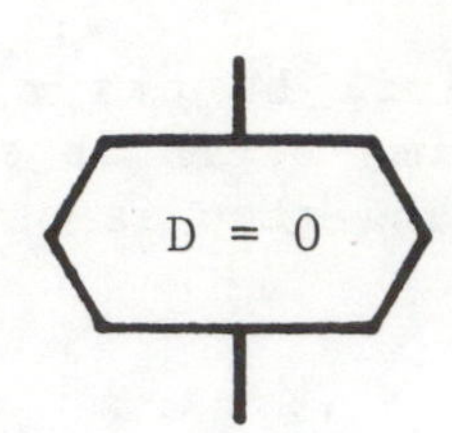

Connectors are used to keep the flowchart from being cluttered with flowlines. They are used mainly for looping back to the top of the page.

Output blocks are used to indicate data output to either a video screen or printer. Output may be a data listing, a graph, or a histogram.

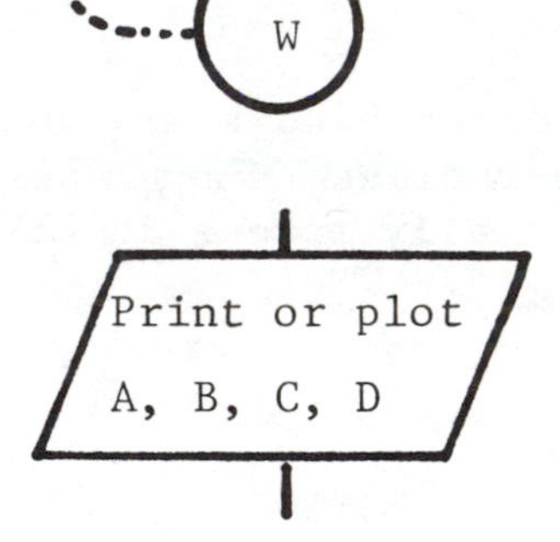

Direction of flow is indicated by flowlines. Arrowheads are not needed if direction of flow is down or to the right.

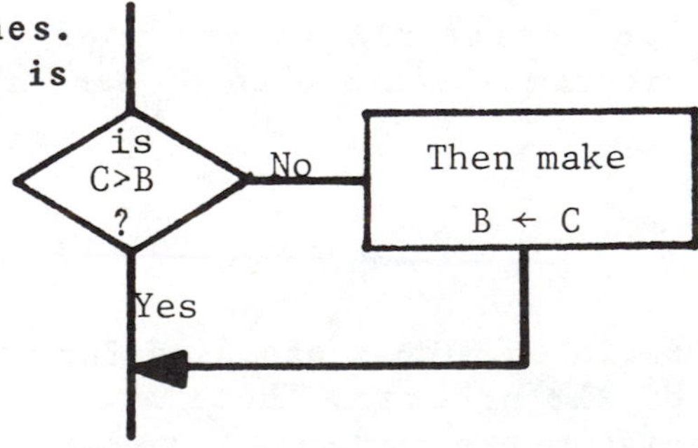

4.4 Examples of Simple Flowcharts

In order to gain some facility in the use of flowcharts, the student should review the following examples taken from exercises in the first three chapters:

a) Population Growth Flowchart
   The flowchart for Exercise 1.1, involving the programming of the population growth Equation 1.13, is given in Figure 4.1.

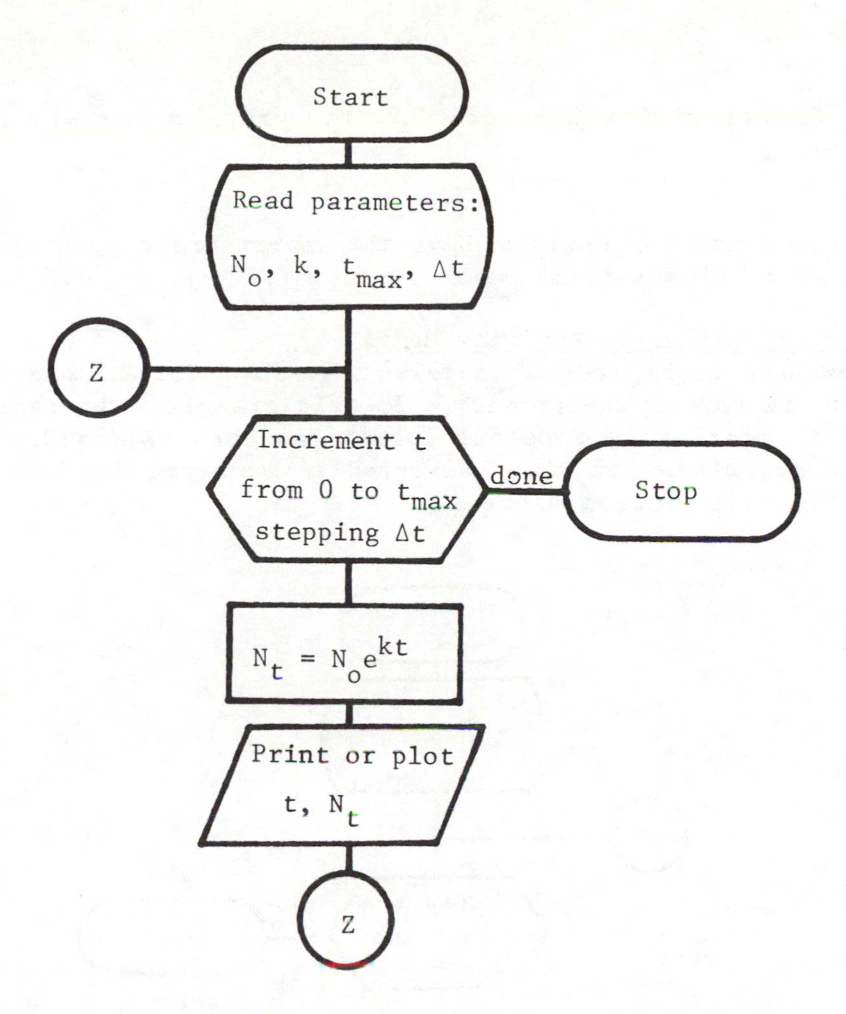

Figure 4.1. Flowchart of population growth program.

Figure 4.1 contains the basic steps important to most flowcharts:

1. A terminal block outline indicates the start.
2. The read block is used to indicate the list of constants or variables which need to be defined in order for the program to operate.
3. The Z connector permits reentry into the loop from below.
4. The incrementing block shows that time is to be incremented from 0 to t_{max} with Δt as the increment size.
5. When t_{max} is exceeded, the program exits to the stop terminal.
6. If t_{max} is not exceeded, the program proceeds to the process block and calculates P according to the equation given.
7. The resulting data are printed or plotted and control is returned to point Z by the connector.

Exercise 4.1
Draw up the flowchart for Exercise 1.2, the program for the simulation of radioactive decay.

Exercise 4.2
Draw up the flowchart for Exercise 1.7, the invertebrate population density at various distances below a point source.

b) Flowchart of pH/Enzyme Activity Model
The flowchart in Figure 4.2 is taken from Exercise 2.3 and deals with the effect of pH on enzyme activity. In this example, the same basic input, increment, process, and output format has been employed. In this case, pH is incremented, and then converted to hydrogen ion concentration for use in the main process block.

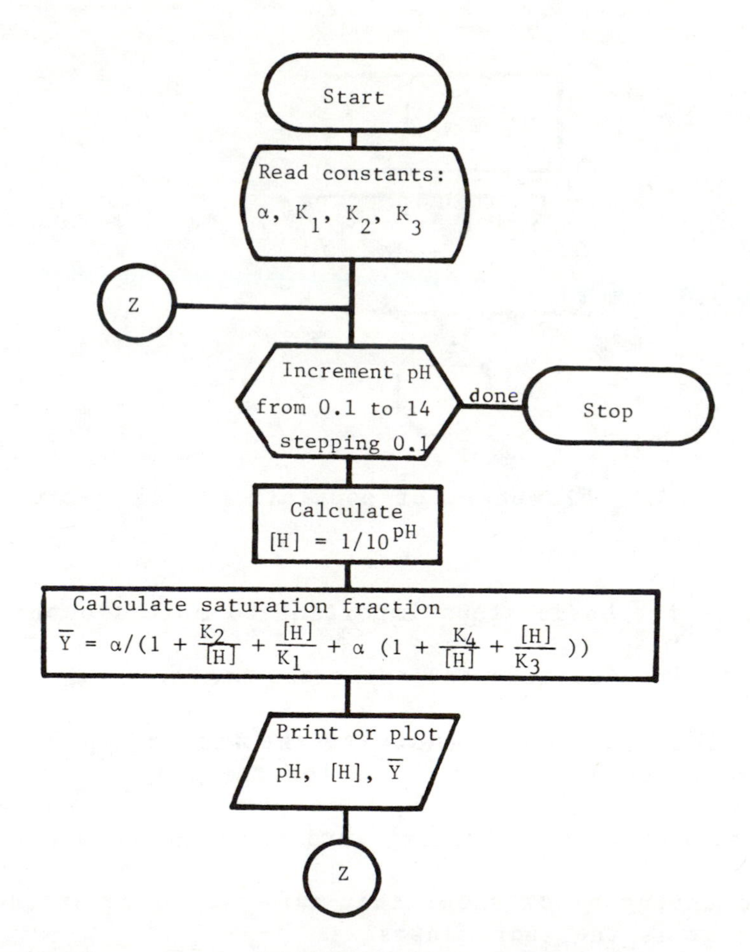

Figure 4.2. Flowchart for the pH/Enzyme Activity Model. K_1, K_2, K_3, K_4, are equilibrium constants and α is the substrate concentration relative to K_s.

Exercise 4.3
Draw up the flowchart for Exercise 2.8, the allosteric enzyme model.

Conclusion

The technique of flowcharting has been very briefly introduced in this chapter. It is assumed that, as more complicated flowcharts are introduced, the student will readily adapt to the system, and become quite comfortable reading flowcharts, and creating his own. For further information on the technique of flowcharting, consult Boillot, et al. (1974), or Edwards and Broadwell (1974).

CHAPTER 5
NUMERICAL SOLUTION OF RATE EQUATIONS

Chapter 1 provided a short review of the analytical solution of rate equations (differential equations) using the rules of integral calculus. Because of their non-linear nature, most differential equations of biological interest are not readily solved by such analytical techniques. Thus, most simulations depend on numerical solution of rate equations, or more precisely, numerical integration. It is this technique more than any other which makes simulation dependent upon the digital computer. Therefore, this chapter is the key to understanding and working exercises in most of the chapters that follow. Try to master it completely before going on to the next section.

Numerical integration, in its simplest form, is very easily understood. It employs the very straightforward technique of calculating an incremental change in a variable and adding it to the old value to obtain the new value of the variable. For example, if one is calculating the population of organisms at time $t + \Delta t$, one would add the change in population that occurs during time increment, Δt, to the population size at time t. The computer is ideally designed to perform repetitive operations of this nature and to print or plot the results for any number of time intervals. Numerical integration can be a highly accurate technique. The results approach those of the analytical solution to varying degrees, depending upon the method employed. In any case, the small errors involved are not serious, since in simulation, one is more interested in the behavioral modes exhibited by a model than in precise duplication of real data. Excellent review of various numerical integration methods are given by Patten (1971) and by Benyon (1968). The discussion which follows emphasizes simple procedures which are easily adapted to the microcomputer. However, the user will be surprised at the accuracy which these techniques are able to achieve.

James D. Spain, BASIC Microcomputer Models in Biology ISBN 0-201-10678-7

5.1 The Finite Difference Approach**

All numerical integration techniques are based upon the fact that for small incremental changes in x (the independent variable), the difference quotient, $\Delta y/\Delta x$, approximates the derivative of the differential equation, dy/dx. You will probably remember that dy/dx is actually defined as follows:

$$\frac{dy}{dx} = \lim_{\Delta x \to 0} \frac{\Delta y}{\Delta x}$$

Therefore if

$$\frac{dy}{dx} = f(y)$$

then

$$\frac{\Delta y}{\Delta x} \cong f(y)$$

and

$$\Delta y \cong f(y) \cdot \Delta x \qquad (5.11)$$

This leads to the general equation upon which all numerical integration techniques are based.

$$y_{(x + \Delta x)} = y_x + \Delta y \qquad (5.12)$$

Equation 5.11 defines the change in y for a finite change in x, and 5.12 calculates the new value of y after the change has occurred.

5.2 The Euler Technique**

Equations 5.11 and 5.12 actually define what has been called the Euler Integration Technique. In order to see how this method would apply to a specific example, let us examine the differential equation for exponential growth.

$$\frac{dN}{dt} = kN \qquad (5.21)$$

where N is population density and k is the growth rate constant. Like most differential equations of interest to modelers, the independent variable is time, t. The differential equation is approximated by the following difference equation:

$$\frac{\Delta N}{\Delta t} \cong kN \qquad (5.22)$$

or

$$\Delta N \cong kN\Delta t \qquad (5.23)$$

and the new value of N is obtained by the equation:

$$N_{(t + \Delta t)} = N_t + \Delta N \qquad (5.24)$$

To carry out an Euler integration, increment time from one to t_{max} and at each time interval calculate the change in N, and the new value of the population density, $N_{(t + \Delta t)}$. This new value then becomes the N_t for the following time interval.

Although this equation is simple enough to be calculated by a single combined equation, it is better to develop the habit of first calculating the change using an equation like 5.23 and then updating the variable with an equation such as 5.24. This procedure is referred to as the 'two stage' approach.

Exercise 5.1**
Write difference equations analogous to 5.23 and 5.24 for the numerical integration of each of the following differential equations from Chapter 1: a) exponential decay, Equation 1.22, b) diffusion across a boundary, Equation 1.51, c) fish growth, Equation 1.61, and d) inhibited growth, Equation 1.81.

Select one of the above models and write a computer program to implement the equations you have derived. Use constants similar to those given in Chapter 1. When programming a numerical integration, always start incrementing the time loop at one rather than zero.

Compare the general shape of the curve obtained by the Euler numerical integration with that obtained from the equation that was derived by the analytical integration. Note that although there is probably not a one to one correspondence of points, the shapes of the curves are generally the same.

5.2 The Time Increment as a Factor in Numerical Integration**

The farther Δt departs from the limiting value of zero, the farther the results of the finite difference approach depart from the true value of the integral. This may be seen when one compares the analytical solution of Equation 1.11 with the results of Euler integration, using two different values for Δt. This is shown graphically in Figure 5.1.

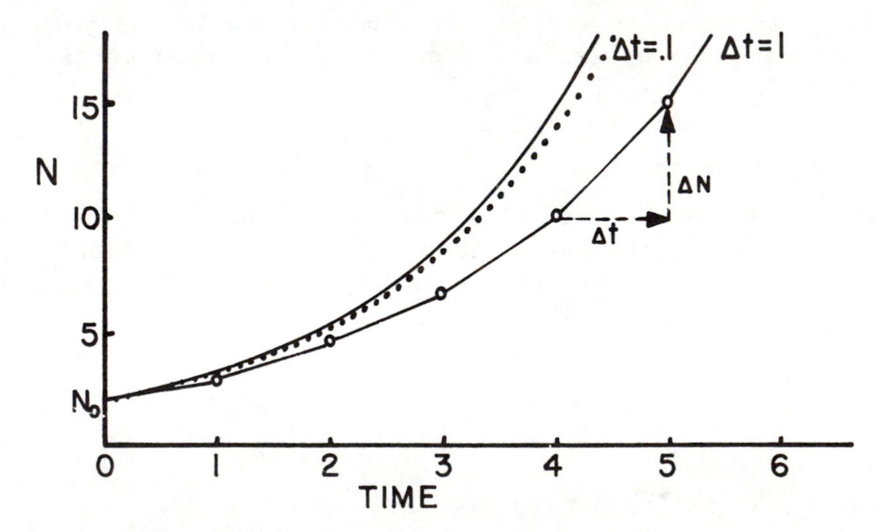

Figure 5.1. The effect of the time increment on an Euler Integration.

The error for a single step is directly proportional to the size of Δt. In addition, the error is cumulative, since each new value of P is dependent upon the last value rather than the original value as in the analytical solution.

Exercise 5.3**
Write a computer program which will allow you to compare three methods for the integration of the decay equation (Equation 1.21). These are

a) The analytical solution,
b) Euler integration with $\Delta t = 1$, and
c) Euler integration with $\Delta t = 0.1$.

Use the two stage approach.

Let $C_0 = 100$, increment t from 0 to 30, and let $k = 0.1$.

Have the computer print out the following:

a) time,
b) the true value of C_t, based on the analytical solution,
c) C_t when $\Delta t = 1$,
d) C_t when $\Delta t = 0.1$,
e) the error between b) and c), and
f) the error between b) and d).

This exercise is most conveniently programmed using nested FOR-NEXT loops, one for the major time units, and an inner loop for smaller time increments. After you have studied the effect of $\Delta t = 0.1$, modify the program and run it again with $\Delta t = 0.01$.

5.4 The Improved Euler or Second Order Runge-Kutta Method

A reduction in time increment decreases the error but is accompanied by a proportional increase in computer time required to accomplish a given integration. Hence, there is definite value in finding methods which decrease the error by other means. The error in the simple Euler method arises from the fact that it assumes that the slope in the interval Δt has a constant value which is a function of N_t, whereas it should be changing continuously between N_t and $N_{t+\Delta t}$. See Figure 5.1.

Methods of the Runge-Kutta type are based on an improved estimate of the slope within the interval, Δt. In the second order Runge-Kutta method, this estimate is obtained from the slope at y_t and an estimate of the slope at $y_{t+\Delta t}$. The slope between these two points is assumed to be the simple average of the two. The equations which accomplish this are as follows:

$$\Delta y_t = f(y_t) \cdot \Delta t \tag{5.41}$$

$$y'_{t+\Delta t} = y_t + \Delta y_t \tag{5.42}$$

where $y'_{t+\Delta t}$ is the first estimate of $y_{t+\Delta t}$. This estimate is used in Equation 5.41 to calculate the change, $\Delta y'_{t+\Delta t}$, which is used to calculate the improved estimate of $y_{t+\Delta t}$:

$$y_{t+\Delta t} = y_t + (\Delta y_t + \Delta y'_{t+\Delta t})/2 \tag{5.43}$$

The analogous equations for exponential growth (1.11) would be:

$$\Delta N_t = k \cdot N_t \cdot \Delta t \tag{5.44}$$

$$N'_{t+\Delta t} = N_t + \Delta N_t \tag{5.45}$$

$$\Delta N'_{t+\Delta t} = k \cdot N'_{t+\Delta t} \cdot \Delta t \tag{5.46}$$

$$N_{t+\Delta t} = N_t + (\Delta N_t + \Delta N'_{t+\Delta t})/2 \tag{5.47}$$

This technique is illustrated graphically in Figure 5.2.

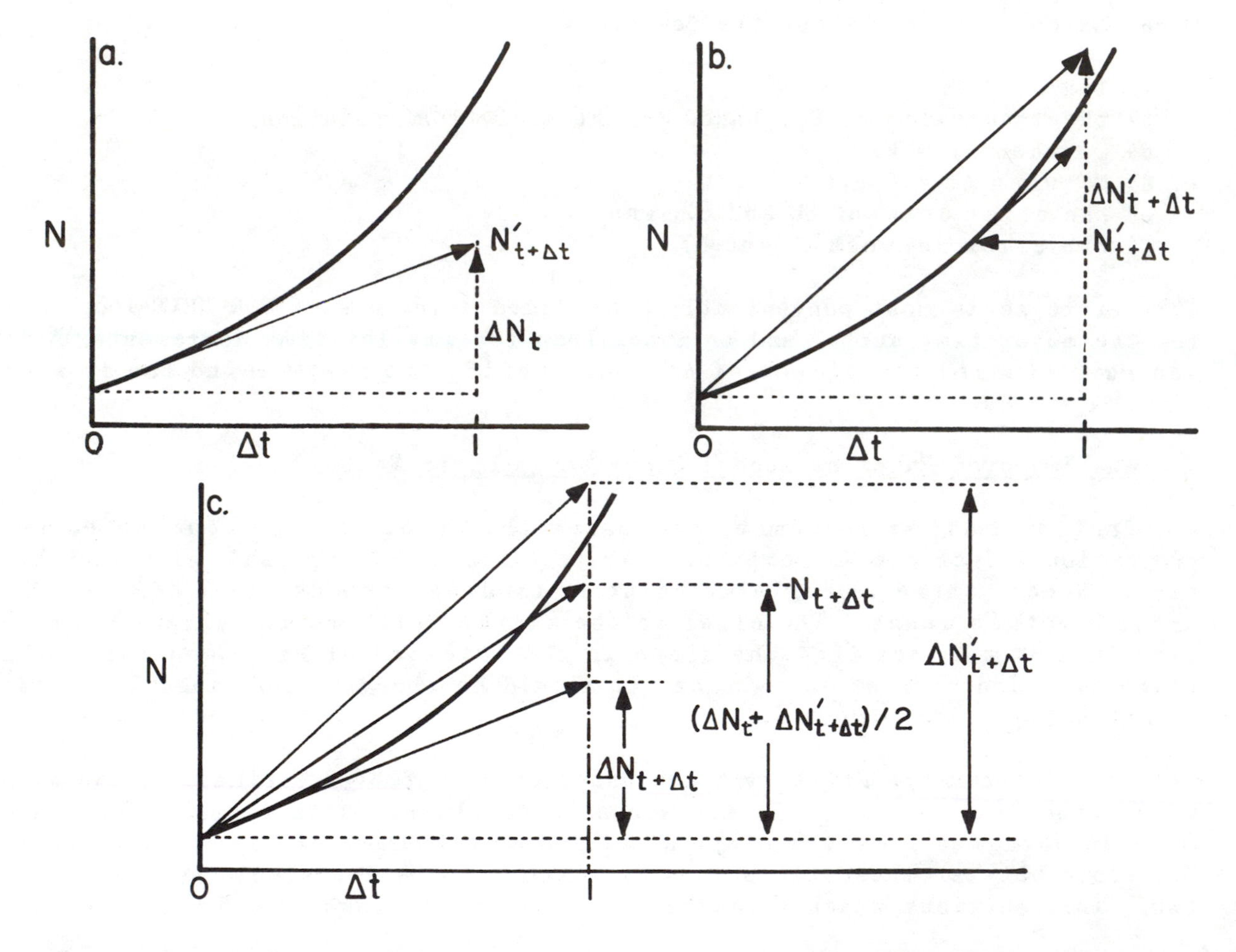

Figure 5.2. Graphical description of the second order Runge-Kutta Method.

Exercise 5.3
Write the equations required for the second order Runge-Kutta integration of the exponential decay curve (Equation 1.21).

Exercise 5.4
Write a program which will allow you to compare the results of the three methods for integration of Equation 1.21: a) The analytical solution, b) the

simple Euler integration, and c) the second order Runge-Kutta or improved Euler integration. Use the same parameters as in Exercise 5.3, and print out time, the values of C from method a), b), and c), and the errors between a) and c).

5.5 An Iterated Second Order Runge-Kutta Method

Since one can improve the estimate of y_t by basing the change, Δy, on both the value of y_t and an estimate of $y_{t+\Delta t}$, it is logical that the accuracy of $y_{t+\Delta t}$ would be further improved by repeating the process using the improved estimate of $y_{t+\Delta t}$ to calculate the change. Thus, it is possible to use an iterative process to further improve the second order Runge-Kutta integration. The following coding sequence uses subscripted variables to accomplish this for the integration of the exponential growth model:

```
100 REM EXPONENTIAL GROWTH MODEL
110 REM ITERATED SECOND ORDER RUNGE-KUTTA SOLUTION
120 N = 2 : K = 0.3 : DT = 0.1
130 IM = INT(1/DT + 0.1)
140 FOR T = 1 TO 20
150 FOR I = 1 TO IM
160 N(1) = N
170 FOR J = 1 TO 3
180 DN(J) = K * N(J) * DT
190 N(J + 1) = N + (DN(J) + DN(1))/2
200 NEXT J
210 N = N(4)
220 NEXT I
230 PRINT T, N
240 NEXT T
250 STOP
```

The first iteration through the J-loop calculates the first estimate of $N_{t+\Delta t}$, designated N(2). The next iteration (J = 2) uses the value of N(2) and the original value N(1) to make the first improved estimate of $N_{t+\Delta t}$, called N(3). The third iteration calculates the final value for $N_{t+\Delta t}$ based on the average of the ΔN_t, i.e. DN(1), and the best estimate of $\Delta N_{t+\Delta t}$, i.e. DN(3). Further iterations of this loop do not seem to make significant improvements.

Exercise 5.5

Write a program to compare the iterated second order Runge-Kutta with both the analytical and the simple Euler solution. Use the limited growth model, Equation 1.81, to test the three methods. Find out how much longer it takes to obtain the same accuracy by varying the size of Δt using the simple Euler method compared with the iterated second order Runge-Kutta method with Δt set at 0.1.

Conclusions

This chapter has described three simple numerical methods for the solution of differential equations. The exercises have been designed to give you some idea of the degree of accuracy one is able to obtain from each method. Benyon (1968) investigated the accuracy of these and several more sophisticated methods and found that in most cases the second order Runge-Kutta integration achieved an acceptable level of accuracy faster than any of the others. It would therefore seem to be the method of choice where accuracy is important. However, as indicated previously, most simulations are concerned more with the particular behavioral modes exhibited by a model than with precise duplication of experimental data. For this purpose, the simple Euler technique is usually adequate. If a higher degree of accuracy is required, it can always be obtained without altering the computer program by simply shortening the time increment.

To permit comparison of the two techniques, this chapter has been concerned with equations which lend themselves to analytical solution. However, through the remainder of the text, we will be concerned largely with differential equations of a nonlinear nature, and the value of numerical methods should become much more obvious.

PART II
MULTICOMPONENT SYSTEMS MODELS

In the second part of this book we will be concerned with models that involve the interaction of several equations in a concerted way to simulate multi-component systems. These models are based on what can be called the summation principle of modeling.

The summation principle states that if a single equation gives verifiable output when implemented independently, then it may be combined with other such equations to effectively model a multicomponent system. Although there is a belief that the behavior of a multicomponent system is not necessarily the sum of its parts, it may be argued that those instances where the principle is not upheld represent situations where unrecognized non-linearities, and/or feed-back interactions are in effect. In other words, the supporters of the summation principle would contend that when we fully understand any given system, this principle will always prove to be true. It is through the process of simulation that unrecognized interactions such as these can become more fully understood. The general concept of the replacement principle acting in conjunction with the summation principle is illustrated in Figure 1.

The system models described in this section are all of the deterministic type. A deterministic model is one in which random perturbations of the system are assumed to be absent. In such models, the state of a given variable at any time is entirely determined by previous states of that variable as well as the other variables upon which it is dependent. It must be understood that deterministic models can depart considerably from the real system, depending on the importance of the random component. This fact must be accepted as one of the basic assumptions associated with deterministic models.

Stochastic or probabilistic models which take into account random effects on the system are the subject of Part Three of this book.

James D. Spain, BASIC Microcomputer Models in Biology

ISBN 0-201-10678-7

a. CONCEPTUAL MODEL OF THE REAL SYSTEM

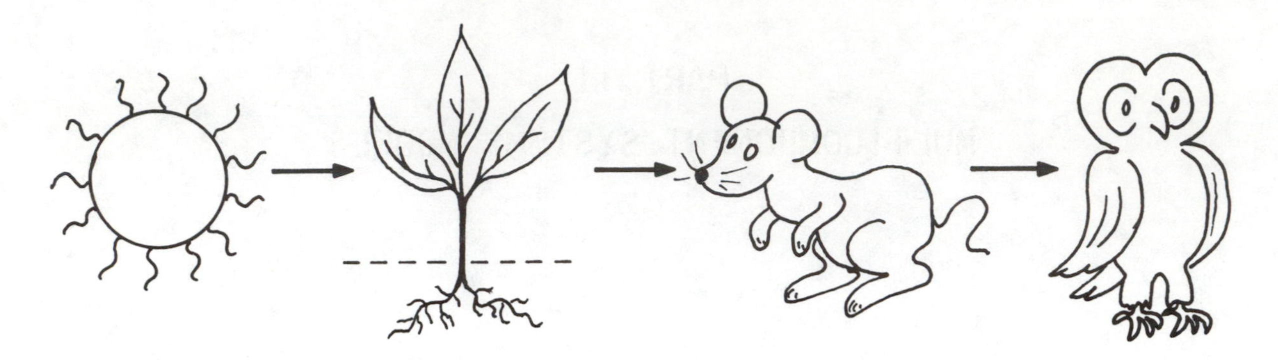

b. RESPONSE DATA IN GRAPHICAL FORM

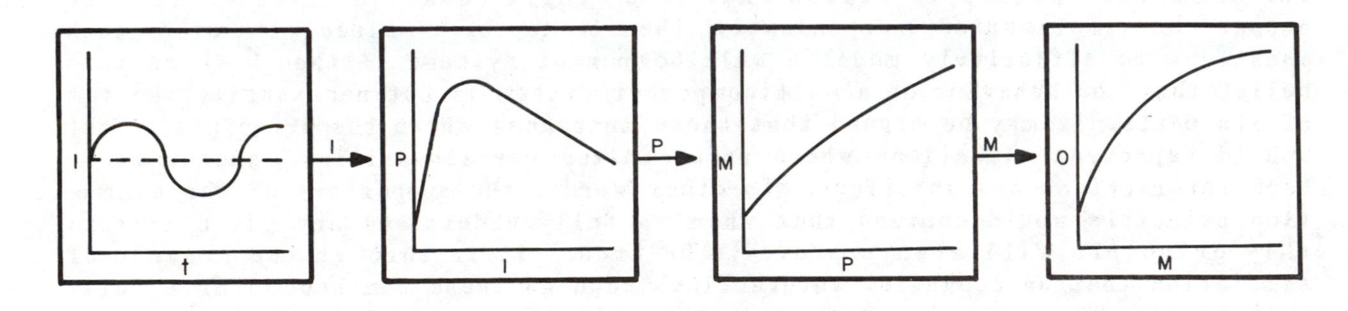

c. BLOCK DIAGRAM OF THE MATHEMATICAL MODEL

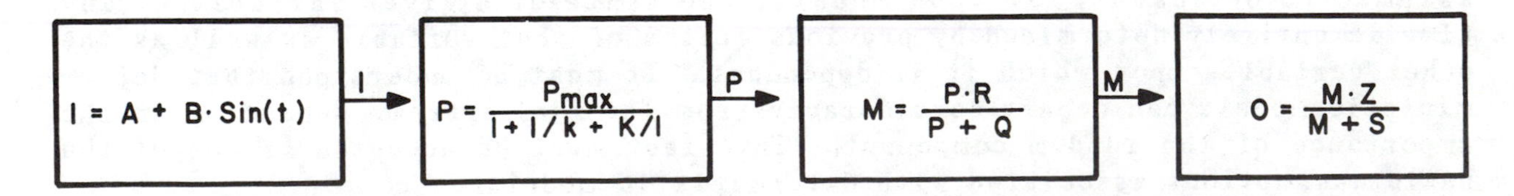

Figure 1. Stepwise replacement of the components in a single hypothetical biological system by modeling equations to form a multicomponent system model.

Your major objective in this section should be to understand the general numerical approach and how it may be applied to the solution of complex biological problems. Other more specific objectives include

1) learning how to solve several differential equations simultaneously by the technique of numerical integration,

2) understanding the holistic nature of biology and the integrating role that mathematics plays,

3) learning to use the compartment model concept,

4) learning to apply matrix algebra to the implementation of multicomponent systems models, and

5) applying modeling techniques to at least one area which directly interests you.

CHAPTER 6

BIOCHEMICAL REACTION KINETICS

Models of the deterministic type will conform to the real system to varying degrees depending upon the size of the populations being modeled. The larger the populations, the more the random component will be averaged out, and the more the deterministic simulation data will conform to real data, assuming the underlying model is itself valid. Biochemical reaction kinetics has been selected for discussion at this time for several reasons:

1. In chemical reactions, one is dealing with enormous populations of molecules. Therefore, the systems are about as deterministic as one is likely to encounter. In such cases, simulation data should match almost exactly with real data, thus providing a simple means of validating the models.

2. Chemical reaction kinetics have been studied for a very long time and the mathematics is very well established.

3. Many of the present-day concepts about modeling in the areas of physiology and ecology have evolved from the theory of chemical reaction kinetics and the law of mass action. This discussion should therefore provide an excellent point of departure for deterministic simulation in general. See Lotka (1956), Garfinkel (1962, 1964), and Chance (1960).

4. Simulations dealing with chemical reaction kinetics are quite straightforward, often permitting both analytical and numerical solution. The exercises should therefore give you practice in the use of numerical integration techniques on equations which may be verified by analytical methods.

James D. Spain, BASIC Microcomputer Models in Biology ISBN 0-201-10678-7

If your primary interest is in population dynamics or ecosystem analysis, you may be tempted to skip over this section. That would be a mistake as this chapter illustrates some very fundamental principles which are not easily understood without first seeing how they apply to molecular interactions.

6.1 Kinetics of Bimolecular Reactions**

In Chapter 1, you had an opportunity to examine the kinetics of bimolecular reactions using model equations derived through the analytical process. In this section, you will be using the more direct approach of numerical integration. The results should be almost the same, except that the more powerful numerical method allows us to assume any initial concentrations of the two reactants, A and B. Because of this, we may observe a complete spectrum of reaction orders between the extremes of first order and second order.

Assume we are dealing again with the reaction:

$$A + B \xrightarrow{k} P$$

The rate of formation of product, P, will be dependent upon the concentrations of the reactants, A and B, and is expressed as follows:

$$\frac{d[P]}{dt} = k[A][B] \qquad (6.11)$$

In order to model such a reaction, the differential equations may be integrated numerically using the simple Euler Method. The following set of equations would be employed:

$$\Delta[B] = \Delta[A] = -k[B][A]\Delta t \qquad (6.12)$$

$$\Delta[P] = +k[B][A]\Delta t \qquad (6.13)$$

$$[B]_{t+\Delta t} = [B]_t + \Delta[B] \qquad (6.14)$$

$$[A]_{t+\Delta t} = [A]_t + \Delta[A] \qquad (6.15)$$

$$[P]_{t+\Delta t} = [P]_t + \Delta[P] \qquad (6.16)$$

Note that whenever there are several differential equations being solved simultaneously, as in the present example, it is necessary to perform numerical integration in two stages. First, each of the differences, ΔA, ΔB, etc., must be defined by one or more equations. Then, in a second stage, each is added to the appropriate state variable to define the value of that variable at time $t+\Delta t$. Typically, initial concentrations of B and A would be entered into the computer, and then the computer would print out the concentrations of B and A for each time increment. To validate the model by determining reaction order, the values for 1/[B], and ln[B] should also be obtained. By plotting these values, the student should be able to show that when the two reactants B and A are approximately equal in concentration, the rate will follow second order kinetics, i.e. will be dependent on the square of the concentration of either reactant.

The data should conform to the following equation from Chapter 1:

$$\frac{1}{[B]} = kt + \frac{1}{[B]_0} \tag{1.94}$$

Second order kinetics are characterized by a straight line when 1/[B] is plotted as a function of t. See Section 1.9 for theory. Some typical simulation data are graphed in Figure 6.1.

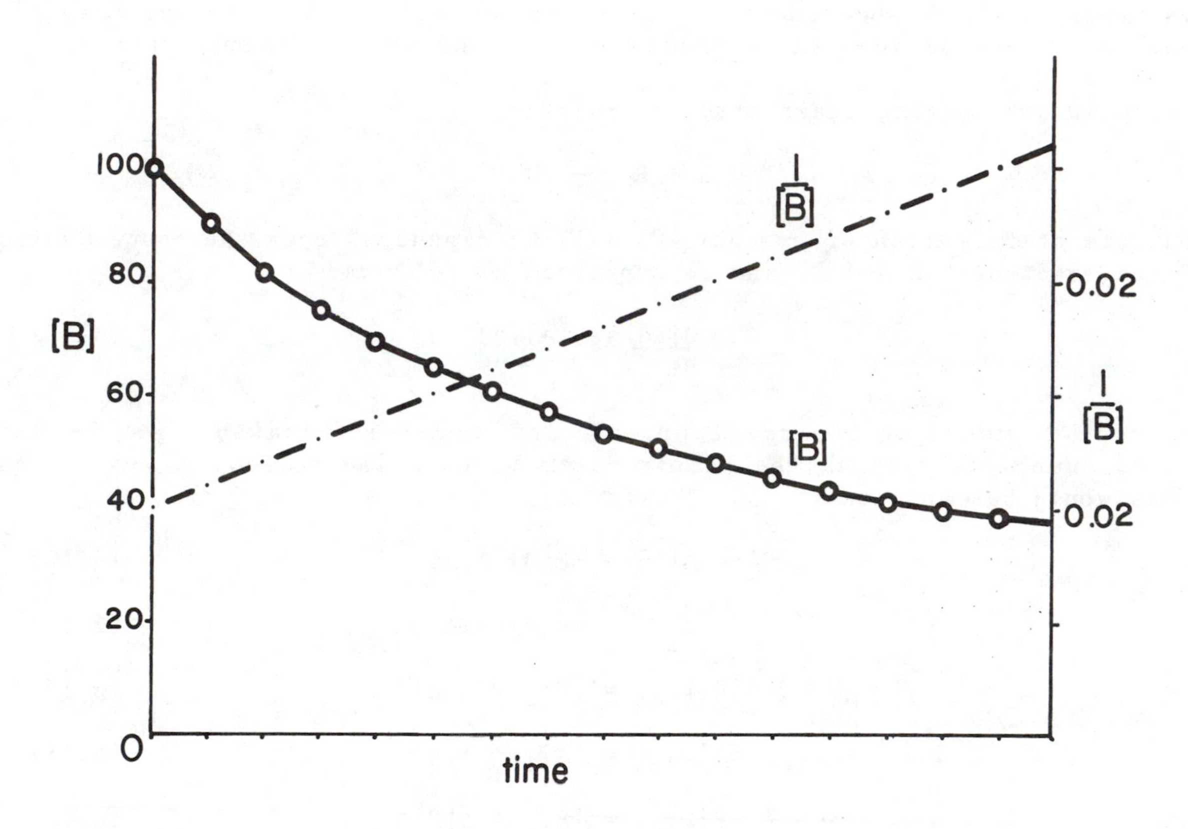

Figure 6.1. Second order kinetics bimolecular reaction simulation $[B]_0 = 100$, $[A]_0 = 100$, $k = .001$.

When the concentration of one reactant greatly exceeds the other, the rate will follow first order kinetics. This results from the fact that the concentration of the reactant in excess will remain essentially constant as the reaction proceeds. The relevant equation is as follows:

$$\ln[B] = -k't + C \tag{1.97}$$

First order reaction kinetics is characterized by a straight line when ln[B] is plotted as a function of t.

Exercise 6.1**
Write a program to simulate bimolecular reactions. Test your simulation with the following input parameters: $\Delta t = .1$, [B] = 100, [A] = 100, k = .0001. All exercises in this chapter employ arbitrary units of concentration and time because they model reaction types rather than real examples. Repeat the simulation with [B] = 10 and [A] = 1000. In each case, plot 1/[B] and ln[B] as a function of time in order to determine reaction order. Compare results with those obtained in Chapter 1.

6.2 Chemical Equilibrium Model**

This is a model designed to demonstrate the relationship between rate constants and the equilibrium constant for a simple reaction of the type:

$$A + B \underset{k_2}{\overset{k_1}{\rightleftharpoons}} C + D$$

The equilibrium constant is defined as follows:

$$K_{eq} = \frac{[C]_{eq} \cdot [D]_{eq}}{[A]_{eq} \cdot [B]_{eq}}$$

The computer may be programmed to determine changes in concentrations of A, B, C and D by first calculating the flux equations associated with the forward and reverse reactions:

$$F_1 = k_1[A][B]\Delta t \qquad (6.21)$$

$$F_2 = k_2[C][D]\Delta t \qquad (6.22)$$

This constitutes stage one of the numerical integration process. Stage two is then accomplished by updating the four state variables or reaction concentrations for time $t+\Delta t$

$$[A]_{t+\Delta t} = [A]_t - F_1 + F_2 \qquad (6.23)$$

$$[B]_{t+\Delta t} = [B]_t - F_1 + F_2 \qquad (6.24)$$

$$[C]_{t+\Delta t} = [C]_t + F_1 - F_2 \qquad (6.25)$$

$$[D]_{t+\Delta t} = [D]_t + F_1 - F_2 \qquad (6.26)$$

Note that since each variable is dependent upon the other three, a two stage numerical integration is necessary for simultaneous solution.

Exercise 6.2**
Write the flowchart and a computer program for the chemical equilibrium model. Set it up so that any initial concentrations of A, B, C, and D may be entered along with the rate constants k_1 and k_2. The computer should be programmed to print out time, the net forward reaction rate, the ratio [C][D]/[A][B], the

concentration of A, B, C, and D, and, to facilitate data analysis, ln([C][D]/[A][B]). Assume the following initial conditions for Exercise 6.2:

$$[A] = [B] = 100, \ [C] = [D] = 0, \ k_1 = .002, \text{ and } k_2 = .001.$$

A reasonable approach to equilibrium should be obtained after about 30 time intervals. The true ratio $[D]_{eq}[C]_{eq}/[B]_{eq}[A]_{eq}$ may then be obtained by plotting [D][C]/[B][A] or ln([D][C]/[B][A]) as a function of net forward reaction rate, $F_1 - F_2$, and extrapolating to zero net rate. See Figure 6.2 for an illustration of this technique.

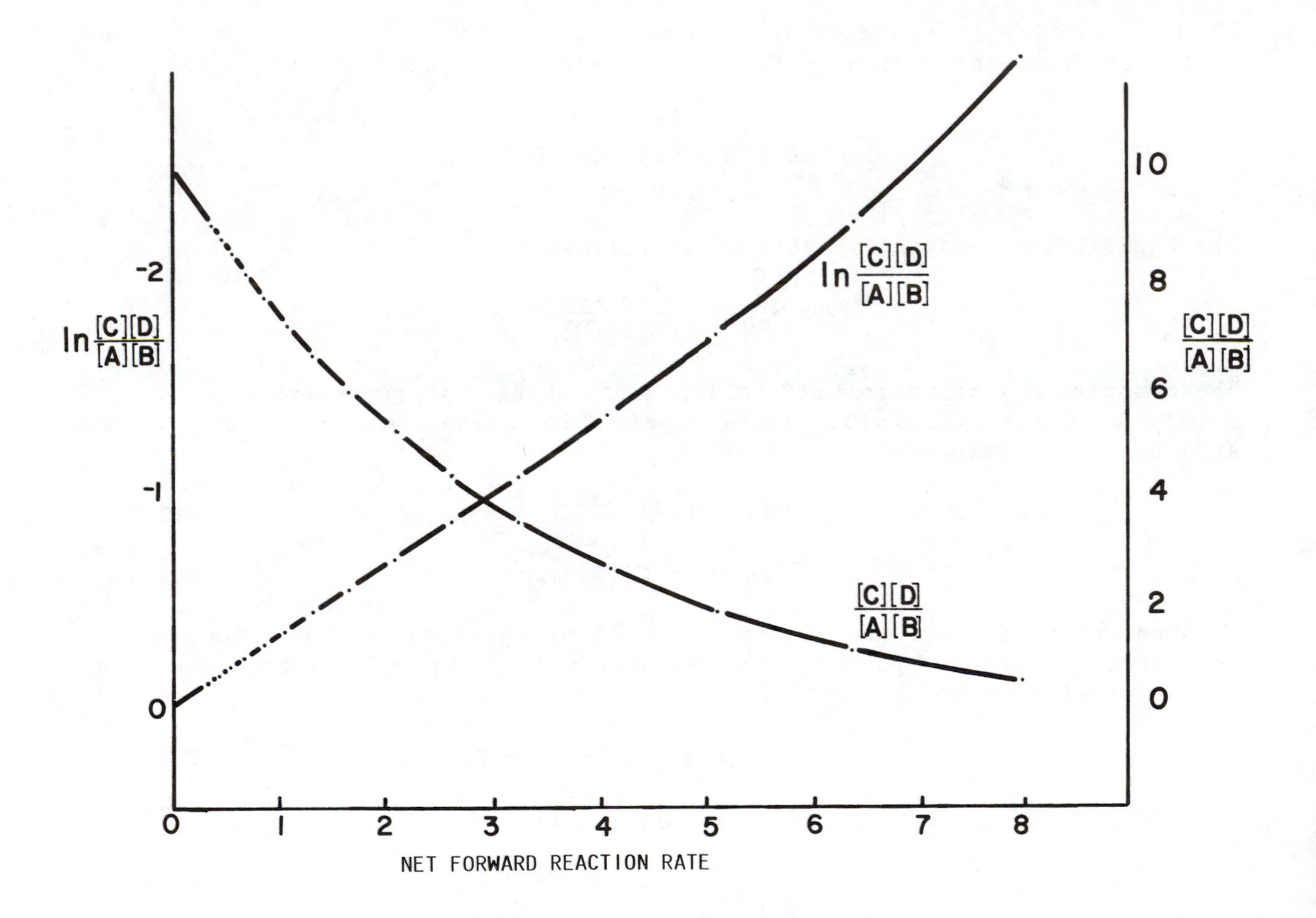

Figure 6.2. Chemical equilibrium simulation $k_1 = k_2 = 0.001$
$k_1/k_2 = K_{eq} = 1$.

The following well known properties of chemical equilibria may be used to validate the model you have programmed:

a) $K_{eq} = k_1/k_2$.

b) The same K_{eq} should be obtained irrespective of starting concentrations of reactants and products.

c) The net reaction rate should be reduced as equilibrium is approached.

d) The ln([C][D]/[A][B]) should vary in direct relation to net reaction rate.

6.3 Kinetics of a Sequential Reaction

This is an example of a mathematical model of a chemical reaction in which the product of one reaction is the reactant of a second.

Assume a reaction in which B plus C yields an intermediate D which then breaks down to yield a product E, as follows:

$$B + C \underset{k_2}{\overset{k_1}{\rightleftharpoons}} D \overset{k_3}{\longrightarrow} E$$

The kinetics are described by the following differential equations:

$$\frac{d[B]}{dt} = \frac{d[C]}{dt} = -k_1[B][C] + k_2[D] \quad (6.30)$$

$$\frac{d[D]}{dt} = k_1[B][C] - k_2[D] - k_3[D] \quad (6.31)$$

$$\frac{d[E]}{dt} = k_1[B][C] - k_3[D] \quad (6.32)$$

These equations can be numerically integrated in two stages as follows:

First we define the reaction fluxes for each of the rate governed processes:

$$F_1 = k_1[B][C]\Delta t \quad (6.33)$$

$$F_2 = k_2[D]\Delta t \quad (6.34)$$

$$F_3 = k_3[D]\Delta t \quad (6.35)$$

Then define the new values of each of the concentrations involved:

$$[B]_{t+\Delta t} = [B]_t - F_1 + F_2 \quad (6.36)$$

$$[C]_{t+\Delta t} = [C]_t - F_1 + F_2 \quad (6.37)$$

$$[D]_{t+\Delta t} = [D]_t + F_1 - F_2 - F_3 \quad (6.38)$$

$$[E]_{t+\Delta t} = [E]_t + F_3 \quad (6.39)$$

Exercise 6.3
Write a program to model the kinetics of sequential reactions. Use the following parameters to test your program:

$$[B]_0 = 100,\ [C]_0 = 100,\ [D]_0 = 0,\ [E]_0 = 0$$
$$k_1 = .001,\ k_2 = .05,\ k_3 = .1,\ \Delta t = 1.$$

Plot concentrations as a function of time, and identify the point where reactant D is in steady-state.

6.4 The Chance-Cleland Model for Enzyme-Substrate Interaction**

The numerical solution of the classical Michaelis-Menten Enzyme Model was first reported by Chance (1960), and has also been attributed to Cleland by several authors including Mahler and Cordes (1966). The Chance-Cleland model uses techniques similar to those described above to simulate the interaction of enzyme and substrate to produce product. The Michaelis-Menten model was described in detail in Section 2.1. The assumed reaction is as follows:

$$E + S \underset{k_2}{\overset{k_1}{\rightleftharpoons}} C \xrightarrow{k_3} E + P$$

where S is the substrate, E is the enzyme, C is the enzyme-substrate compound, and P is the product. k_1, k_2, and k_3 are rate constants for the reactions indicated. The kinetic equations are:

$$\frac{d[S]}{dt} = -k_1[S][E] + k_2[C] \qquad (6.41)$$

$$\frac{d[C]}{dt} = k_1[S][E] - (k_2 + k_3)[C] = -\frac{d[E]}{dt} \qquad (6.42)$$

$$\frac{d[P]}{dt} = k_3[C] \qquad (6.43)$$

Exercise 6.4**
Write equations for the numerical integration of these four differential equations using the two stage approach described above. Write up the flowchart and the computer program. Initial concentrations of S, C, P, and E should be entered into the computer along with constants k_1, k_2, k_3. Let the program generate values for concentrations of S, C, P, and E at successive time intervals, as well as the rate of product formation, $k_3[C]$.

Typical initial values might be as follows:

$$k_1 = 0.005,\ k_2 = 0.005,\ k_3 = 0.1$$
$$[S]_0 = 100,\ [E]_0 = 10,\ [C]_0 = [P]_0 = 0.$$

Your data should conform to the classical curve which appears in Figure 6.3.

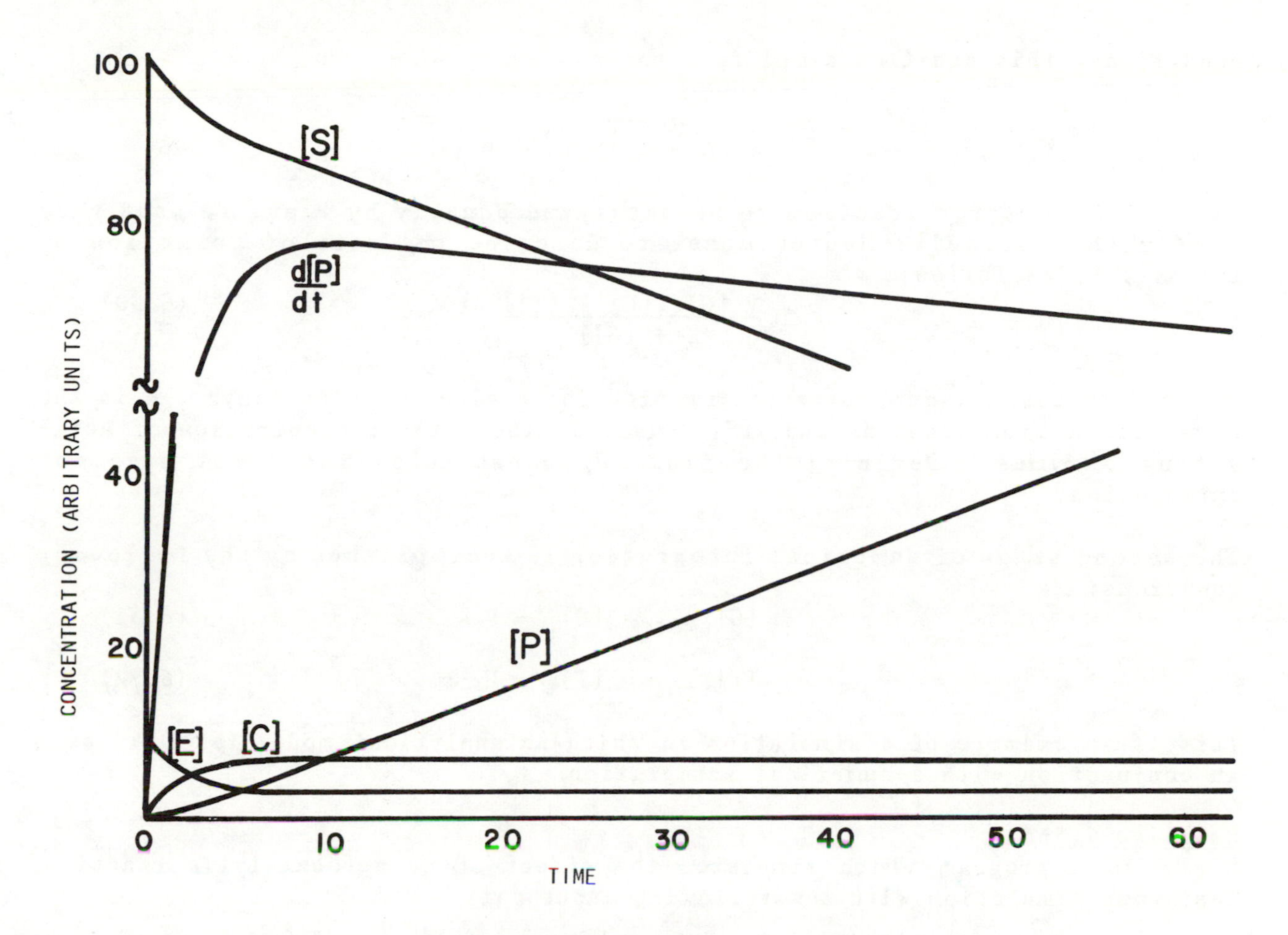

Figure 6.3. Simulation of enzyme kinetics

6.5 Autocatalytic Reactions**

Autocatalytic reactions are those which catalyze the formation of the catalyst itself! This property causes reactions in this class to exhibit positive feedback characteristics almost identical to the growth of organisms or populations of organisms. Autocatalytic reactions have therefore been used as models for growth of living organisms.

Classically these reactions are exemplified by the activation of trypsin in the gastrointestinal tract. Trypsin is formed in the intestine from the product of the pancreatic cells, trypsinogen. This proenzyme or zymogen is converted to the active form by enterokinase of the small intestine. Trypsin, however, itself is capable of activating trypsinogen in the same way. The activation is therefore autocatalytic.

An equation which may be employed to describe this process is the following:

$$G + E \rightleftarrows C \longrightarrow T + E$$

where G represents trypsinogen, E is the activating enzyme, enterokinase, T is trypsin and C is the enzyme-substrate complex. Assuming steady-state conditions, this equation simplifies to:

$$G \xrightarrow{E + T} T$$

We will assume the reaction to be catalyzed equally by E and T, and then employ the Michaelis-Menten model to describe the rate of formation of trypsin, F, as follows:

$$F = \frac{V \cdot [G] \cdot ([E] + [T]) \cdot \Delta t}{(K + [G])} \tag{6.50}$$

where V is the maximum rate of formation for a given unit of enzyme, K is the Michaelis-Menten constant and ([E] + [T]) is the total concentration of activating enzymes. Defining the flux, F, constitutes the first stage of integration.

The second stage of numerical integration is accomplished by the following equations:

$$[G]_{t+\Delta t} = [G]_t - F \tag{6.51}$$

$$[T]_{t+\Delta t} = [T]_t + F \tag{6.52}$$

This is an example of a simulation in which an analytical model (6.50) is used in conjunction with a numerical integration.

<u>Exercise 6.5</u>**

Write up a program which simulates the effect of an autocatalytic reaction. Test your simulation with the following input data:

$$[G]_0 = 100, \; [E]_0 = 1, \; K = 10, \; V = .2, \; [T]_0 = 0.$$

Let Δt = .1 and obtain a print out of [G] and [T] for about 30 seconds. Plot the concentration of G and T as functions of time. Compare the results of the autocatalytic reaction model with those of the population growth model (1.8).

<u>Exercise 6.6</u>

Modify the Chance-Cleland model (6.4) to include rate constants k_4, k_5 and k_6 involving the interaction of product with enzyme to produce an enzyme-product compound which then converts to an enzyme-substrate compound as follows:

$$B + E \underset{k_2}{\overset{k_1}{\rightleftarrows}} C \underset{k_6}{\overset{k_5}{\rightleftarrows}} D \underset{k_4}{\overset{k_3}{\rightleftarrows}} P + E$$

where C is enzyme-substrate compound and D is enzyme-product compound. Assume the following constants in addition to those used in Exercise 6.4:

$$k_4 = .001, \; k_5 = .1, \; k_6 = .1.$$

Generate enough data to compare this model with the simpler one described in Exercise 6.4.

Exercise 6.7

Modify the Chance-Cleland model to take into account a second substrate, A, reacting with substrate B in an ordered fashion with the enzyme to produce intermediates C_1, C_2, C_3 which break down to products F and D. Assume the following overall reaction takes place:

$$A + E \underset{k_2}{\overset{k_1}{\rightleftharpoons}} C_1 + B \underset{k_4}{\overset{k_3}{\rightleftharpoons}} C_2 \underset{k_6}{\overset{k_5}{\rightleftharpoons}} F + C_3 \underset{k_8}{\overset{k_7}{\rightleftharpoons}} D + E$$

Use Exercises 6.4 and 6.6 as a guide to assume initial concentrations for reactants and rate constants. Compare your results with those in an advanced biochemistry text such as Mahler and Cordes (1966) or Lehninger (1975).

Exercise 6.8 Deterministic Model of Linear Diffusion Gradient

This simulation provides a means of studying the pattern which results when solutes diffuse from an initial sharp boundary. Assume that a diffusion cell is represented by a series of 20 subcompartments of unit volume as follows:

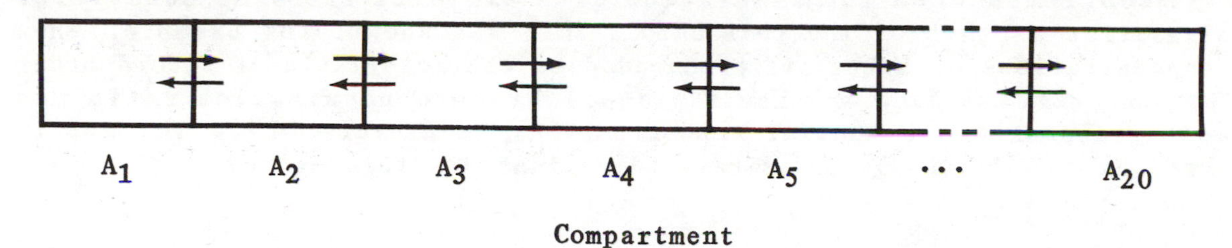

The net flux from compartment A_n to compartment A_{n+1} is described by the equation:

$$F_n = k(A_n - A_{n+1})\Delta t$$

where k is the diffusion rate constant, A_n is the concentration in compartment A_n, and A_{n+1} is the concentration in compartment A_{n+1}. If the compartments are defined as having unit volume, then the concentration of any two compartments are altered as follows:

$$(A_n)_{t+\Delta t} = (A_n)_t - F_n$$

$$(A_{n+1})_{t+\Delta t} = (A_{n+1})_t + F_n$$

The model is designed to simulate the changes which occur in a concentration gradient at various times after a sharp boundary is formed in a linear diffusion cell such as might be employed for molecular weight determination of proteins. You may best observe the output by plotting concentration as a function of distance (compartment number) at various times. Other curves of interest would be concentration difference between compartments as a function of distance, and log of concentration as a function of distance. This program requires the use of an array of subscripted variables. It should not be attempted unless you are a fairly competent programmer. A suggested value for k is 0.1.

Exercise 6.9

Modify one of the exercises in this chapter to incorporate the equations for improved Euler integration (second order Runge- Kutta) instead of simple Euler as implied in the description of the exercise. Include a flowchart to assist in documenting your program.

Conclusion

This chapter has presented examples of systems which lend themselves to deterministic type simulations involving the technique of numerical integration. Exercise 6.1 could have been solved analytically. However, the other exercises would have been very difficult, if not impossible, to solve by this means. Thus, they provide excellent examples of the tremendous power and simplicity of the basic numerical integration technique. Although you were not given any real data with which to compare, the examples are sufficiently classical that most of you have been exposed to them previously in one course or another. This is a good time to remind you that simulations are not intended to duplicate real data point for point. Generally the validity of a model is confirmed when the simulation data exhibits the same behavioral characteristics as that of the real data. This was shown, for example, when equal concentrations of reactants produced data characteristic of second order kinetics in Exercise 6.1, or when the equilibrium concentration ratio was produced irrespective of initial concentrations in Exercise 6.2. The models in future chapters will not be as easy to validate in this manner.

CHAPTER 7

DYNAMICS OF HOMOGENEOUS POPULATIONS OF ORGANISMS

In the last chapter we found that reactions between homogeneous populations of molecules were effectively simulated by deterministic models based on numerical integration of the reaction kinetics equations. When the first attempts were made to model homogeneous populations of organisms, it seemed logical to use some of the same principles, including the law of mass action. This law states that the rate of interaction is a direct function of the product of the concentrations of the interacting molecular species. The relationship of population dynamics equations to the law of mass action has been discussed by Lotka (1956), and by Garfinkel (1962 and 1964).

Homogeneous populations of organisms are those in which fine structural details such as differences in age, sex, genotype, and behavior have been ignored. The population is assumed to be a single homogeneous mass of average organisms evenly distributed within the environment, and described by a single state variable representing population density. Typically, the state variable has units such as organisms per hectare or organisms per cubic meter, and is directly analogous to chemical concentration in moles per liter, the state variable employed in reaction kinetics. By analogy to the law of mass action, we may assume, as a first approximation that the rate which populations interact with each other by competition, predation, or parasitism is a direct function of the product of the population densities involved.

7.1 The Verhulst-Pearl Equation**

In Chapter 1 we discussed two differential equations which provide much of the basis for models dealing with population dynamics. They were the uninhibited growth model (1.11) and the inhibited growth model (1.81). If an arbitrary limit is placed on population density N, the uninhibited growth model gives the typical J-shaped growth curve observed in certain examples of algal bloom, plant growth, and insect infestation (Odum, 1971). The inhibited growth

---

James D. Spain, BASIC Microcomputer Models in Biology

ISBN 0-201-10678-7

curve, on the other hand, has a sigmoid or S-shape and is typical of most other populations existing in a limited environment. The equation for inhibited growth was first proposed by Verhulst in 1838 and developed by Pearl in 1920. It is therefore called the Verhulst-Pearl equation or the logistic growth curve. The form of the equation given in Chapter 1 was as follows:

$$\frac{dN}{dt} = kN(L - N) \qquad (1.81)$$

A more typical form of the equation is

$$\frac{dN}{dt} = rN\left(\frac{K - N}{K}\right) \qquad (7.11)$$

where N is the population density, r is the intrinsic growth rate constant which includes the net effect of both birth and death, and K is the carrying capacity of the environment. K is the same as L in Equation 1.81. This means that r in Equation 7.11 is equal to k·L. This equation may also have the form

$$\frac{dN}{dt} = rN\left(1 - \frac{N}{K}\right) \qquad (7.12)$$

or

$$\frac{dN}{dt} = rN - \frac{rN^2}{K} \qquad (7.13)$$

Using Equation 7.12, it is seen that if N is small relative to K, then $1 - N/K$ is only slightly less than 1. Under these conditions, growth is exponential ($dN/dt \cong rN$). But when $N \cong K$ then $1 - N/K \cong 0$ and growth is $\cong 0$. Hence, the population grows asymptotically to the limit value, K. When $N > K$, the term $1 - N/K$ becomes negative, net growth becomes negative, and the population declines to the carrying capacity.

Equation 7.13 provides a different insight into this model. The first term, rN, is the same as that for unlimited growth, and is positive, hence adding to the population, while the second term is negative and is a function of N x N (a constant times N^2). As such, it may be thought of as a loss to the population resulting from the crowding interaction of organisms, one on another. The population density approaches a steady-state condition in which the increase is exactly balanced by the density dependent decrease or negative feedback term. This equation is the first example of the application of the law of mass action to population dynamics. The crowding effect may be thought of either as an increased death rate or a decreased birth rate. It would be a mistake, however, to attach too much biological significance to this term because of the homogeneity assumptions described in the introduction to this chapter.

Exercise 7.1**

Use the following data set taken from Pearl (1927) to evaluate r and K in the Verhulst-Pearl equation. These constants may be approximated by graphical methods, but are best obtained by fitting the data to the linear form of Equation 1.82 which is similar to the exponential (sigmoid CURFIT Equation 7). Although the data is expressed in biomass, we can assume that this is proportional to population density. Remember, k in Equation 1.82 is equal to r/K in the Verhulst-Pearl model.

Time (Hrs.)	0	1	2	3	4	5	6	7	8	9	10	12	14	18
Yeast Biomass (mg/100 ml)	4	7	12	19	28	48	70	103	140	176	205	238	256	265

After evaluating the constants, use numerical integration on Equation 7.12 or 7.13 to simulate this growth process. Plot the curve obtained from the simulation data along with the actual data. It will probably be necessary to employ the improved Euler integration in order to obtain a good fit to the data.

7.2 Oscillation and Overshoot in Population Density**

The Verhulst-Pearl equation results in a stable population level near K, the carrying capacity for the environment. The stability of Equation 7.13 is the result of the interaction of the positive feedback characteristics of the rN term and the negative feedback characteristics of the rN^2/K term. One of the basic principles of control theory (see Chapter 15) suggests that the oscillation or overshoot commonly associated with population growth of higher organisms is the result of a lag in the information employed in the negative feedback loop. Typically this would result from the delay which occurs between the process of reproduction, and the process of resource utilization. For example, in humans, there is a lag of about 10 years before a child begins to make a significant impact on food resources.

Although details of this nature are best handled by age class models such as those discussed in the next chapter, they may be approximated by building a time lag into the equations. For example, Equation 7.12 could be modified to

$$\frac{dN_t}{dt} = rN_t \left(1 - \frac{N_{t-d\Delta t}}{K}\right) \qquad (7.21)$$

where $N_{t-d\Delta t}$ refers to the population that existed d time intervals previous to t.

$$N_{t-d\Delta t} \cdots \longrightarrow N_{t-2\Delta t} \longrightarrow N_{t-1\Delta t} \longrightarrow N_t$$

In the computer program this technique requires one to retain several previous population densities so that $N_{t-d\Delta t}$ is available when needed.

Another type of time lag is the lag in reaching reproductive age. This lag would appear in the rN term of the equation. The lag could be simulated by using $rN_{t-d\Delta t}$ instead of rN_t. Here d represents the number of time intervals required to reach reproductive age. Both of these time lags are discussed by Wangersky and Cunningham (1957).

Exercise 7.2**

Write a computer program to simulate population growth in which there is a time lag in resource utilization amounting to three time intervals. The flowchart in Figure 7.1 will assist you. Assume the carrying capacity, K, is 1000, $N_0 = 2$, $r = 0.5$. Allow the simulation to proceed for 50 time intervals.

Notice that it is necessary to initialize N_t, N_{t-1}, N_{t-2}, N_{t-3} with the same value, N_0, in order to get the simulation started. Graph the data for N_t as a function of time. An interesting addition to this exercise is obtained when the rate of growth is also plotted as a function of time. The axes on the graph must be scaled to allow for the negative values which result.

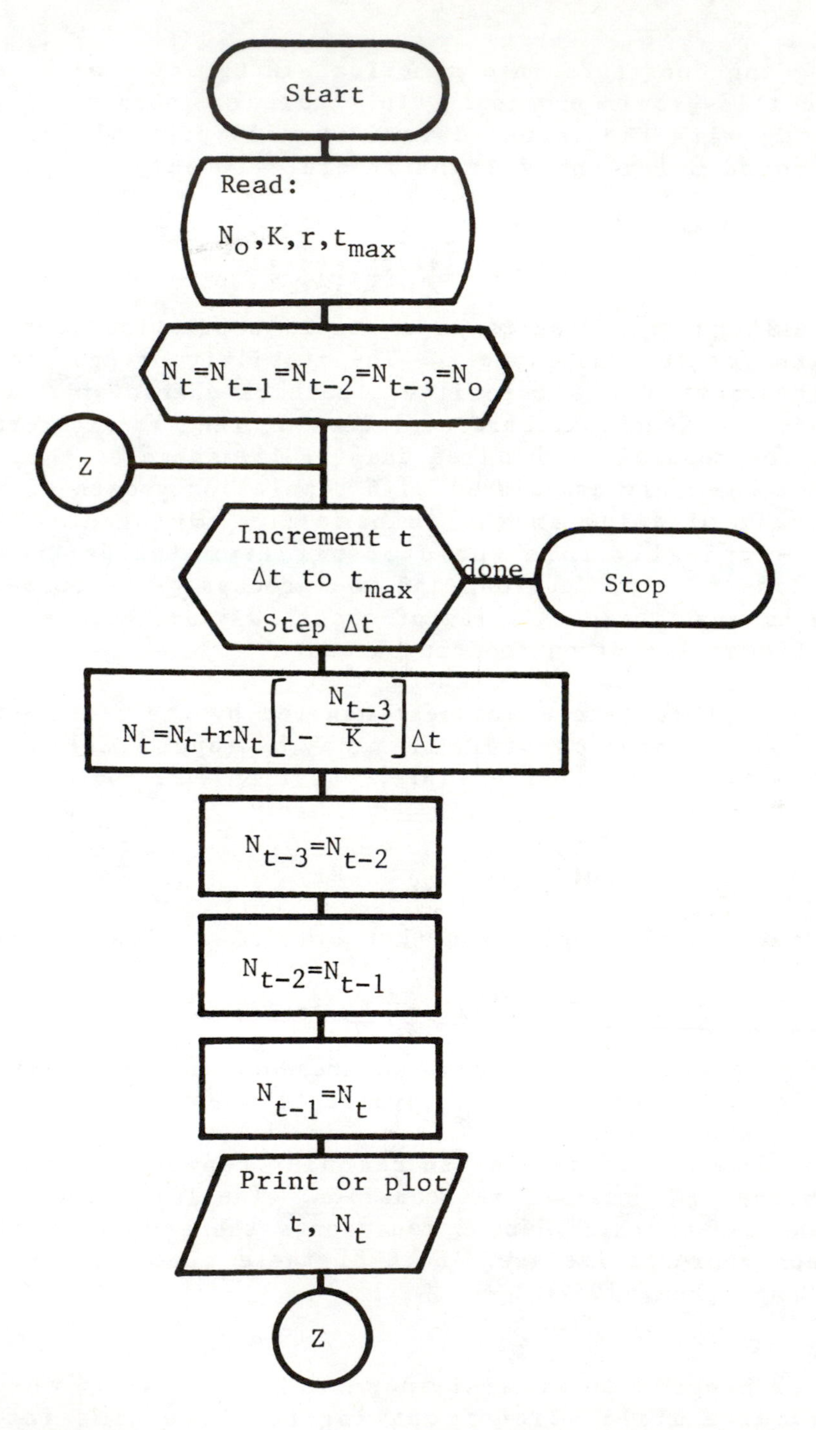

Figure 7.1. Flowchart of model which employs the Verhulst-Pearl Equation with time lag.

```
10   REM   LOGISTIC GROWTH WITH LAG
20   REM   EXERCISE 7.2
30   REM   SUBMITTED BY JOHN DOE
35   REM   10/10/81
40   POKE 232,0: POKE 233,64: SCALE= 1: HCOLOR= 3: ROT= 0
50   PRINT  CHR$ (4)"BLOAD SMALL CHARACTERS,A$4000"
100   REM  N IS POPULATION DENSITY AT TIME T
120   REM  N1,N2,N3 ARE POPULATION DENSITIES 1,2,3 TIME INTERVALS BEFORE T
130   REM  K IS THE CARRYING CAPACITY
140   REM  DN IS THE CHANGE IN POPULATION DENSITY
150   REM  R IS THE INTRISIC GROWTH RATE
200   READ N,K,R
205   HGR
210 N1 = N:N2 = N:N3 = N:N4 = N:N5 = N
215 DT = 1
220 X$ = "TIME":Y$ = "DENSITY":YM = 800:XM = 100: GOSUB 3000: REM   DRAW A
    XES
225 ZF = 1:ZG = 0
230  FOR T = 1 TO 100
240 DN = R * N * (1 - N5 / K) * DT
250 N = N + DN
255 N5 = N4
260 N4 = N3
265 N3 = N2
270 N2 = N1
280 N1 = N
290 X = T:Y = N: GOSUB 4000: REM  PLOT POINT
300  NEXT T
400  DATA  2,500,.3
500  END
3000  REM   GRAPH SUBROUTINE WHICH DRAWS AND LABELS AXES THEN PLOTS POINTS

3010  REM   DEVELOPED BY J. SPAIN, MICHIGAN TECHNOLOGICAL UNIV., AND B. WI
    NKEL, ROSE-HULMAN INST.
3020  REM  X$=VARIABLE PLOTTED ON X AXIS
3030  REM  Y$=VARIABLE PLOTTED ON Y AXIS
3040  REM  YM =MAXIMUM UNITS ON THE Y AXIS
3042  REM  YN = MINIMUM UNITS ON THE Y AXIS
3045  REM XM = MAXIMUM UN ITS ON THE X AXIS
3047  REM  MN = MINIMUM UNITS ON X-AXIS
3050  HCOLOR= 3: SCALE= 1:SC = 1: ROT= 0
3055  REM     LIST OF RESERVED VARIABLES:X,X0,XM,X$,Y,Y0,YM,YN,Y$,Z,ZF,ZG,
    ZP,Z0,I,L$,SC
```

7.3 Variable Carrying Capacity

Although the carrying capacity in the Verhulst-Pearl equation is normally considered to be constant, there is no reason why it may not be made to vary in some orderly fashion, as would be the case if it were simulating annual growth and decline of vegetation upon which an herbivore population was dependent. On the other hand it might represent an herbivore population on which a carnivore population was dependent. The carrying capacity usually varies in an annual cycle which might be represented by either a sine curve or a sawtooth-type curve. These two possibilities could be modeled by the following equations:

1) An annual cycle may be based on a sine function as shown in Figure 7.2.

$$K = k \sin\left(\frac{2\pi \cdot t}{t_\lambda}\right) + C \qquad (7.31)$$

where C is the annual mean value for K, k is equal to $K_{max} - C$, t_λ is the total time in one annual cycle (52 weeks, 365 days, etc). For additional information on this equation, see Chapter 10. See May (1976) for further discussion on variable carrying capacity.

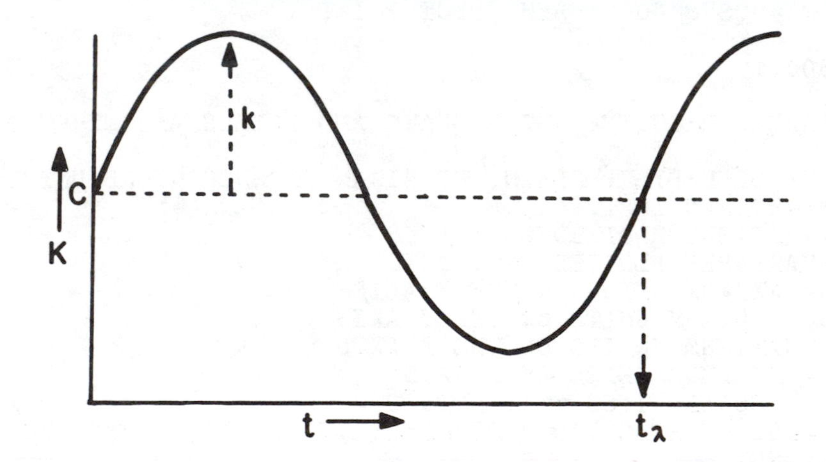

Figure 7.2. Curve for sinusoidally varying carrying capacity.

2) An annual cycle based on sawtooth or ramp function as shown in Figure 7.3.

$$K = \left| kt + C \right|_{t_0}^{t_\lambda} \qquad (7.32)$$

In this equation k is the slope, and C is the annual minimum. The equation would be evaluated between t_0 and t_λ, at which point it would start over at t_0.

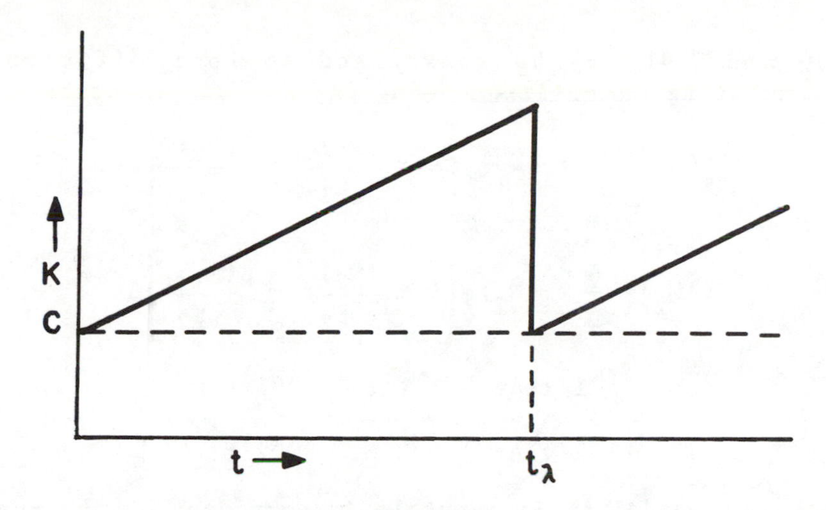

Figure 7.3. Sawtooth or ramp-like variation of carrying capacity.

Exercise 7.3
Use the above information to write a program simulating population growth in an environment in which the carrying capacity is varying in an annual sawtooth pattern. Assume $C = 30$, $N_0 = 2$, $r = 0.5$, $t_\lambda = 52$, $k = 6$, and $t_{max} = 100$. Use a Δt of 0.1 or less, otherwise populations will vary in an unrealistic fashion (go negative) when carrying capacity drops. Plot both N_t and K_t as a function of time.

7.4 Gause's Model for Competition Between Two Species**

Models dealing with competition between two species are generally based on a pair of logistic type equations first proposed by Lotka and Volterra, and subsequently developed in the classical paper by Gause and Witt (1935). The basic differential equations involved are

$$\frac{dN_1}{dt} = r_1N_1 - \frac{r_1N_1^2}{K_1} - \frac{r_1\alpha N_1N_2}{K_1} \tag{7.40}$$

and

$$\frac{dN_2}{dt} = r_2N_2 - \frac{r_2N_2^2}{K_2} - \frac{r_2\beta N_2N_1}{K_2} \tag{7.41}$$

where N_1 and N_2 are the population densities of species 1 and 2 respectively, r_1 and r_2 are the growth rate constants for the two species, K_1 and K_2 are the carrying capacities of the environment for each species when growing alone, α is a constant relating the effect of species 2 on species 1, and β is a constant relating the effect of species 1 on species 2. Basically these differential equations are saying that the increase in population density is equal to the unlimited growth rate less the self crowding effects and less the effects of species interaction. It is only this latter term which distinguishes this equation from the simple logistic growth equation. It should also be noted that the N_1N_2 term represents another application of the law of mass action to populations of organisms.

Equations 7.40 and 7.41 may be rearranged to form difference equations suitable for simulating competition using the two stage approach.

Stage 1 is

$$(\Delta N_1)_t = r_1(N_1)_t\left[1 - \frac{(N_1)_t}{K_1} - \frac{\alpha(N_2)_t}{K_1}\right]\Delta t \qquad (7.42)$$

and

$$(\Delta N_2)_t = r_2(N_2)_t\left[1 - \frac{(N_2)_t}{K_2} - \frac{\beta(N_1)_t}{K_2}\right]\Delta t \qquad (7.43)$$

Stage 2 is

$$(N_1)_{t+\Delta t} = (N_1)_t + (\Delta N_1)_t \qquad (7.44)$$

and

$$(N_2)_{t+\Delta t} = (N_2)_t + (\Delta N_2)_t \qquad (7.45)$$

With this simulation model, it is possible to vary each of the parameters so as to determine which conditions produce stable populations of the two species, and which conditions lead to the exclusion of one species by the other.

Equation 7.41 may be rearranged to obtain a new insight into the utilization of carrying capacity by the two species as follows:

$$\frac{dN}{dt} = \frac{r_1N_1}{K_1}(K_1 - N_1 - \alpha N_2) \qquad (7.46)$$

Note how the carrying capacity, K_1, is reduced by the terms N_1 and αN_2.

The constant α expresses the effect of an individual of species 2 on the carrying capacity of species 1, for instance, if α is 0.5, then one individual of species 2 is equivalent to 0.5 individuals of species 1. β expresses the equivalent effect of one individual of species 1 on the carrying capacity of species 2. Values greater than 1 indicate that the competing species is taking the place of more than one individual of the other species. Negative values have the effect of adding to the carrying capacity of the other species, as would be the case with prey organisms.

To analyze the results of any given experiment, it is helpful to plot the data with N_1 on the x axis and N_2 on the y axis to produce a phase space diagram. On such phase space diagrams it is always helpful to include isocline lines showing where zero growth occurs for N_1 or N_2. The equations for these two lines are obtained by setting growth rate equal to zero in Equations 7.40 and 7.41. They are then solved for N_2 by factoring out common terms and simplifying. The N_1 isocline is described by the equation

$$N_2 = \frac{K_1}{\alpha} - \frac{N_1}{\alpha} \qquad (7.47)$$

This line has an intercept of K_1/α on the N_2 axis and an intercept of K_1 on the N_1 axis.

The N_2 isocline is described by the equation

$$N_2 = K_2 - \beta N_1 \quad (7.48)$$

This line has an intercept of K_2 on the N_2 axis and an intercept of K_2/β on the N_1 axis. These isoclines are easily drawn on the phase diagram by connecting the intercepts with straight lines. The isoclines are very useful for predicting the direction and final destination of population trajectories originating anywhere on the phase space diagram. The N_1 population will increase anywhere in the region to the left of the N_1 isocline and decrease anywhere to the right of it. In the same manner, the N_2 population will increase anywhere below the N_2 isocline and decrease anywhere above it. The isoclines thus pinpoint the location of maxima or minima in population trajectories.

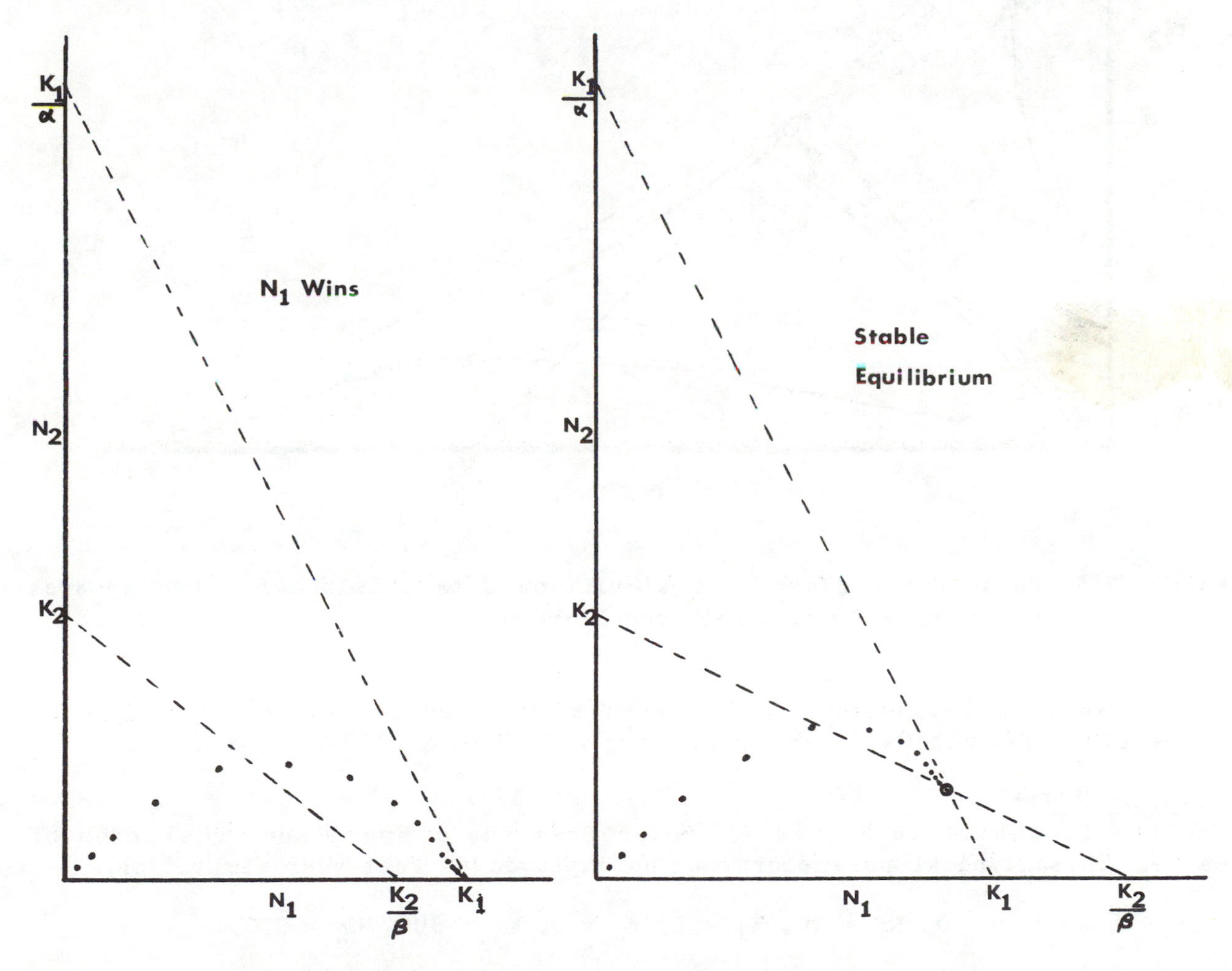

Figure 7.4. Phase space diagrams of simulation data illustrating two types of competitive interaction.

Some typical simulated phase space diagrams resulting from plotting population densities of species N_2 as a function of species N_1 appear in Figures 7.4 and 7.5. Note how the resulting trajectories are dependent on system parameters.

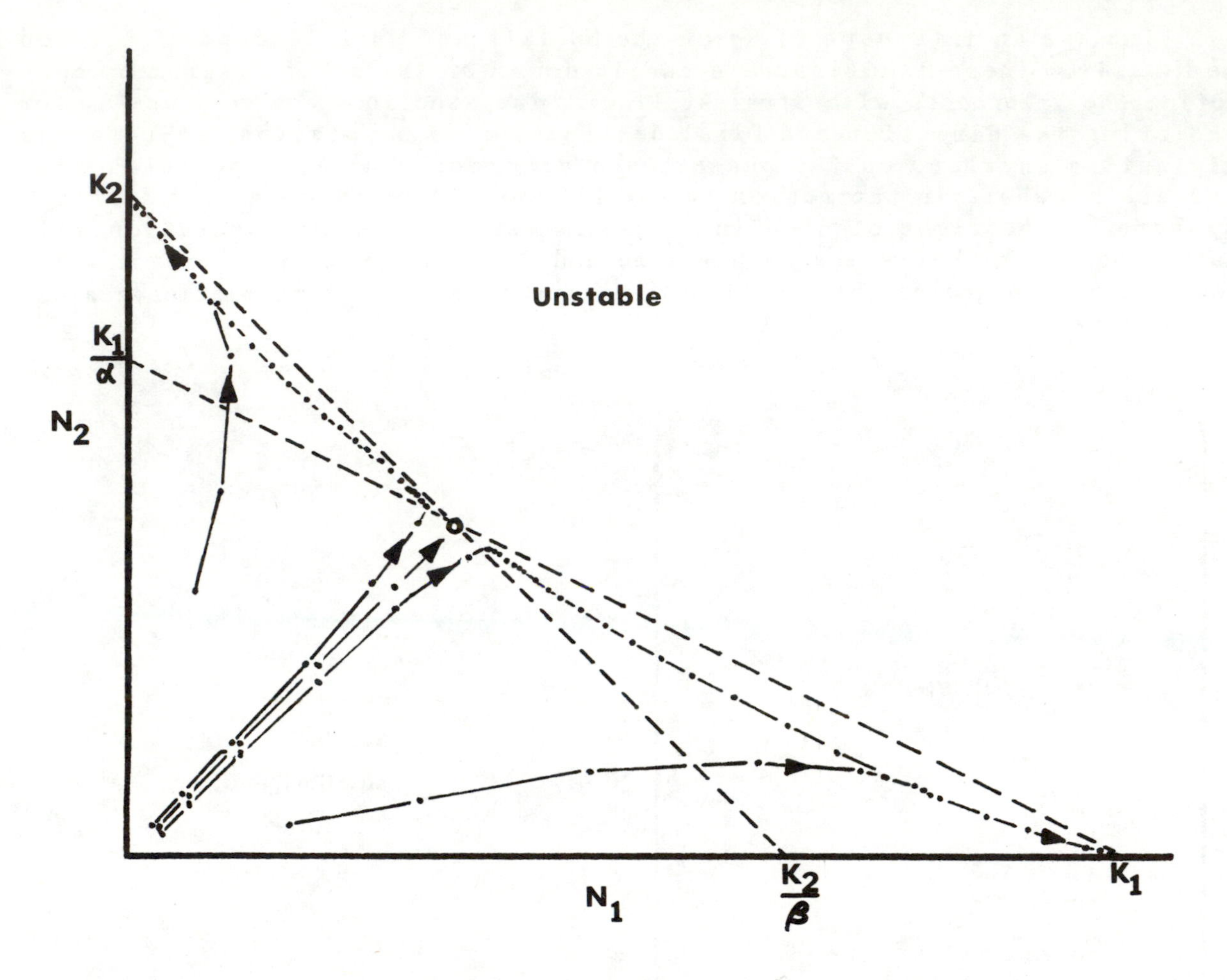

Figure 7.5. Phase space diagram of simulation data obtained for two species which exist in unstable equilibrium.

For a complete discussion of the results and significance of this type of simulation, consult Gause and Witt (1935), or Pielou (1977).

<u>Exercise 7.4</u>**

Use the flowchart in Figure 7.6 to program the competition simulation of Gause. These typical parameters may be employed to test your simulation.

$$N_1 = 10,\ N_2 = 10,\ r_1 = 1,\ r_2 = 1,\ K_1 = 300,\ K_2 = 300,$$

$$\alpha = 0.5,\ \beta = 0.5,\ \Delta t = 1,\ t_{max} = 30.$$

Plot data in the form of a phase space diagram as illustrated in Figures 7.4 and 7.5. Make at least two additional runs of the simulation using parameters which illustrate different types of population interaction.

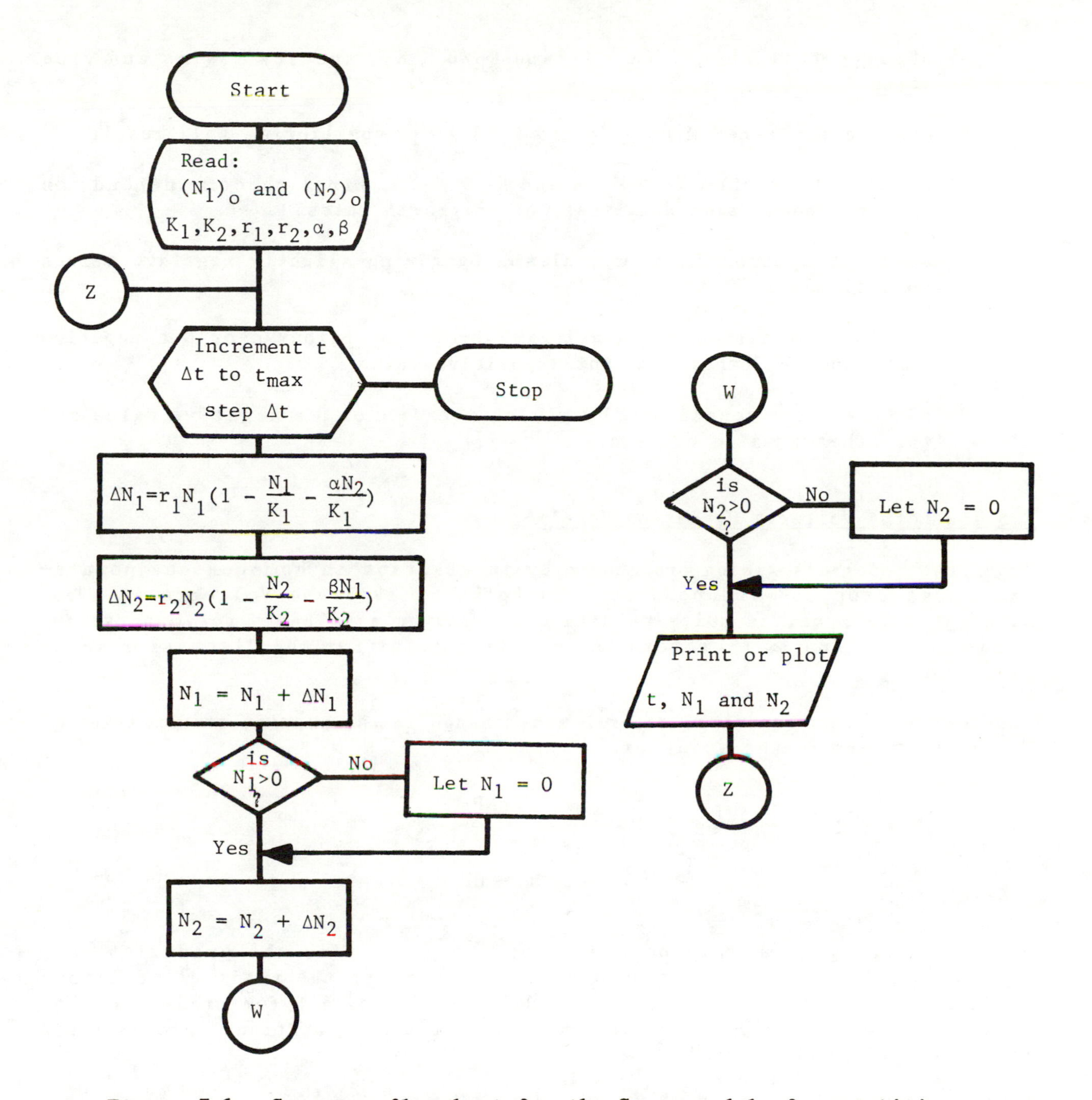

Figure 7.6. Computer flowchart for the Gause model of competition.

Using the competition models described, one may

1) observe the effect of different growth rates when the values of α and β are kept below 1,

2) observe the effect of different carrying capacities,

3) observe the effect of different α and β values,

4) observe that when $K_1/\alpha > K_2$ and $K_2/\beta < K_1$, species 1 wins and vice versa,

5) observe that when $K_1/\alpha > K_2$ and $K_2/\beta > K_1$, equilibrium will result,

6) observe that when $K_2 > K_1/\alpha$ and $K_1 > K_2/\beta$, final outcome depends on original population densities, or on growth rates,

7) simulate symbiosis, or mutualism, by giving slightly negative values to both α and β,

8) simulate predation or parasitism, by giving either α or β a negative value and the other constant a positive value,

9) simulate commensalism, by giving either α or β a negative value and the other a value of zero.

7.5 The Lotka-Volterra Predation Model**

The first model describing predator-prey interaction in homogeneous populations was proposed independently by Lotka in 1923 and Volterra in 1931. Although this model is quite primitive, it serves as a basis for many subsequent models, and is therefore the starting point for the discussion about predation.

The following two equations describe the change in the population densities of N, the prey, and P, the predator.

$$\frac{dN}{dt} = rN - pNP \qquad (7.51)$$

$$\frac{dP}{dt} = \alpha pNP - dP \qquad (7.52)$$

Here r is the growth rate constant for the prey, p is the predation rate constant, α is the assimilation constant expressing the ratio of predators produced to prey killed, and d is the death rate constant for predators. The cross product term, pNP, is a characteristic of these equations, and is based on the law of mass action.

One of the difficulties encountered in simulations based upon the simple cross product term for predation is that the simple Euler integration of Equation 7.51 can result in values of N which are less than zero. Ordinarily, this can be prevented by using the more precise improved Euler integration or by shortening the time interval.

As with the competition model, it is useful and instructive to display results from this model in the form of a phase diagram in which predator density, P, is plotted as a function of prey density, N. The result in the case of the Lotka-Volterra model is an elliptical trajectory which moves in a counterclockwise direction with time. Trajectories of this type are characteristic

Lotka-Volterra model is an elliptical trajectory which moves in a counter-clockwise direction with time. Trajectories of this type arc characteristic of most predator-prey systems. However, the Lotka-Volterra model is unique in that it results in a neutral cycle with constant amplitude determined solely by initial predator and prey populations for a given set of parameters. See Rosenzweig and MacArthur (1963).

As with the competition model it is useful to determine the prey and predator isoclines. The prey isocline is obtained by setting the prey growth rate equation (7.51) equal to zero and solving for predator density. This results in the equation:

$$P = \frac{r}{p} \tag{7.53}$$

The prey isocline is therefore a straight line parallel to the prey axis.

The predator isocline is similarly obtained by setting the predator growth rate equation equal to zero and solving for prey density. This results in the equation:

$$N = \frac{d}{\alpha p} \tag{7.54}$$

Thus, the predator isocline is a vertical line which crosses the prey isocline at $N = d/\alpha p$. As before, these isoclines are useful in predicting maxima and minima in the predator-prey phase space diagrams.

<u>Exercise 7.5**</u>
Write a computer program for the numerical integration of the Lotka-Volterra equations. These equations are a particularly good test of any numerical integration technique. They will require either very short time intervals or more advanced methods such as the Runge-Kutta. The results should be presented both as a plot of predator and prey densities as a function of time and as a phase space diagram of predator density as a function of prey density.

If the numerical integration has been performed correctly, the amplitude of oscillations for both predator and prey should remain constant through time.

The following parameters may be used to demonstrate the behavior of this model:

$$N_0 = 1500,\ P_0 = 50,\ r = 0.1,\ p = 0.002,\ d = 0.2,\ \alpha = 0.1.$$

7.6 Modifications of the Predation Rate Term**

One of the most serious criticisms of the Lotka-Volterra Model is that the cross product expression, pNP, is unrealistic in that it causes predation to be directly proportional to prey density no matter how low the predator population densities. In other words, there is no saturation or satiation of predators as one would observe in the real world. This can be seen if the equation for predation is rearranged and solved for prey killed per predator per time interval.

$$\frac{\Delta N_p}{\Delta t} = -\ pPN \qquad (7.61)$$

and

$$-\ \frac{\Delta N_p}{P\cdot\Delta t} = +\ pN \qquad (7.62)$$

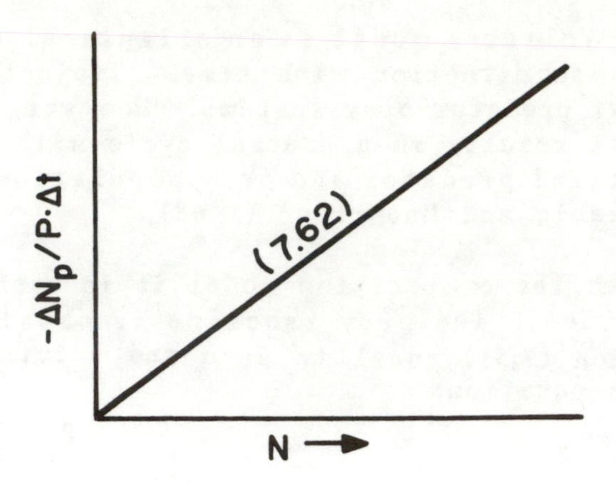

where ΔNp is change in prey density due to predation. A more realistic form of the curve would be the saturating or hyperbolic type. This has the following form:

$$-\ \frac{\Delta N_p}{P\cdot\Delta t} = +\frac{aN}{k + N} \qquad (7.63)$$

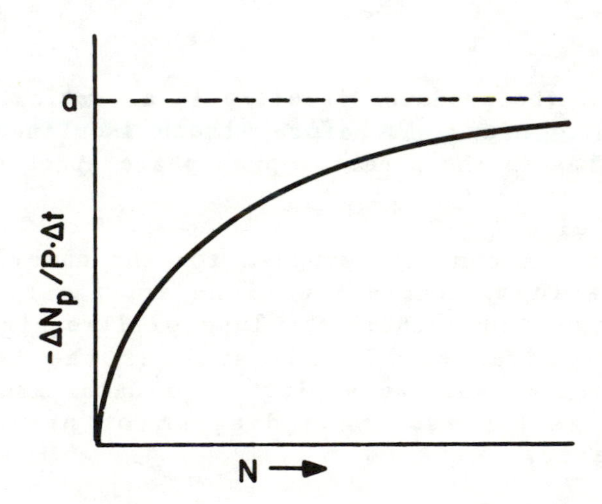

In this equation, a is a constant representing the maximum predation rate under satiating conditions. Equations of this form were employed by Holling (1959) and Tanner (1975). A steady-state derivation of this equation was discussed in Section 2.3.

O'Neill, et al. (1972) employed an equation of a similar type, but used the variable P (predator density) in place of the constant, k.

$$-\ \frac{\Delta N_p}{P\Delta t} = +\ \frac{SN}{(P + N)} \qquad (7.64)$$

The O'Neill equation has the advantage of being effective in modeling predation under essentially all possible population densities of both predators and prey. For further discussion, see DeAngelis, et al. (1975).
Still another equation which results in curves of the saturating type was

prey. For further discussion, see DeAngelis, et al. (1975).
Still another equation which results in curves of the saturating type was devised by Watt (1959) and Ivlev (1961). It has the following form:

$$-\frac{\Delta N_p}{P\Delta t} = S(1 - e^{-kN}) \qquad (7.65)$$

In this equation S would refer to the satiation level of kills per predator per time interval. The constant, k, expresses how rapidly this satiation level is achieved but in a different manner than the constant used in Equation 7.63.

In each of the above models, the effect is to make predation rate proportional to predator density when prey density is high, and proportional to both prey and predator density when prey density is low, thus preventing predation rates from becoming unrealistically high when prey is abundant.

Exercise 7.6**
Modify Equations 7.51 and 7.52 to employ the O'Neill Predation Model in place of the simple cross product term. Compare the results of this model with those of 7.5.

7.7 Modification of the Growth Terms**

Another criticism of the simple Lotka-Volterra equations is that neither equation takes into account that growth is limited by the carrying capacity. Therefore, Leslie and Gower (1960) proposed the following logistic type equations to describe the process:

$$\frac{dN}{dt} = rN(1 - \frac{N}{K} - \alpha P) \qquad (7.71)$$

$$\frac{dP}{dt} = sP(1 - \frac{P}{\beta N}) \qquad (7.72)$$

In these equations, growth of prey is limited by both the carrying capacity, K, and the predation term, αP. The growth of predators is a function of the intrinsic growth rate constant, s, and the number of prey, N. The constant β relates prey numbers to predator carrying capacity. Note that the death term of the Lotka-Volterra equation has now been incorporated into the logistic term $(1 - P/\beta N)$.

Exercise 7.7**
Write a flowchart for the predator-prey system which is described by the following growth rate equations which employ both Holling-Tanner and Leslie-Gower expressions:

$$\frac{dN}{dt} = rN(1 - \frac{N}{K}) - \frac{aPN}{k+N} \qquad (7.73)$$

$$\frac{dP}{dt} = sP(1 - \frac{P}{\alpha N}) \qquad (7.74)$$

Program the simulation and generate some simulation data. The following parameters are suggested:

$$N_0 = 25, P_0 = 5, r = 0.4, K = 900, a = 1, k = 200, s = 0.1, \alpha = 0.5.$$

Exercise 7.8
Use the growth rate equations in Exercise 7.7 to calculate the predator and prey isoclines. Draw these isoclines on the phase diagram and plot the data from Exercise 7.7. Repeat the simulation with the following set of data:

$$N_0 = 60, P_0 = 10, r = 0.3, K = 1000, s = 0.03, \alpha = 0.45, a = 1, k = 200.$$

Conclusion

This chapter has been concerned with the classical techniques for modeling the basic dynamics involving interactions between homogeneous populations. We have considered equations based upon the simple mass action law. We found that this simple cross product approach lacks in realism for certain kinds of interaction. Such simplifying approximations have caused many biologists to question the validity of modeling. It should not be forgotten, however, that simple approximations such as this have provided a means to get started with the difficult task of mathematically describing complex biological systems. Such simplifications were never intended to provide the final answer. As illustrated in this chapter, the cross product term has led to several more realistic ways for describing such processes as predation. The approach has great value because of its simplicity and limited theoretical basis. For this reason, you will see the cross product term used frequently in other parts of this text and elsewhere in the literature.

Several excellent reviews of population dynamics are available, should you wish to pursue this subject further. These include Gazis, Montroll, and Ryniker (1973), May (1976), Pielou (1977), Rosen (1970), Maynard Smith (1974), and Tanner (1975).

CHAPTER 8

SIMULATION MODELS BASED ON AGE-CLASSES AND LIFE TABLES

In the last chapter, we studied homogeneous populations in which birth and death processes were assumed to be independent of age and sex. For more realistic modeling, it is necessary to take into account the fact that in higher organisms only certain age-classes participate in reproduction, that natural death rates are usually a function of age, and that the processes of predation, parasitism, disease, and cannibalism are exerted to different degrees on different age-classes of a given organism. Models which divide populations into age-classes are described as having higher resolution than those which ignore age-classes.

Because of the mass of data involved in manipulation of age classes, age-specific survival rates and natality rates, matrix algebra is often employed for such models. However, this chapter will use a more easily understood stepwise approach. The application of matrix algebra to age-class models will be reserved for Chapter 17.

Before discussing the age-class model, let us first examine, from a qualitative standpoint, how age affects reproductive rate and natural death.

8.1 Age Specific Natality Rates**

In general, higher organisms must reach an age of sexual maturity before they are able to reproduce. Many higher organisms achieve their maximum reproductive rate shortly after sexual maturity, then the reproductive rate slowly declines with increasing age. One model which meets these requirements is a simple triangular shaped ramp function suggested by Lewontin (1965) to describe age specific fecundity. (See curve I, Figure 8.1)

Other organisms show a continuous increase in fecundity with increasing age

---

James D. Spain, BASIC Microcomputer Models in Biology

ISBN 0-201-10678-7

and size. This is a characteristic of many fishes, where fecundity would have the form of curve II in Figure 8.1, perhaps as a function of body weight. Finally, there are organisms which reproduce only once during their lifetime.

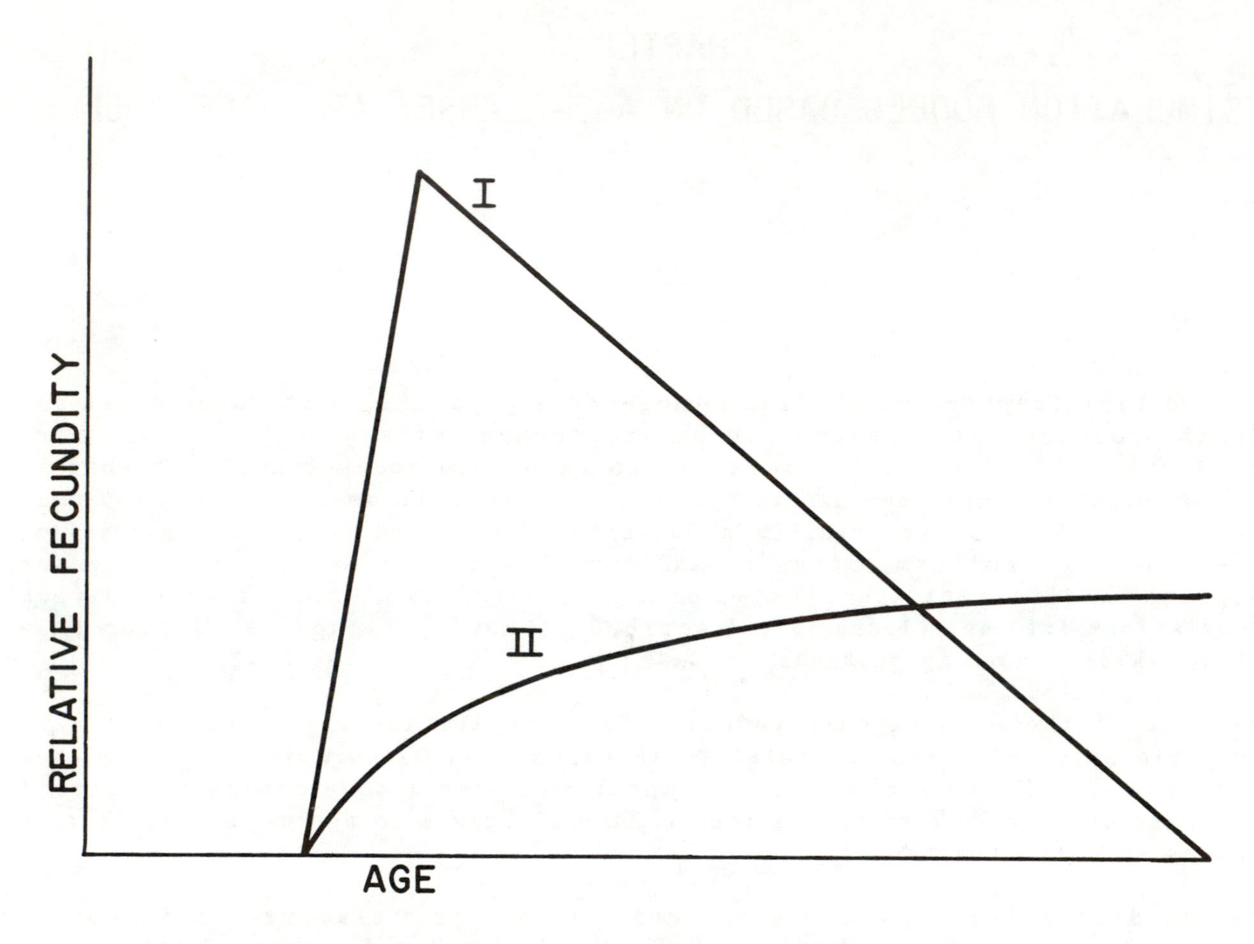

Figure 8.1. Variation in age-specific natality.

Because of the wide variation in age-specific natality patterns, the practical approach is simply to use a table of age-specific natality coefficients, based on observed reproductive rates, to calculate total natality. Since we are usually dealing with a limited number of discrete age classes the amount of data involved is not unreasonable, and the use of an equation as a function generator is hardly justified.

8.2 Age-Specific Survival Rates**

Survivorship curves for various organisms usually take one of three basic forms which were described by Pearl and Miner (1935). In such cases, a cohort, or group of individuals all born at the same time, is followed through its entire lifespan. The number of individuals surviving each time interval is plotted as a function of time. Units may be normalized to 100 in order to compare the survivorship curves for different organisms.

In Figure 8.2, it is seen that type I involves a species in which a cohort dies off more or less simultaneously near the end of the lifespan. Type II is shown by organisms which have a constant mortality rate throughout their lifespan. Type III is characteristic of organisms which have a very high mortality shortly after birth, but a relatively long life for the survivors. Various marine invertebrates, such as oysters and most plants tend to show a type III pattern. Many organisms are intermediate between I and II. Some organisms, including some human populations, show a slight variation of the type I curve because of a relatively higher mortality during the first years of life. This pattern is shown as type IV.

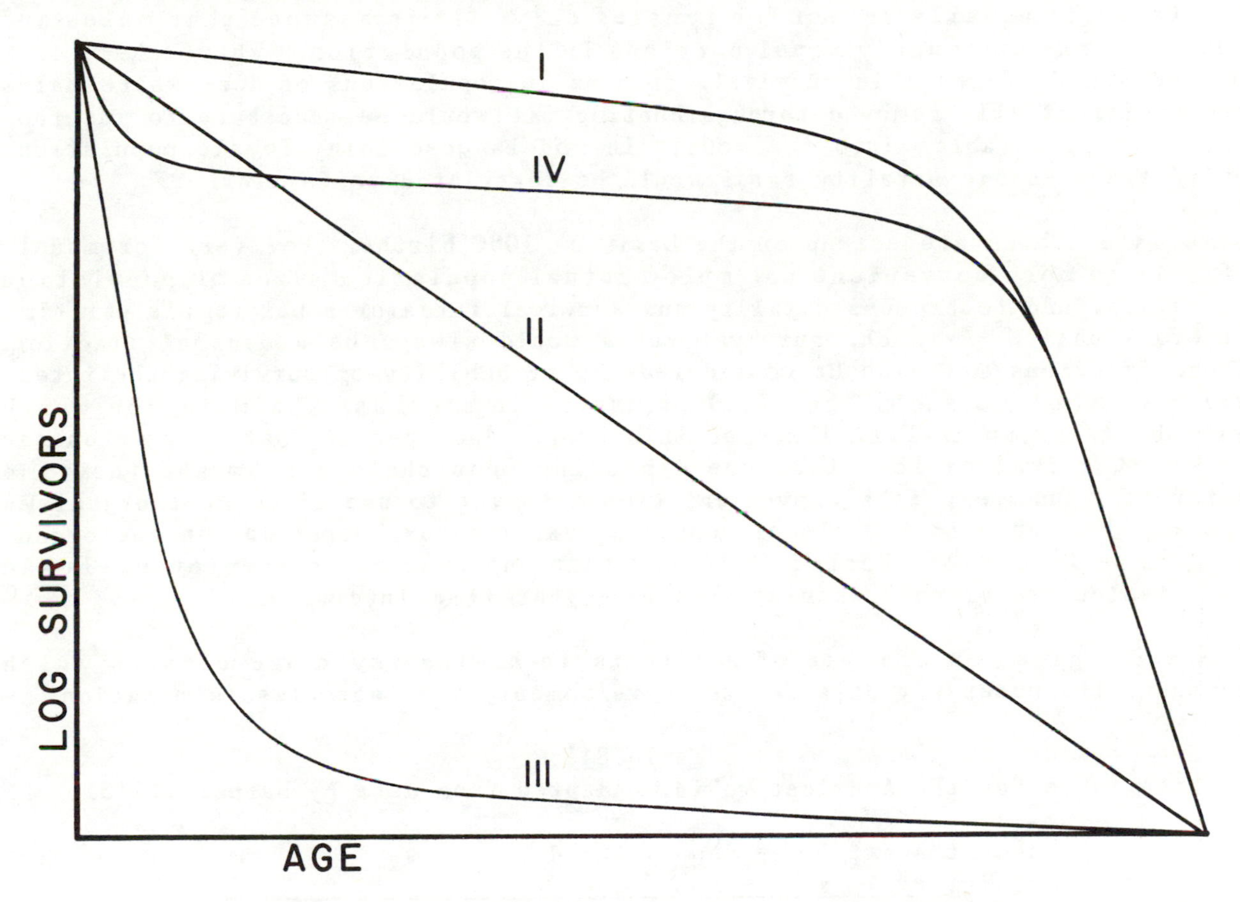

Figure 8.2. Four types of survivorship curves. Based on Pearl and Miner (1935). See text for explanation of types.

The direct approach to modeling survival for age-class simulations would be to store the survival rates for each age-class or else to incorporate survival rates as constants in each equation where needed. Another approach is to assume no natural death up to a point, then either truncate the life cycle completely or apply a fixed survivorship value for the last age-class. This last procedure departs rather markedly from reality, so that it would be employed only when the overall model was relatively insensitive to the survival sub-model.

8.3 Life Table**

The basic life table usually incorporates the following information about a population: the beginning of the age interval, x, the number of survivors at age x, here designated as N_x, the number of deaths in the age interval x to x+1, d_x, the proportion of survivors or age-specific survival rate for the interval x to x+1, s_x, and the age-specific natality rate for the age interval x to x+1, m_x. Other parameters are sometimes included, depending on the objective of the life tables.

The table is usually set up for females only. It is assumed that males and females make up roughly equal portions in the population. Where there is a marked sex difference in survival, such as in populations of deer where males are preferentially removed through hunting, it would be necessary to incorporate two life tables into the model. In models describing female populations only, age-specific natality rates would be restricted to females.

Some life tables are set up on the basis of 1000 births. However, for modeling, it is more convenient to employ actual population sizes or population densities, and to express natality and survival rates on a per female per time interval basis. As such, survival rates would always be a decimal fraction. These fractions may also be considered the probability of surviving the interval x to x+1. As such, they will be useful in stochastic simulation which will be taken up in Part Three of this book. Age specific natality rates can be greater than or less than one depending upon the organism and the time interval. One year is a convenient time interval to use since most organisms are keyed to an annual cycle of reproduction. However, depending on the organism, time intervals shorter or longer than one year may be employed. Human life tables are often broken up into five-year time intervals.

Table 8.1 gives an example of a life table having seven age-classes which contains the necessary data for the development of an age-class simulation.

Table 8.1
Life table for the American robin. Adapted from data by Farner (1945).

Age at start of interval, x	N_x	d_x	s_x	m_x
0	1400	903	.355	0
1	497	268	.461	1.4
2	229	130	.433	2.1
3	99	63	.364	1.6
4	36	26	.300	1.4
5	10	4	.200	1.1
6	6	5	.001	1.0

Age class 6 refers to six years and older.

The reproductive equation for the basic age-class model would be

$$\Delta N = \Sigma N_x m_x \tag{8.30}$$

$$= N_0 m_0 + N_1 m_1 + N_2 m_2 + N_3 m_3 + N_4 m_4 + N_5 m_5 + N_6 m_6 \tag{8.31}$$

The update equations describing the survival of each age-class must be solved in the following order:

$$N_6 \leftarrow N_5 s_5 + N_6 s_6 \quad (8.32)$$

$$N_5 \leftarrow N_4 s_4 \quad (8.33)$$

$$N_{x+1} \leftarrow N_x s_x \quad (8.34)$$

$$N_1 \leftarrow N_0 s_0 \quad (8.35)$$

$$N_0 \leftarrow \Delta N \quad (8.36)$$

$$N_t = N_0 + N_1 + N_2 + N_3 + N_4 + N_5 + N_6 \quad (8.37)$$

Note that the total natality is computed first, and saved as a dummy variable, ΔN. The value of ΔN is assigned to N_0 only after the previous value of N_0 has been employed to calculate N_1. It is also useful to have the total population, N_t, calculated and printed for each time interval. In this example, age-class six represents all individuals surviving past age-class five. It therefore includes survivors of age-class six, as well as survivors of age-class five.

Exercise 8.1**
Write a program for the simple age-class simulation of an American robin population. Implement the simulation with the steady-state data which have been provided in Table 8.1, and generate about 10 years of simulation data. Program your simulation to print out the number of individuals in each age class, as well as the total population. See Figure 8.3 for a flowchart of this model.

Exercise 8.2**
Equation 8.31 of the robin simulation may be modified as follows so that reproduction is limited by the carrying capacity in a logistic fashion.

$$\Delta N = (N_0 m_0 + N_1 m_1 + N_2 m_2 + N_3 m_3 + N_4 m_4 + N_5 m_5 + N_6 m_6) \cdot (2 - N_t/K) \quad (8.38)$$

Substitute this equation in the above program, and implement the simulation with the steady-state data, and a carrying capacity, $K = 2277$. After 10 time intervals, perturb the system by letting $N_0 = 1000$ and generate an additional 10 time intervals of data. Plot N_t and ΔN as a function of time to show the effect of the perturbation on the steady-state system.

The logistic term in this equation differs from the $(1 - N_t/K)$ term used in the Verhulst-Pearl model because in that model the intrinsic growth constant, r, describes the net effect of both birth and death processes. In this model, the $(2 - N_t/K)$ term is used to modify only total natality without affecting survival. The latter is assumed to be a density independent factor. Note that when $N_t = K$, this logistic term, $(2 - N_t/K)$, is equal to one, and natality is at the steady-state level, $\Sigma N_x m_x$. Thus, age-specific natality rates in this model represent half-maximum natality rates, as they may be increased by a factor of two as N_t approaches zero.

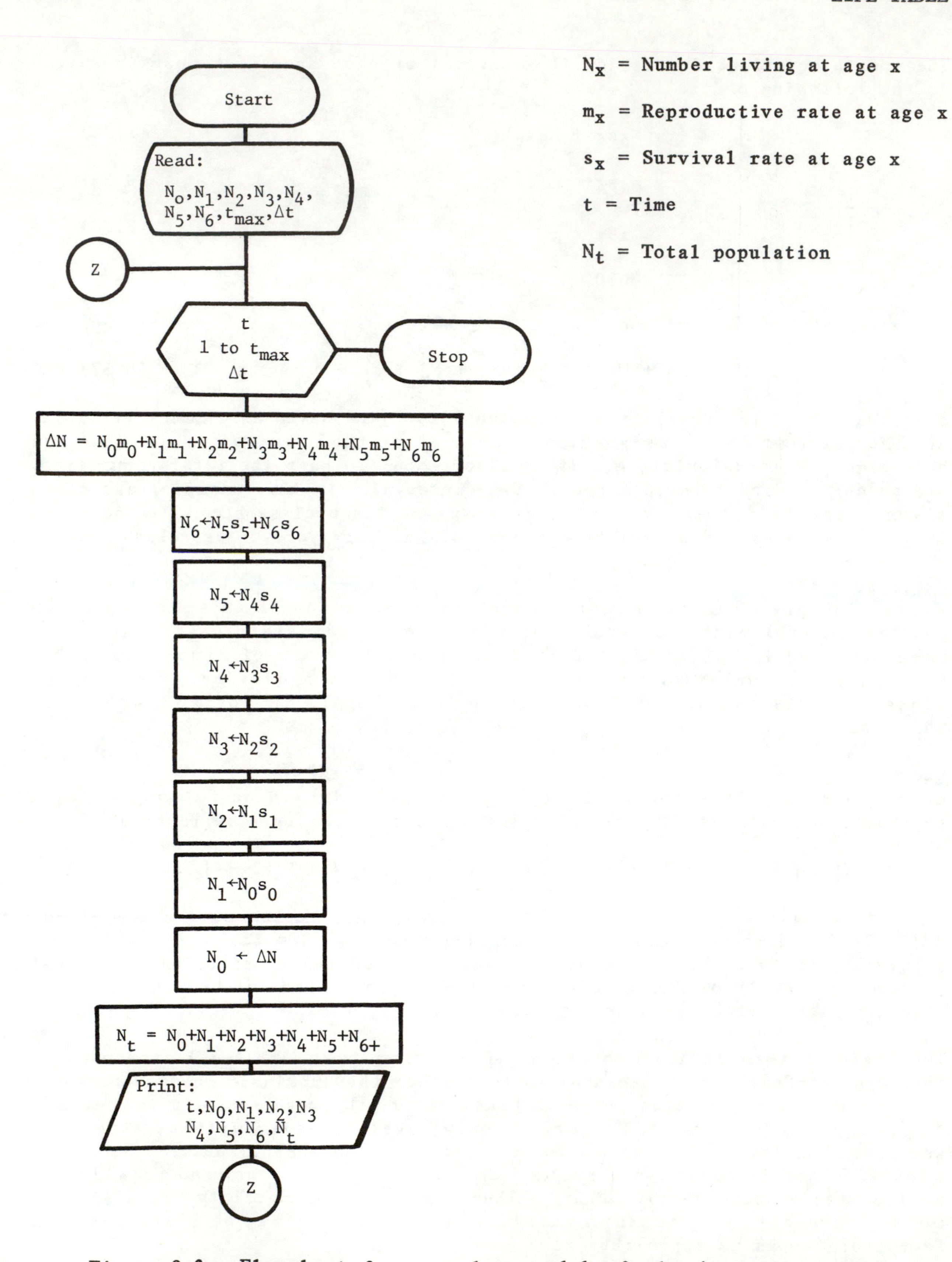

Figure 8.3. Flowchart for age-class model of the American robin.

Exercise 8.3**
Use the following data describing the U.S. population in 1965 to develop a demographic simulation. Program the simulation to calculate the predicted population table for each five-year interval until the year 2000. In addition to population structure, calculate and print out the total population and the average annual growth rate after each five-year interval. As a modification, you might include a means for proportional reduction in fertility of all age classes in order to investigate this effect on the population structure at year 2000. Survival rates and natality rates are based on five year intervals.

Table 8.2

Age-Class	Age	No. Females in Millions	Survival	Natality
1	0-5	9.397	0.978	-
2	5-10	10.268	.988	-
3	10-15	9.784	.998	-
4	15-20	8.786	.997	0.457
5	20-25	7.531	.996	.249
6	25-30	6.083	.996	.164
7	30-35	5.538	.994	.092
8	35-40	5.897	.991	.045
9	40-45	6.341	.986	.015
10	45-50	6.097	.979	.005
11	50-55	5.556	.969	-
12	55-60	4.952	.954	-
13	60-65	4.250	.932	-
14	65-70	3.543	.889	-
15	70-75	2.941	.810	-
16	75-80	2.198	.710	-
17	80-85	1.286	.500	-
18	85-90	0.727	.250	-
19	90-95	0.325	.120	-
20	95-100	0.175	.050	-

Data based on that of Justice (1968).

8.4 Use of Subscripted Variables in Age-Class Models

Programming large age-class models by the direct approach involves many equations of almost identical form. Whenever this occurs in writing computer programs, it is always a good idea to consider using subscripted variables in order to reduce the number of programming instructions. The age-class model is an excellent example where this type of program simplification may be used. The following list of program instructions employ subscripted variables to accomplish the same calculations as equation sequence 8.30 to 8.37.

Although this list of instructions is about the same length as the one using the direct approach, with slight modifications, it could also be used for models having 100 age-classes. For larger models, the use of subscripted variables is always more efficient. Be on the lookout for other opportunities to use this efficient programming technique!

```
100  REM AGE CLASS MODEL USING SUBSCRIPTED VARIABLES
110  DN = 0
120  FOR YX = 0 TO 6
130  DN = DN + N(YX) * M(YX)
140  NEXT YX
150  NT = 0
160  FOR YX = 6 TO 0 STEP - 1
170  N(YX + 1) = N(YX) * S(YX)
180  NT = NT + N(YX + 1)
190  NEXT YX
200  N(6) = N(6) + N(7)
210  N(0) = DN
220  NT = NT + N(0)
```

8.5 Age-Class Model with Sex Differentiated Survival and Reproduction

In certain populations, survival is a function of both age and sex. This phenomenon normally occurs in wildlife populations which are subject to heavy hunting pressure. The age-class model of the white-tailed deer population, selected to illustrate this approach, is based on the life history information presented in Table 8.3.

Table 8.3 Life history information, white-tailed deer

x	s_x	m_x	F_x	M_x	P_x^*
0	.62	0.00	1000	1000	0.00
1	.87	.50	620	694	.80
2	.88	.60	539	543	.89
3	.89	.63	474	406	.92
4	.90	.66	422	289	.95
5	.88	.68	380	195	.98
6	.86	.70	334	120	.99
7	.78	.68	287	62	.99
8	.62	.60	224	29	.99
9	.30	.50	139	11	.99
10	.10	.40	42	2	.99
11	0.00	.30	4	0	.99

Based on data from Dahlberg and Guettinger (1956).

*Hypothetical hunter preference values are based on a weighted average of .893. See explanation below.

In Table 8.3, x refers to the age in years at the beginning of the year, s_x is the natural survival per year, m_x is total fawn production per female per year based on steady-state natality, F_x and M_x refer to the population of females and males in each age-class. P_x is an assigned hunter preference which is a function of the value of the buck and its ability to be recognized. M_t and F_t are the totals of male and female deer in reproductive age-classes 1 through 11. M_t from the table is 2351 and F_t is 3465. The area involved is assumed to be 200 square miles.

For this model, one must retain two sets of population data, one for the males and one for the females. Both are assumed to be subject to the same natural survival rates, but in addition, the males are subject to hunting mortality. The hunting mortality sub-model is based on the idea that hunting success for each age-class is a function of the number of hunters, H, the age-specific preference coefficient, P_x, and density of adult male deer, M_t.

These variables are incorporated in an equation similar to the O'Neill predation model (Equation 7.64).

$$(\Delta M_x)_h = -M_x\left(\frac{H \cdot P_x}{H + M_t}\right) \tag{8.50}$$

where $(\Delta M_x)_h$ is the hunting loss incurred by males of age-class x during a single mating season. The equation for male survival from all causes would be the following:

$$M_{x+1} = M_x \cdot s_x\left(1 - \frac{H \cdot P_x}{H + M_t}\right) \tag{8.51}$$

This model assumes no hunting loss to the doe population.

The average value for P_x may be approximated from data taken by Jenkins and Bartlett (1959) for the deer herd in the northern portion of the lower peninsula of Michigan. The herd contains approximately 50,000 bucks. During a single hunting season, 284,000 hunters took 38,000 deer from an area of approximately 10,000 sq. miles. Under these conditions, the average value for P_x is .893. As the equation is set up, P_x is independent of area. The age specific values for P_x are based on reasonable expectations.

Populations in which the number of males is small relative to the number of females suffer from the additional problem that some females may not become fertilized during the mating season. A sub-model describing mating success should therefore be included in the overall simulation model. It would seem logical to make reproductive success a function of the number of reproducing males and to employ a saturating equation similar in form to 7.64. In this equation, reproductive success, R, is expressed in terms of the fraction of females becoming fertilized during one mating season.

$$R = \frac{M_t}{k + M_t} \tag{8.52}$$

where k is a constant describing how rapidly the female population is saturated with males. Assuming that one male is required for each 6 females to give 90 percent fertilization, the resulting value for k would be 261 bucks/200 sq. miles.

This sub-model would be combined with an overall reproduction equation similar to Equation 8.31 to give

$$\Delta N = (F_1m_1 + F_2m_2 + F_3m_3 + \ldots + F_{11}m_{11})\left(\frac{M_t}{k + M_t}\right) \tag{8.53}$$

The differential sex ratio at birth may be utilized to calculate the fraction of male births (.528) and female births (.472). These two factors are employed to calculate F_0 and M_0 in equations analogous to 8.36.

$$F_0 = .472(\Delta N) \quad (8.54)$$

$$M_0 = .528(\Delta N) \quad (8.55)$$

To maintain the population within reasonable limits, it will also be necessary to multiply Equation 8.53 by the logistic term, $(2 - (M_t + F_t)/5816)$, where 5816 is the carrying capacity for the system.

Exercise 8.4
Construct a flowchart for the white tailed deer and hunter system which is described above. Set up the model so that one may vary the number of hunters, and observe the effect on the total population of deer, their reproductive capacity, and the proportion of adult males in the population.

Exercise 8.5
Program the deer-hunter management model above, and carry out some simulated experiments to observe the effects of hunting pressure on total population, its reproductive capacity, and the adult sex ratio. Note that the ratio given in Table 8.3 is the result of some hunting pressure. Use appropriate diagrams to express your results. To simplify programming this model, you may limit it to 10 age-classes (0-9) for both males and females. Vary hunters from 0 to 10,000 in steps of 1000.

8.6 Fisheries Model Based on Age-Classes

Age-class models are also important to fisheries management because of the age-specific nature of both fecundity and fishing pressure on fish populations. Most fish also have a Type III natural mortality pattern resulting from a high mortality of eggs and immature stages. Since this mortality is quite often density dependent, there is an interesting relationship between the population density of the parental stock, and the number of recruits which are added to the parental stock. The density dependent stock-recruitment model, first proposed by Ricker (1954) is shown in Figure 8.4.

The equation which describes this curve is

$$F = P \cdot e^{(P_r - P)/P_m} \quad (8.60)$$

where F is the density of recruited spawners, P is parental stock density, P_r is the parental stock density which exactly replaces itself, and P_m is the parental stock needed to produce maximum recruitment. This model is mainly designed for use with homogeneous populations of fish, but, with modifications, it may be adapted to age-class models.

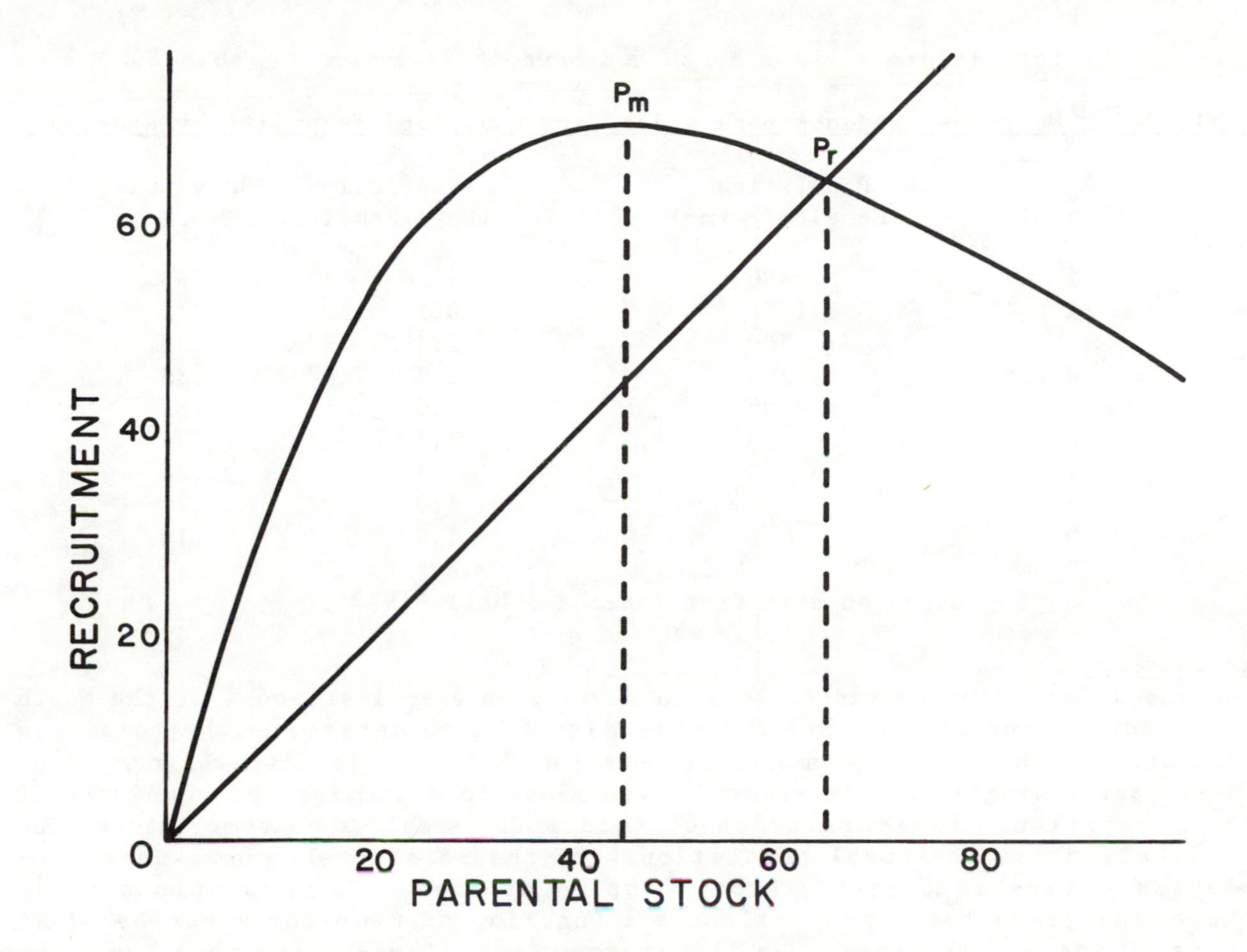

Figure 8.4. Recruitment of new adults as a function of parental stock density. P_r is replacement density, P_m is maximum recruitment density.

Jones and Hall (1973) employed a model of this type to relate recruitment of one year old haddock to egg production by the spawning stock. Their curve had a form similar to Figure 8.4. However, no equation was given. To model this response, Equation 8.60 was modified as follows:

$$N_1 = F \cdot E \cdot e^{(E_r - E)/E_m} \tag{8.61}$$

where N_1 is the number of one year old female recruits/cohort, F is a factor relating surviving eggs and larvae to female fish, E is the total eggs/cohort, E_r is the number of eggs required for a replacement or equilibrium population level, and E_m is the number of eggs required for maximum recruitment. All values are based on 3 square kilometers of pelagic area. The following conform approximately with the data of Jones and Hall:

$$F = 19.2 \times 10^{-6} \text{ female fish/egg}, \quad E_r = 130 \times 10^6 \text{ eggs/3 km}^2,$$

$$E_m = 80 \times 10^6 \text{ eggs/3 km}^2.$$

The life history information on North Sea haddock is given in Table 8.4.

Table 8.4. North Sea haddock population, survival, and fecundity by age-class

Age (Years)	Population (Females/3 km^2)	Fecundity (Eggs/Female)	Survival (S_x)
1	2500	--	.65
2	1630	0.58×10^5	.42
3	700	1.4 ''	.35
4	250	2.2 ''	.33
5	90	3.0 ''	.33
6	30	3.8 ''	.33
7	10	4.4 ''	.30
8	5	5.0 ''	.20
9	2	5.5 ''	.10

Based on data from Jones and Hall (1973).

Exercise 8.6
Use the information provided above to develop an age-class model of the North Sea haddock population. Use the fecundity data to determine the total egg production. Then, employ empirical Equation 8.61 to determine the number of first year recruits, and the survival equations to determine the remainder of the population. Implementation of this model should show some interesting oscillations in the total population. It should also show dominant age-classes moving as cohorts through the population. To demonstrate these phenomena graph total population as a function of time for a period of at least 30 years, and generate a histogram of age-classes each year for a ten year period.

8.7 Epidemic Simulation

Although an epidemic simulation does not involve age-classes, it does involve subdividing a population of organisms into sub-classes of a similar nature. The approach is basically that of Kermack and McKendricks (1932).

Assume that N represents the population density of animals which lack immunity to a particular disease. Assume that recruits of non-immune individuals, R, are continuously being added to this population through birth and immigration processes. Non-immune animals have a certain probability of becoming infected when they come in contact with contagious animals, C. On becoming infected, assume that these animals themselves become contagious for one time interval. During the following time interval they remain sick, but no longer contagious, perhaps because of isolation from the population as a whole. This stage is designated E.

Following this stage, the organisms either recover as immune individuals, I, or leave the population as dead individuals, D, depending on a survival constant, S. Some non-immune individuals may also result from immune individuals losing their immunity as a function of a first order coefficient, L. The diagram for this model is presented in Figure 8.5.

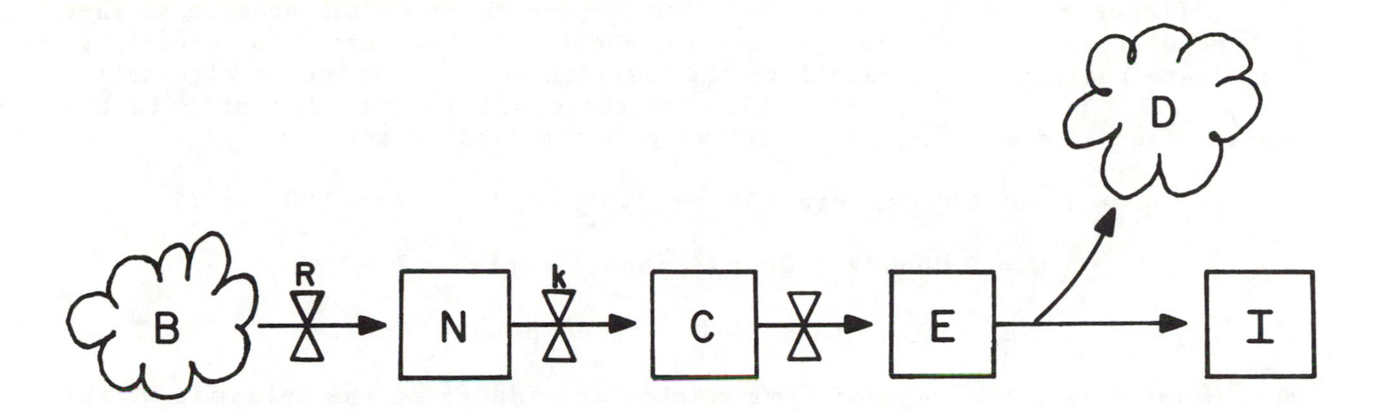

Figure 8.5. Diagram of typical epidemic model.

The conversion of N to C is described by the differential equation

$$-\frac{dN}{dt} = \frac{dC}{dt} = kNC \tag{8.70}$$

where k is the infectivity constant which is a function of the infectivity of the pathogen, and the probability of contact between N and C. The update equations for Euler integration of this process are:

$$\Delta N_c = kN \cdot C\Delta t \tag{8.71}$$

$$I \leftarrow I + E \cdot S \tag{8.72}$$

$$D \leftarrow D + E(1 - S) \tag{8.73}$$

$$E \leftarrow C \tag{8.74}$$

$$N \leftarrow N + R - \Delta N_c + L \cdot I \tag{8.75}$$

$$C \leftarrow \Delta N_c \tag{8.76}$$

The time interval, Δt, is equal to the time an individual spends in the contagious stage, C, and to simplify the model, also the time spent in the sick but not contagious stage, E. To retain proper continuity, the equations must be solved in the order listed.

The model demonstrates the threshold density effect of Kermack and McKendrick (1932) in that the population density of susceptible animals must be above a particular threshold in order for an active epidemic to be initiated by a single contagious individual. Mathematically this is the reciprocal of the infectivity constant.

Exercise 8.7**
Draw a flowchart for this model, and then prepare the computer program so that each population sub-class is recorded for about 150 time intervals (weeks). To facilitate analysis, also calculate the fraction of the population with active disease, $F = (C + E)/(I + N + C + E)$. Use the simulation to demonstrate the course of an epidemic using the following hypothetical data:

$$N_0 = 1{,}000{,}000 \text{ people/100 mi}^2, \quad C_0 = 1 \text{ person/100 mi}^2,$$

$$k = 0.000002 \text{ (100 mi}^2\text{/people}\cdot\text{weeks)}, \quad S = .9,$$

$$E_0 = D_0 = I_0 = 0, \quad R = 5{,}000 \text{ people/week.}$$

Plot N and F as a function of time to show the course of the epidemic. After you have the basic epidemic model running, implement the immunity loss pathway by having a certain fraction of immune individuals lose their immunity each week. A typical value for the loss coefficient, L, might be 0.03 per week.

8.8 Intraspecific Action: Cannibalism

The following background information on flour beetles was obtained from a classic paper by Thomas Park (1948). The mean length of various stages in the life cycle is given as follows in days: egg stage, 4.6, larval stage, 22.6, pupal stage, 6.6, and adult, 120. The oviposition rate averages about 12.5 eggs per female per day. Experiments were performed in which the beetles were allowed to grow and reproduce in a limited volume of whole wheat flour and brewer's yeast at constant temperature and humidity. The flour was sifted at 30 day intervals, the larvae, pupae and adults counted and returned to a fresh supply of media. The following is typical of the data obtained:

Age days	Larvae, pupae number/gram	Adults number/gram
30	24	3
60	3	24
90	1	24
120	0.7	23
150	0.7	22
180	2	20
210	2.5	14
240	4	10
480	9	5
960	10	5

From examination of these data, it seems clear that intensive cannibalism occurs at two time periods. Initially, there are few adults, and there is very little loss in going from larval or pupal stage to adult stage (24 larvae and pupae become 24 adults). However, later in the experiment, there is a very low carry-over from the larval or pupal stage to the adult stage (8-10 larvae and pupae become about 5 adults, most of which were probably old surviving adults). Therefore, cannibalism appears to be occurring during the pupal stage. The other period of cannibalism apparently occurs during the egg stage. For example, 20 adults should be laying about 250 eggs per day, and over a 30 day period this should result in about 7500 eggs, and a short period later, 7500 larvae and pupae. However, after day 60, less than one larva and pupa per gram are observed. A rather intense rate of cannibalism appears to be taking place. Park, et al. (1965) indicate that cannibalism is the primary mechanism for limiting the population of flour beetles under these conditions. The mathematical model presented below is based on simple assumptions concerning reproduction, natural survival, and cannibalism.

The following assumptions are made:

1) The life cycle of the beetle is divided into four stages. The first three, egg, larval, and pupal stages are assumed to be of equal length, one time interval. The adult stage is assumed to be three time intervals for all individuals. After the third interval, only a fraction continue to live. This survival fraction, S, is multiplied by the number of adults which have completed the third time interval.

2) The oviposition rate is assumed to be constant throughout the adult stage. The number of eggs laid per unit time interval would be described by:

$$\frac{\Delta E}{\Delta t} = R \cdot A \tag{8.80}$$

here ΔE is the number of eggs laid per time interval (Δt), R is the oviposition rate per adult, and A is the total number of adults. (Note that R would have a value of about 50 if the time interval is assumed to be 10 days.)

3) Cannibalism is assumed to occur both at the egg stage, and at the end of the pupal stage. The rate of cannibalism is assumed to be a function of both the density of adults and the density of eggs or pupae being cannibalized. Hence

$$-\frac{dE}{dt} = K \cdot E \cdot A \text{ and } -\frac{dP}{dt} = k \cdot P \cdot A \tag{8.81}$$

where K is a cannibalism constant for eggs, and k is the cannibalism constant for pupae. E is the number of eggs, A is the total number of adults, and P is the number of pupae at the end of the pupal stage.

If we assume that the number of adults is relatively constant during one time interval, then these equations may be integrated between the limits of t and t+Δt as follows:

$$\frac{dE}{dt} = -K \cdot E \cdot A$$

$$\int_{E_t}^{E_{t+\Delta t}} \frac{dE}{E} = \int_{t}^{t+\Delta t} -KA\,dt$$

Then $$E_{t+\Delta t} = E_t e^{-KA\Delta t} \qquad (8.82)$$

and by analogy $$P_{t+\Delta t} = P_t e^{-kA\Delta t} \qquad (8.83)$$

The following equations therefore describe this system:

$$A = A_3 + A_2 + A_1 \qquad (8.84)$$

$$A_3 = A_2 + A_3 \cdot S \qquad (8.85)$$

$$A_2 = A_1 \qquad (8.86)$$

$$A_1 = Pe^{-kA\Delta t} \qquad (8.87)$$

$$P = L \qquad (8.88)$$

$$L = Ee^{-KA\Delta t} \qquad (8.89)$$

$$E = RA \qquad (8.90)$$

The model assumes that eggs and pupae are preferred food of adult beetles, but not necessary to their survival.

<u>Exercise 8.8</u>
Diagram a flowchart and write a computer program to simulate flour beetle cannibalism. Implement the model by assuming two adult beetles have been placed in a flour can containing one kilogram of flour. Oviposition rate, R, is assumed to be 100 eggs/adult·10 days, K = 0.30 gram·(10 day)/adult, k = 0.03 gram·(10 day)/adult, and S = 0.8.

The model should show that when the adult population is high, the larval and pupal population is low, and vice versa. The consequence is that the system tends to oscillate. This oscillation is rapidly damped out with survival factors of 0.8–0.9 (mean life of adult of about 6–10 time intervals), but with a survival factor of 0.5 (mean adult lifetime of 4 time units), the system oscillates rather wildly. A typical simulation is shown in Figure 8.6. The constants for cannibalism and oviposition rate may be varied in order to match the model more perfectly to the experimental data.

If one puts in a value of zero for the cannibalism constant, this limit on population is removed and there is an interesting result. You should realize that other limitations not included in the model would come into play in a real population.

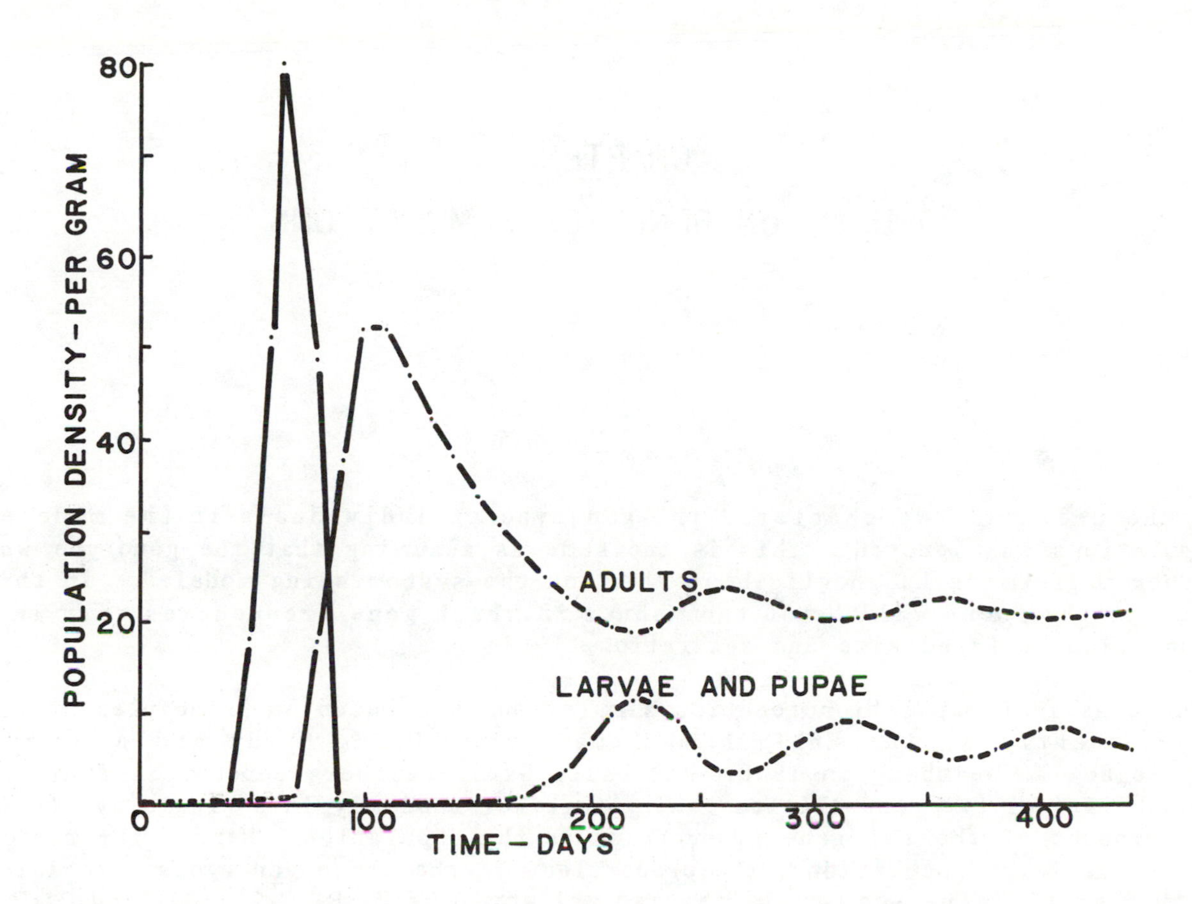

Figure 8.6. Simulation of intraspecific interaction in flour beetles.

<u>Conclusion</u>

This chapter has discussed seven different models which are based on the principle of dividing a population of organisms into sub-classes. Such models are more realistic than homogeneous population models because of the higher level of resolution which this technique provides. The exercises were chosen to illustrate the advantages of using age-class simulations. You should be warned, however, that this technique adds a degree of complexity to the simulation which may, in some cases, cloud the central issue being modeled. As a general rule, models should be no more complex than is required to account for the basic behavior of the system being modeled. This is simply a restatement of the principle that "entities should not be multiplied beyond necessity," made by the medieval scholastic, William of Occam. For an excellent discussion of this philosophy, see Hutchinson (1978).

CHAPTER 9

POPULATION GENETICS SIMULATIONS

In the previous two chapters, the genotype of individuals in the modeled populations was ignored. This is the same as assuming that the genotype was either uniform or had negligible effect on the system being modeled. In this chapter the focus will be on the manner in which gene frequencies vary in a population of fixed size and sex ratio.

Many classical population genetics simulations are based on principles stated independently by the English mathematician, G. H. Hardy and a German physician, W. Weinberg in 1908. The basic Hardy-Weinberg concept is that the probability of two gene alleles coming together randomly in a single zygote is the product of the two gene frequencies in the population. Hence, for random mating in large populations, the proportions of the three genotypes associated with a single gene locus (AA, Aa, and aa) would be $P_A \cdot P_A$, $2P_A \cdot P_a$, and $P_a \cdot P_a$, where P_a and P_A are the probabilities or proportions of the gene alleles a and A in the population as a whole. For further discussion of the Hardy-Weinberg principle, see Gardner (1973) or any other genetics text.

The simple population genetics simulations which are discussed in this chapter are for the most part based on an approach developed by Dahlberg (1948) and Baum (1967).

9.1 The Effect of Selection on Random Populations**

The following population genetics model simulates the effect of various levels of selection on populations of diploid organisms containing different proportions of dominant and recessive genes for a single characteristic.

Assume a characteristic which is determined by a single pair of alleles. Let a denote the recessive allele and A the dominant allele. Assume that genotype

James D. Spain, BASIC Microcomputer Models in Biology

ISBN 0-201-10678-7

aa, the homozygous recessive, results in a phenotype which can respond to selection, either for or against, by participating in reproduction to a greater or lesser extent compared to other phenotypes in the population. As there are only two alleles for the gene in the population, the proportion of a alleles can be represented by a single variable, r, where $P_a = r$ and $P_A = 1 - r$.

Furthermore assume that

1) at a given time, aa, aA, and AA have the respective proportions S, T, and U,
2) these proportions are the same for males and females,
3) mating is random among all individuals who participate in reproduction, and
4) no migration occurs into or out of the area described.

Then, the probability that one partner in a mating will donate gene a to an offspring is

$$r = \frac{0.5T + kS}{kS + T + U} \qquad (9.11)$$

where k is a constant describing the fitness of the homozygous recessive, aa, relative to the other genotypes.

According to the Hardy-Weinberg principle, the proportions in the succeeding generation will be

$$S = r^2 \qquad (9.12)$$

$$T = 2r(1 - r) \qquad (9.13)$$

$$U = (1 - r)^2 \qquad (9.14)$$

Selection may be simulated by calculating Equations 9.11 through 9.14 in a repetitive fashion, printing out the proportions of aa, aA, and AA for each succeeding generation. This simulation may explore any desired level of selection, either for or against genotype aa, relative to others. The flowchart for the basic population genetics simulation is given in Figure 9.1.

This simulation can demonstrate the following principles of population genetics:

1) In the absence of selection, any set of gene frequencies which have a total equal to one will revert to Hardy-Weinberg equilibrium in a single generation.

2) Selection against a recessive gene occurs when a reduced fitness value for the aa genotype is allowed to act over a period of several generations.

3) Ascendance of a dominant mutant is demonstrated by starting with a population having a very small fraction aA, and then simulating the initiation of a dominant mutant A. If A is given a selective advantage

by making the fitness of aa less than one, we may observe the rate at which gene A accumulates in the population. Figure 9.2 is an example of this type of simulation.

4) The limited value of eugenics is demonstrated by the low rate of removal of recessive genes when their proportion is very low, and selection is exerted only on individuals which express the phenotype, i.e. the homozygous recessive.

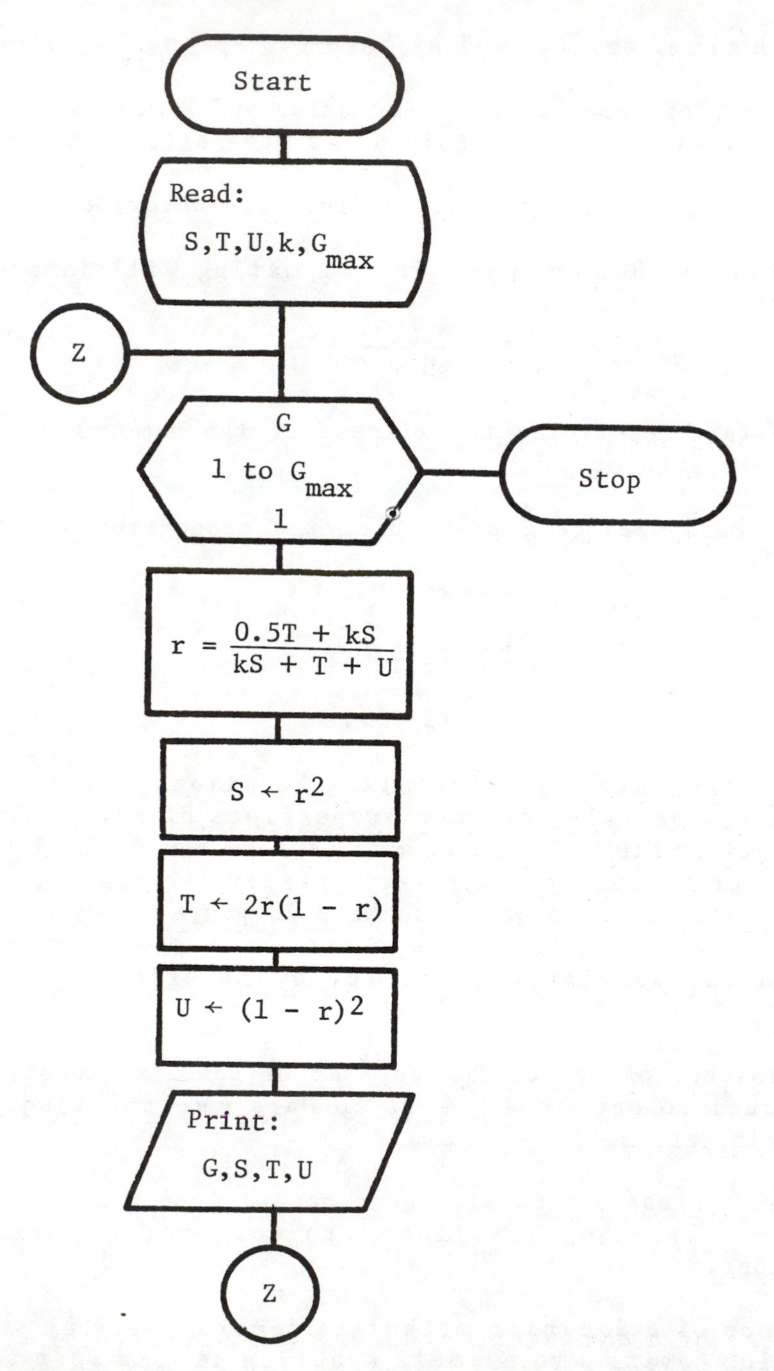

Figure 9.1. Flowchart for model describing selection against recessive genes.

Exercise 9.1**
Program the basic Hardy-Weinberg population genetics simulation. This system may be tested by comparing the simulation data produced when k = 0 with that of Exercise 2.9. After this has been checked out, obtain about 30 generations of simulated data using the following initial parameters:

$$S = .25,\ T = .50,\ U = .25,\ \text{and}\ k = .2.$$

Graph r and S as a function of time to show how selection has affected the frequency of recessive genes in the population.

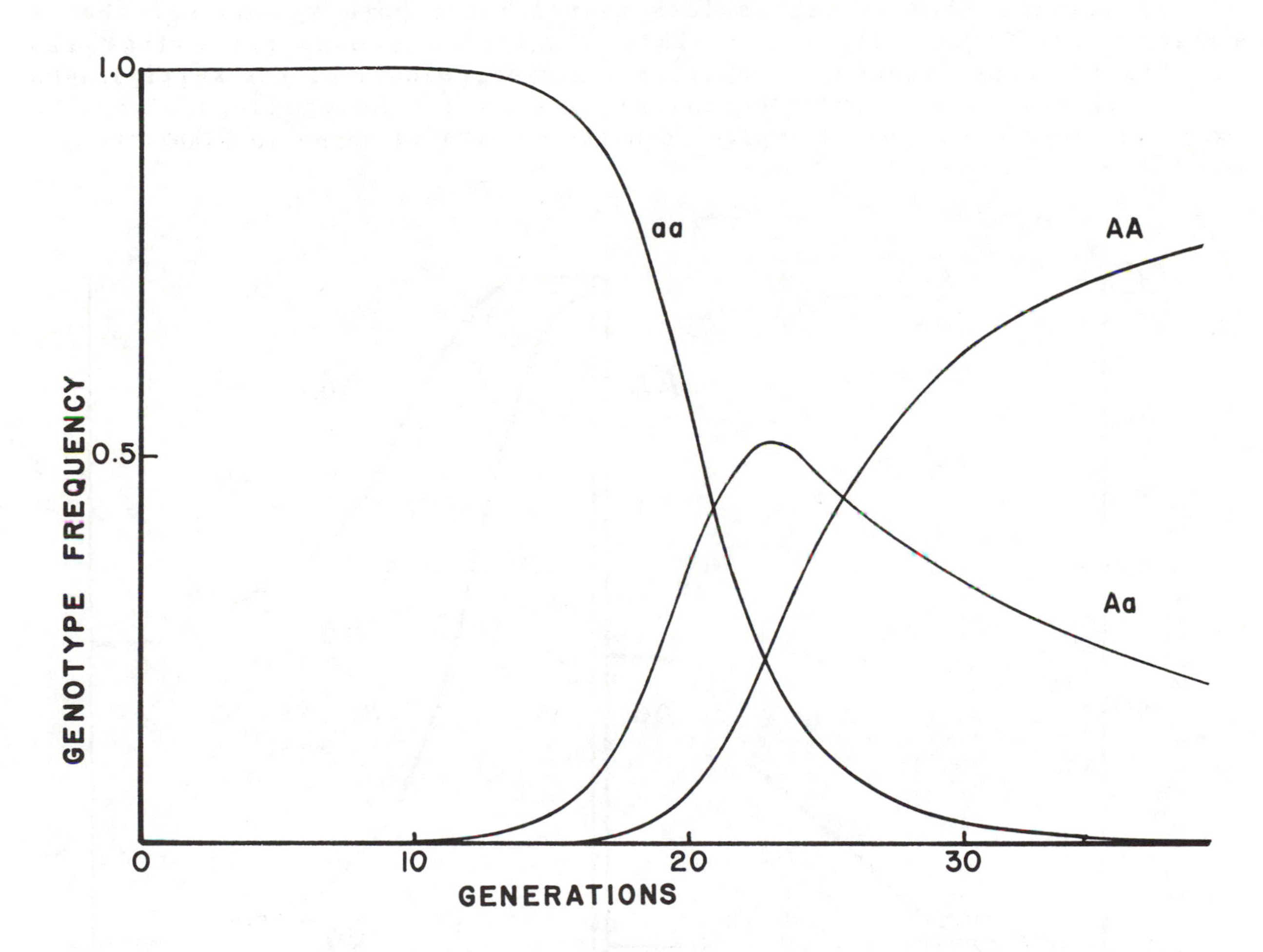

Figure 9.2. Simulated ascendancy of a dominant mutant with twice the fitness of the homozygous recessive.

9.2 Balanced Genetic Load and Balanced Polymorphism

The term balanced load encompasses many phenomena, but the most important is hybrid vigor, the observation that the heterozygote is often more fit than either of the homozygous genotypes. A number of mechanisms for this effect have been proposed, but all result in the population gene pool approaching an equilibrium condition in which there is a balanced gene frequency.

This model employs the same assumptions as in Section 9.1, except that the probability that one partner in a mating will donate an a-allele to an offspring is

$$r = \frac{0.5T + k_a S}{k_a S + T + k_A U} \quad (9.21)$$

where k_a is the fitness of the aa individuals and k_A is the fitness of the AA individuals, both expressed relative to heterozygote fitness which is assumed to be 1.0.

If one assumes fitness values less than 1.0 for both k_A and k_a, then a balanced condition will occur. This simulation demonstrates that the equilibrium gene frequencies obtained are independent of the initial gene frequency when there is selection against both homozygous genotypes. A graphical representation of typical simulation data is shown in Figure 9.3.

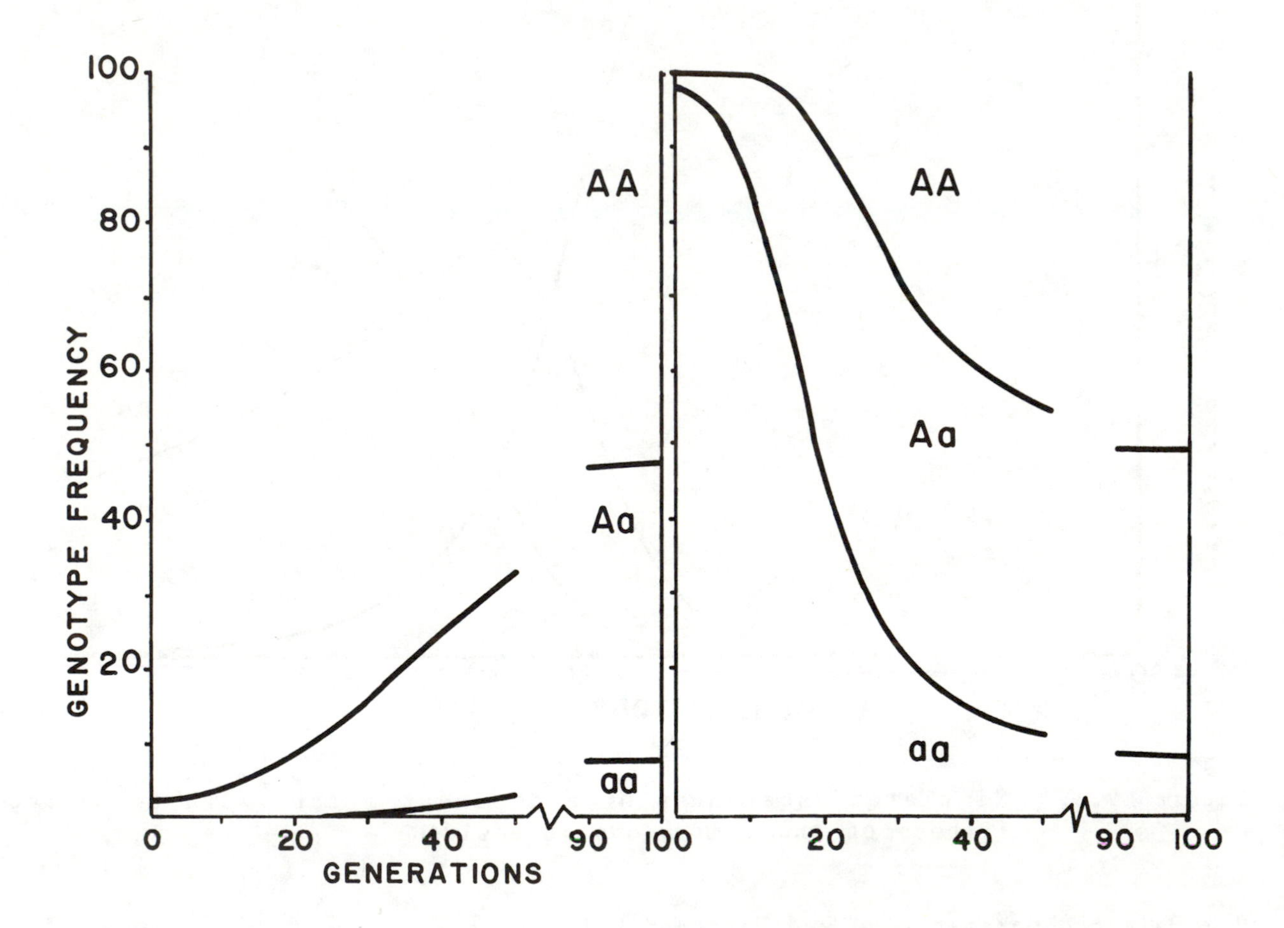

Figure 9.3. Balanced genetic load. Based on data of Wills (1970). Fitness of Aa is 1.00, of AA is 0.92,of aa is 0.80.

A classical example of this type of polymorphism is exhibited by native African populations which contain high frequencies of the gene for sickle-cell anemia. These populations are found in regions where a certain type of malaria is quite prevalent. The sickle-cell gene imparts some resistance to malaria so that heterozygous individuals are more fit than either of the homozygous genotypes. Those who are homozygous with respect to the sickle-cell allele usually die from anemia, while those who are homozygous for the normal allele are much more subject to death or debilitation from malaria. The result is a balanced condition in which the proportion of sickle-cell alleles is much higher than would be expected otherwise.

Exercise 9.2
Construct a flowchart for the balanced genetic load simulation described above. Program this simulation and test it with the following parameters:

$$S = 0.09, \ T = 0.30, \ U = 0.61, \ k_a = 0.2, \ k_A = 0.5.$$

Obtain 30 generations of data and graph the recessive gene frequency, r, as a function of time.

Exercise 9.3
The basic Hardy-Weinberg model, described in Section 9.1, may be modified to simulate the effect of selection on genes which show partial dominance. Find out what would be the result if the homozygous recessive individuals had a fitness of 0.1, while the heterozygotes had a fitness of 0.5 relative to the homozygous dominant. Make the necessary modifications in Equation 9.11, and program the simulation model. Use the initial values of S, T, and U from Exercise 9.1 to test the effect of this type of selection.

9.3 Balanced Mutational Load

Mutations which result in the change of one allele to another will, of course, cause a change in gene frequency. For point mutations, this is basically a chemical process involving the conversion of one base in the DNA code to another, thus causing a change in the genetic message at a particular locus. Because mutations are reversible, the process is not unlike that involved in chemical equilibrium. Assume allele A has a certain rate of mutation to allele a, and vice versa, as indicated in the following diagram

$$A \underset{n}{\overset{m}{\rightleftharpoons}} a$$

where m refers to the rate constant for the forward mutation, and n refers to the rate constant of reverse mutation. If r is the frequency of allele a in the population, then at equilibrium the forward and reverse mutation rates are equal, and may be described by the equation

$$n(r) = m(1 - r) \qquad (9.31)$$

This can be rearranged to give

$$r = \frac{m}{m + n} \qquad (9.32)$$

Since selection may also cause a change in allele frequencies, it is useful to combine the mutation model described in Equation 9.31 with the basic selection model described in Equation 9.11. The result is an equation describing the net effect of mutation and selection.

$$r = \frac{0.5T + kS}{kS + T + U} \cdot (1 - m - n) + m \qquad (9.33)$$

The result of selection and mutation acting on recessive genes within a population of randomly mating organisms should be a balanced or equilibrium gene frequency. If n is negligible compared to m, then according to Gardner (1970) the following equation will apply at equilibrium:

$$r = (m/(1 - k))^{1/2} \qquad (9.34)$$

This means that if $m = 0.00005$, and $k = 0.5$, then the equilibrium or balanced frequency for r should be 0.01.

Exercise 9.4

Program the balanced mutational load simulation and test it with the above parameters to see if an equilibrium frequency of 0.01 is obtained for the recessive genes. Make n very small, for example $n = 0.0000001$, $S = 0.0002$, $T = 0.03$, and $U = 0.97$, and run the simulation until a balanced condition is achieved.

9.4 Selection Against Sex-Linked Recessive Genes

Male organisms of many species have only a single X chromosome and are essentially haploid with respect to X, while females have two X chromosomes. Genes which are carried on the X chromosome are said to be sex-linked. Thus, males show a different pattern for the inheritance of sex-linked recessive genes, characterized by a higher frequency of expression.

To model this system, we will assume a characteristic which is determined by a single pair of sex-linked alleles. Assume that aa, the homozygous recessive genotype in females, and a-, the hemizygous recessive genotype in males, result in a phenotype which can respond to selection.

Furthermore, assume that in females aa, aA, and AA have the respective proportions S, T, and U. Mating is assumed to be random among all individuals who participate in reproduction.

The probability that a female will donate gene a to an offspring is

$$r_f = \frac{0.5T + KS}{KS + T + U} \qquad (9.41)$$

where K is the fitness of aa or a- relative to individuals with aA, AA or A- (hemizygous dominant) genotypes.

Among males, the probability of donating gene a to an offspring is

$$r_m = \frac{(r_f)_{t-1} \cdot K}{(r_f)_{t-1} \cdot K + (1 - (r_f)_{t-1})} \quad (9.42)$$

where r_m is the probability that a male will donate gene a to an offspring and $(r_f)_{t-1}$ is the probability that females of the previous generation will donate gene a. According to the Hardy-Weinberg principle, the proportions among females in the succeeding generation will be

$$S = r_f \cdot r_m \quad (9.43)$$

$$T = r_f(1 - r_m) + r_m(1 - r_f) \quad (9.44)$$

$$U = (1 - r_m)(1 - r_f) \quad (9.45)$$

This simulation demonstrates the principle that sex-linked recessive genes are expressed phenotypically much more frequently among males in a randomly mating population.

Exercise 9.5
Write a program for simulating the effect of selection on sex-linked recessive genes utilizing Equations 9.41 through 9.45. Implement the similation with the following parameters:

$$r_m = 0.5, \ S = 0.25, \ T = 0.5, \ U = 0.25, \ K = 0.3.$$

Produce about 30 generations of simulation data, and graph the frequency of both males and females expressing this recessive phenotype, r_m and S, as a function of generation time.

9.5 Selection against Recessive Genes Involving Two Loci

This model is designed to simulate the effect of selection against recessive genes in two independently assorting gene pairs involved in controlling a single phenotype. Phenotypes, such as size and weight, are often quantitatively related to the total number of dominant alleles. Assume the following designations for genotype frequencies:

S = Proportion of aabb
T = Proportion of aabB or aAbb
U = Proportion of aaBB or aAbB or AAbb
V = Proportion of aABB or AAbB
W = Proportion of AABB

If the aabb genotype is totally selected against, the probability that one partner will donate either gene a or b to a mating will be:

$$r = \frac{0.25V + 0.5U + 0.75T}{T + U + V + W} \quad (9.51)$$

S is not included since it is assumed to have zero reproductive capacity.

The Hardy-Weinberg principle may be used to write the following equations:

$$S = r^4 \tag{9.52}$$

$$T = 4r^3(1 - r) \tag{9.53}$$

$$U = 6r^2(1 - r)^2 \tag{9.54}$$

$$V = 4r(1 - r)^3 \tag{9.55}$$

$$W = (1 - r)^4 \tag{9.56}$$

This model demonstrates the shifting of the composition of the gene pool of the population when two gene pairs are involved.

Exercise 9.6
Write a program to implement the model described in Section 9.5. Select any reasonable input parameters, and graph the resulting simulation data to show how there is a shift in the population distribution for the phenotype involved. Rewrite Equation 9.51 for the condition in which S has a fitness other than zero. How is the phenotype distribution affected if the fitness of S is assumed to be 0.8?

9.6 The Inbreeding Model of Wright

The Hardy-Weinberg model, discussed above, is based on the assumption that populations are large enough that mating is random, and the amount of inbreeding is zero. In populations where significant inbreeding is taking place, the Hardy-Weinberg equations may be modified by inclusion of an inbreeding coefficient first described by Wright (1922). The following equations result:

$$S = r^2(1 - f) + r(f) \tag{9.61}$$

$$T = 2(1 - r)(r)(1 - f) \tag{9.62}$$

$$U = (1 - r)^2(1 - f) + (1 - r)(f) \tag{9.63}$$

where f is the inbreeding coefficient. As inbreeding has no effect on gene frequencies, r is assumed to be constant in this model. Inbreeding does cause a change in genotype frequencies, S, T, and U, and these in turn affect the inbreeding coefficient, f. This coefficient is defined by Gardner (1973) as the probability that two allelic genes, united in a zygote, are both descended from a common ancestor. For any inbreeding system the proportion of common genes increases. Therefore, f follows a pattern of variation which is unique to each system. The recurrence relations for f are obtained from analysis of regular systems of mating. According to Li (1976), the recurrence value for f under conditions of self-fertilization is given by the equation

$$f_t = 0.5(1 + f_{t-1}) \tag{9.64}$$

where t and t-1 are generation times. Assuming one went from random mating (f = 0) to self-fertilization, then $f_t = 0.5$, $f_{t+1} = 0.75$, $f_{t+2} = 0.875$, etc.

For a system of sib-mating, the following recurrence equation pertains:

$$f_t = 0.25(1 + 2f_{t-1} + f_{t-2}) \qquad (9.65)$$

For mating between a fixed sire and his daughter, granddaughter, great-granddaughter, etc., the recurrence equation for f is:

$$f_t = 0.25(1 + 2f_{t-1}) \qquad (9.66)$$

For further equations describing the variation of the inbreeding coefficient under different systems of mating, see Crow and Kimura (1970), or Li (1976).

Exercise 9.7

Write a general program which will allow you to calculate the proportion of homozygous dominants, homozygous recessives, and heterozygotes in populations with various patterns of inbreeding utilizing Equations 9.64-9.66. You should plan your program to retain f_t, f_{t-1}, and f_{t-2} for use in the appropriate recurrence equation.

Note that for a general inbreeding model, the recurrence equation employed will occupy only a single statement in the program. Assume $r = 0.5$, and determine the fraction of heterozygotes in the population at each generation time. Use a graph to show the rate at which homozygosity is approached under each of the systems of mating.

Exercise 9.8

Modify the basic Hardy-Weinberg model described in Section 9.1 to permit you to observe the effect of immigration on the frequency of the recessive alleles in a simple population with an assumed selection pressure. Assume f_i is the fraction of the population that is made up of new immigrants having a set frequency of recessive alleles, r_o.

After you have defined the equations involved, program the model on the computer, and implement it with the following input parameters:

$$S = 0.25,\ T = 0.5,\ U = 0.25,\ k = 0.1,\ f_i = 0.2 \text{ and } r_o = 0.5.$$

Exercise 9.9

The really ambitious student should explore the possibility of combining most of the models described in this chapter into a single interactive simulation which would allow a user to investigate almost any combination of simple phenomena related to population genetics.

Conclusion

The field of genetics, and population genetics in particular, is a very fertile one for the development of simulation models. What has been presented here is only an introduction to some of the most basic methods. For further information, refer to Crow and Kimura (1970) which provides an excellent bibliography including numerous books and papers on simulation and modeling. In the examples given we did not attempt to integrate population genetics models with population growth models. However, there is no reason why this cannot be done as long as one meets the assumptions on which both classes of models are based. The models described in this chapter, like those in the preceding chapters, are deterministic in nature. Genetics is also a particularly fertile ground for stochastic models of various types, some of which will be taken up in Chapters 19 and 21.

CHAPTER 10

LIGHT INTENSITY AND PHOTOSYNTHESIS

In our previous discussion, the models have been concerned primarily with the response of organisms and populations to biotic factors of one kind or another. In the next two chapters, we will be discussing the modeling of light and temperature in relation to living organisms. These are the two most important abiotic factors which are included in biological models.

The variation of light intensity and temperature is largely deterministic in that each follows rather well defined cycles of both an annual and daily periodicity. In the environment, light intensity and temperature also show random fluctuations due to cloud cover and storm fronts. However, this stochastic variation will be reserved for a later chapter.

The models describing the variation of both light intensity and temperature will serve as components of larger models dealing with energy input and transfer that will be taken up in subsequent chapters. As such, these submodels are commonly referred to as forcing functions to multicomponent models.

In this chapter we will be considering the temporal and spatial variations in light intensity and the effect of light intensity on biological activity, with particular emphasis on models dealing with the effect of light on the rate of photosynthesis.

10.1 Variations of Light Intensity on an Annual Cycle**

The amount of light striking the earth's surface is a function of the intensity of light striking the upper atmosphere, the angle of incidence, and the amount of absorption from atmosphere or cloud cover. For cloudless days, however, direct radiation per day depends largely on the angular height of the sun, and the length of day. It is therefore possible to estimate the daily

James D. Spain, BASIC Microcomputer Models in Biology ISBN 0-201-10678-7

light intensity at sea level at various latitudes assuming a mean correction for atmospheric turbidity. The curves seen in Figure 10.1 are based on estimated light intensities in calories per cm^2 per day (Hutchinson, 1957). The general form of each curve is sinusoidal, and it is therefore reasonable to model the intensity of light as a sine function of time. The wavelength will be 52 weeks, 365 days, or 12 months, depending upon the units of time being employed. To model the light intensity at 45° north latitude, we would employ the average light intensity (378) and the variations on each side of the average for that latitude ($\pm$249) to write the following equation:

$$I = 249\sin\left(2\pi\frac{(W - 11)}{52}\right) + 378 \qquad (10.11)$$

In this equation, I is the intensity of direct radiation in calories per cm^2 per day. The time in weeks, W, is expressed relative to the calendar year. The angle, in radians, is given as $2\pi(W - 11)/52$ in order to provide for the necessary shifting of phase. Vernal equinox, where $\sin(t) = 0$, is assumed to occur at week 11.

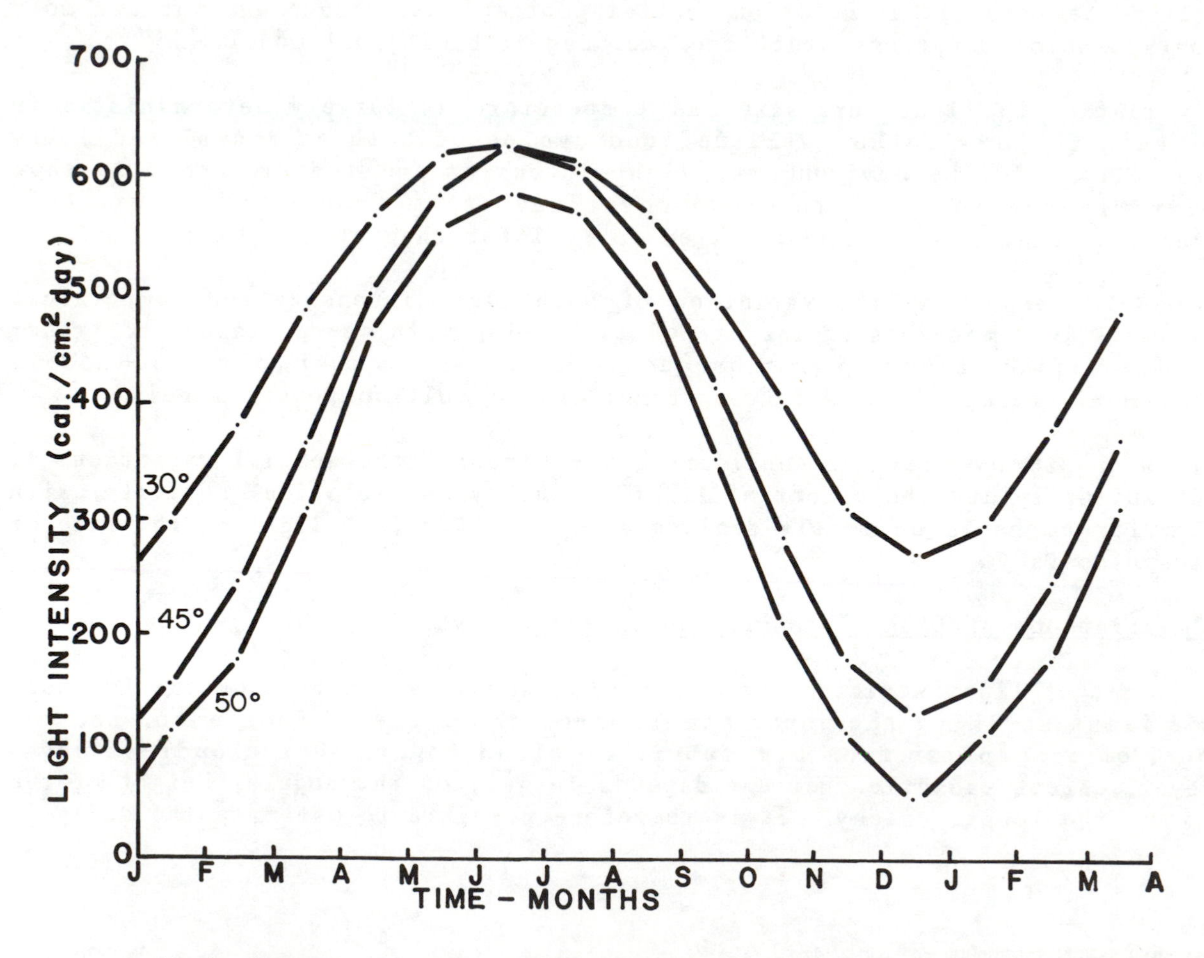

Figure 10.1. Annual variation in light intensity at different latitudes. Based on data from Hutchinson (1957).

The general equation for annual light variation is

$$I = R\sin\left(2\pi\left(\frac{W - 11}{52}\right)\right) + M \qquad (10.12)$$

where M is the mean annual daily intensity, and R is the range on either side of the mean.

Exercise 10.1
Write the model equation for calculating daily light intensities by week at 30° north latitude. Also, use the data given in Figure 10.1 to estimate what the parameters should be for your specific latitude.

Exercise 10.2**
Write a program to calculate daily light intensities by week throughout a single annual cycle. Implement your model with the constants used in Equation 10.11. Graph daily light intensity as a function of time in weeks.

10.2 Modeling Diurnal Variation in Light Intensity

Each day as the sun moves through an angle of 180°, the incident light intensity on each square centimeter of horizontal surface is a sinusoidal function of time. The light intensity at any time may therefore be approximated by a sine curve equation which has been truncated to eliminate negative values.

$$I = I_m\sin\left(2\pi\left(\frac{T - 6}{24}\right)\right) + B \qquad (10.21)$$

where I_m is the mean daily maximum light intensity. The curves described by this equation are given in Figure 10.2.

To model variable day length and maximum (noon) radiation values, a suggested approach is to vary the constant, B. B would be zero under conditions of 12 hours day and 12 hours night, but would have positive values for longer days and more intense noon radiation, and negative values for shorter days and lower noon radiation intensities.

Exercise 10.3**
Employ Equation 10.21 to model the daily light intensity for each hour during a 24 hour period. Program the computer so that it will numerically integrate the total radiation over one day. Assume I_m = 30 cal/cm^2/hr, and a 12 hour photoperiod.

10.3 Effects of Cloud Cover and Backscatter

The preceding equations are based on direct sunlight from clear skies. It is estimated by Gates (1962) that the average cloud cover for the northern hemisphere is about 52 percent. About half of the light hitting clouds is reflected back to space, about 1/5 is absorbed, and the rest is transmitted to the surface of the earth. Taken on an annual basis, averaged over the northern hemisphere, this means that direct radiation should be reduced by 35 percent by multiplying by a factor of 0.65. Of course, if an actual value for average effective radiation is available for the particular locality being modeled, this value should be employed.

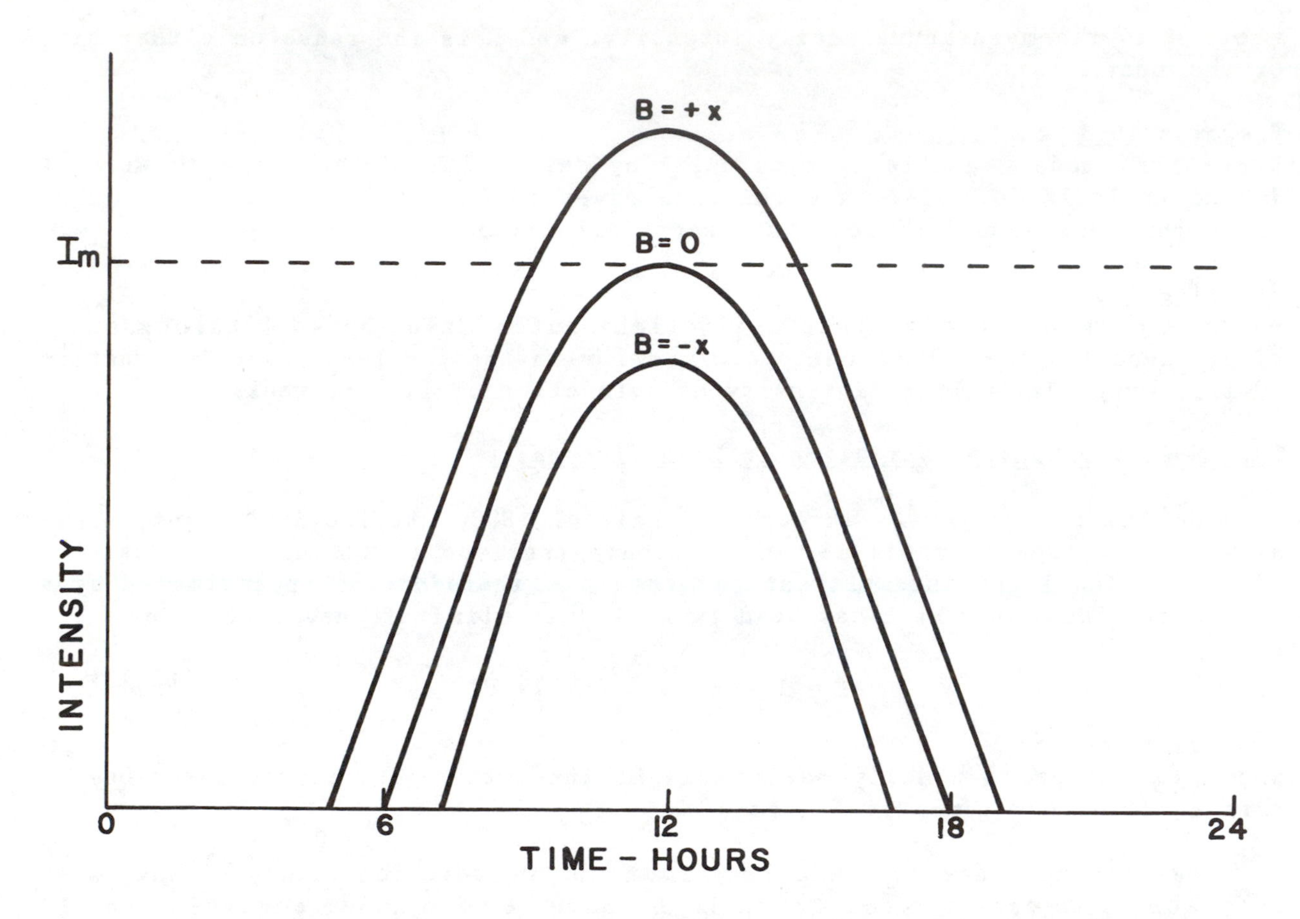

Figure 10.2. Truncated sine curve simulation of diurnal light intensity.

For some areas of the country, cloud cover is highly seasonal. Anderson (1974) developed an elaborate model for light intensity on the coast of southern California which took into account the heavy cloud cover that is experienced during the early months of summer. Cloud cover may be modeled by fitting a polynomial equation to local weather data obtained from the National Oceanic and Atmospheric Administration, Environmental Data Service, Asheville, NC. This empirical equation may then be employed as a correction to the basic sine curve equation described in Section 10.1. This kind of fine tuning is not required for the exercises in this text, however.

10.4 Reflection of Light from Water Surfaces

When light intensities are involved in the modeling of aquatic systems, it may be necessary to take into account the fact that surface light intensity is reduced due to the reflection that occurs at the air/water interface.

Reflection is a function of the incident angle of light, becoming almost total at very low angles. The fraction of light being reflected, F_r, may be obtained from the Fresnel's law equation.

$$F_r = \frac{I_r}{I_i} = \frac{1}{2}\,\frac{\sin^2(\phi_i - \phi_r)}{\sin^2(\phi_i + \phi_r)} + \frac{\tan^2(\phi_i - \phi_r)}{\tan^2(\phi_i + \phi_r)} \tag{10.41}$$

where I_i is the intensity of incoming radiation, and I_r is the intensity of refracted radiation.

The fraction entering the water, F_w, is

$$F_w = 1 - F_r \tag{10.42}$$

The angles involved in Equation 10.41 are defined in Figure 10.3.

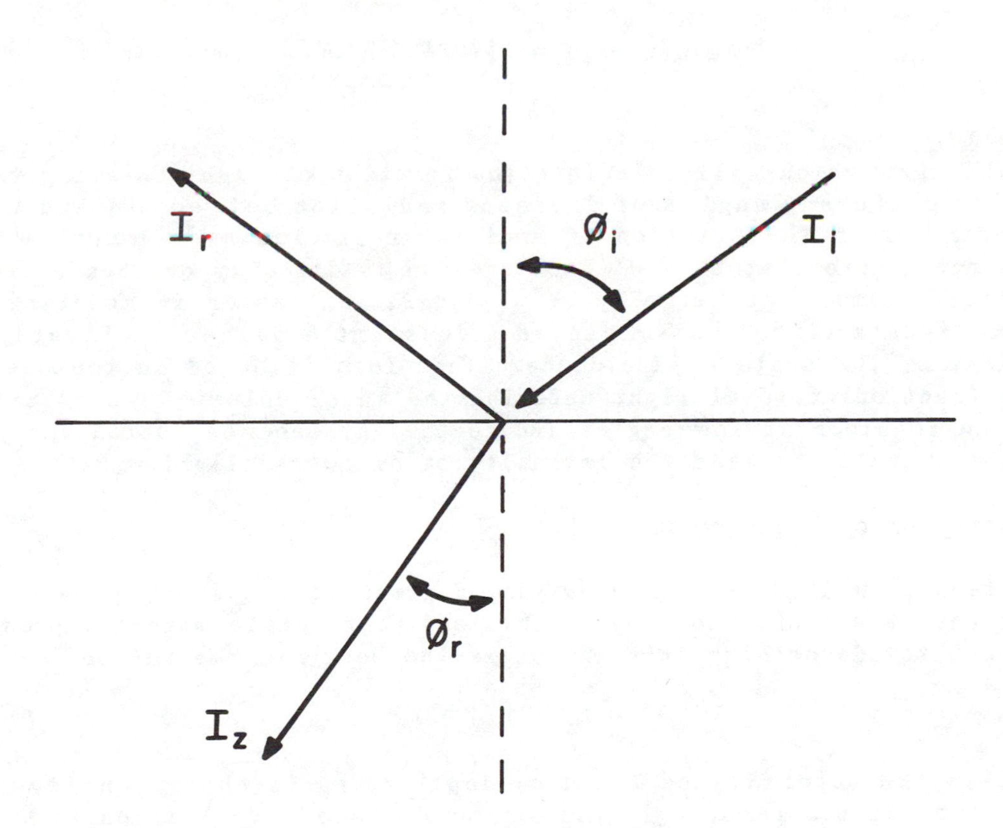

Figure 10.3. Reflection and refraction of light at the surface of water.

Trigonometric functions for the angle of refraction are obtained from the fundamental equation for refraction of light.

$$\sin \phi_r = \frac{\sin \phi_i}{1.33} \tag{10.43}$$

If the arcsin function is not available on your computer to obtain ϕ_r, the following equations may be used to solve Equation 10.41:

$$\cos x = 1 - \sin^2 x \tag{10.44}$$

$$\tan x = \frac{\sin x}{\cos x} \tag{10.45}$$

$$\sin(x + y) = \sin x \cdot \cos y + \cos x \cdot \sin y \tag{10.46}$$

$$\sin(x - y) = \sin x \cdot \cos y - \text{cox } x \cdot \sin y \tag{10.47}$$

$$\tan(x + y) = \frac{\tan x + \tan y}{1 - \tan x \cdot \tan y} \tag{10.48}$$

$$\tan(x - y) = \frac{\tan x - \tan y}{1 + \tan x \cdot \tan y} \tag{10.49}$$

Exercise 10.4

Write a program which will calculate the fraction of light entering the water column for different angles of incident radiation between 0° and 90° in 5° increments. Graph the fraction of incident radiation as a function of angle of incidence. Note that $\phi_i = 0°$ when the sun is directly overhead. Take note that reflection may not actually be a significant factor in modeling aquatic systems. Ryther (1959) has employed a constant 5 percent reflection loss, independent of the angle of incidence. This loss is based on the observation that the fraction of total light entering the water column is almost independent of angle since at low angles indirect light becomes increasingly important and eventually exceeds the intensity of direct sunlight.

10.5 Absorption of Light by the Water Column

Light intensity will decrease to varying degrees in a column of water depending upon the amount of absorbing soluble and particulate material present. The basic model for describing this effect is the Beer's law equation.

$$I_z = I_w e^{-\eta b z} \tag{10.51}$$

where I_z is the intensity of light at depth z, I_w is the intensity of light entering the water column, η is the extinction coefficient which is a function of the amount of absorbing material in the water, and b is a factor relating the length of the actual light path to depth as follows:

$$b = 1/\cos \phi_r \tag{10.52}$$

ϕ_r is the angle of refraction as determined by Equation 10.43.

Exercise 10.5
Assume that a simple truncated sine curve adequately models the light entering the top of the water column. Develop a model which describes the temporal variation in light intensity at a depth of 2 meters. Assume that $\eta = .5$, and 5 percent of the light is reflected at all angles. Draw curves for total light striking the surface, and light intensity at 2m depth for different times of the day.

10.6 Effect of Light on Photosynthesis**

One of the chief reasons we are interested in light intensity is for its effect on the rate of photosynthesis. One of the early models of this relationship was developed by Smith (1936). His equation, in its simplest form, is quite similar to the saturation equation discussed several times before.

$$P = \frac{P_{max} \cdot I}{(K + I)} \tag{10.61}$$

Here, P is the rate of photosynthesis, P_{max} is the maximum rate at saturating light intensities, I is light intensity, and K is the saturation constant ($K = I$ when $P = 1/2\ P_{max}$). A modification of this equation which is preferred by Vollenweider (1965) is as follows:

$$P = P_{max} \cdot \frac{KI}{\sqrt{1 + (KI)^2}} \tag{10.62}$$

The resulting curve appears to fit experimental data better because it has a steeper slope at low light intensities.

Ryther (1956) has observed that most aquatic algae (and perhaps many other plants) show a marked inhibition of photosynthesis at high light intensities. A model suggested by Ryther's data (1959) is embodied in the following equation which was first employed by Steele (1962).

$$P = KI \cdot e^{-kI} \tag{10.63}$$

Here, k is a constant that determines the point of maximum light intensity and the rate of inhibition, and K is a constant that scales the rate of photosynthesis to the proper units. These parameters may be obtained by fitting the maxima function (Equation 10) to the light intensity data. Figure 10.4 was obtained by using such a curve-fitting procedure on the average data taken by Ryther (1959) from 14 species of marine plankton algae.

To account for inhibition, Vollenweider (1965) has modified Equation 10.62 by incorporation of an additional term as follows:

$$P = P_{max} \cdot \frac{KI}{\sqrt{1 + (KI)^2}} \cdot \frac{1}{\sqrt{(1 + (\alpha I)^2)^n}} \tag{10.64}$$

This curve approximates Ryther's data for percent maximum photosynthesis when $P_{max} = 125$ mg $C/m^3 \cdot hr$, $K = 1$, $\alpha = 0.025$, $n = 1$, and I is expressed in thousands of footcandles (kilo-footcandles).

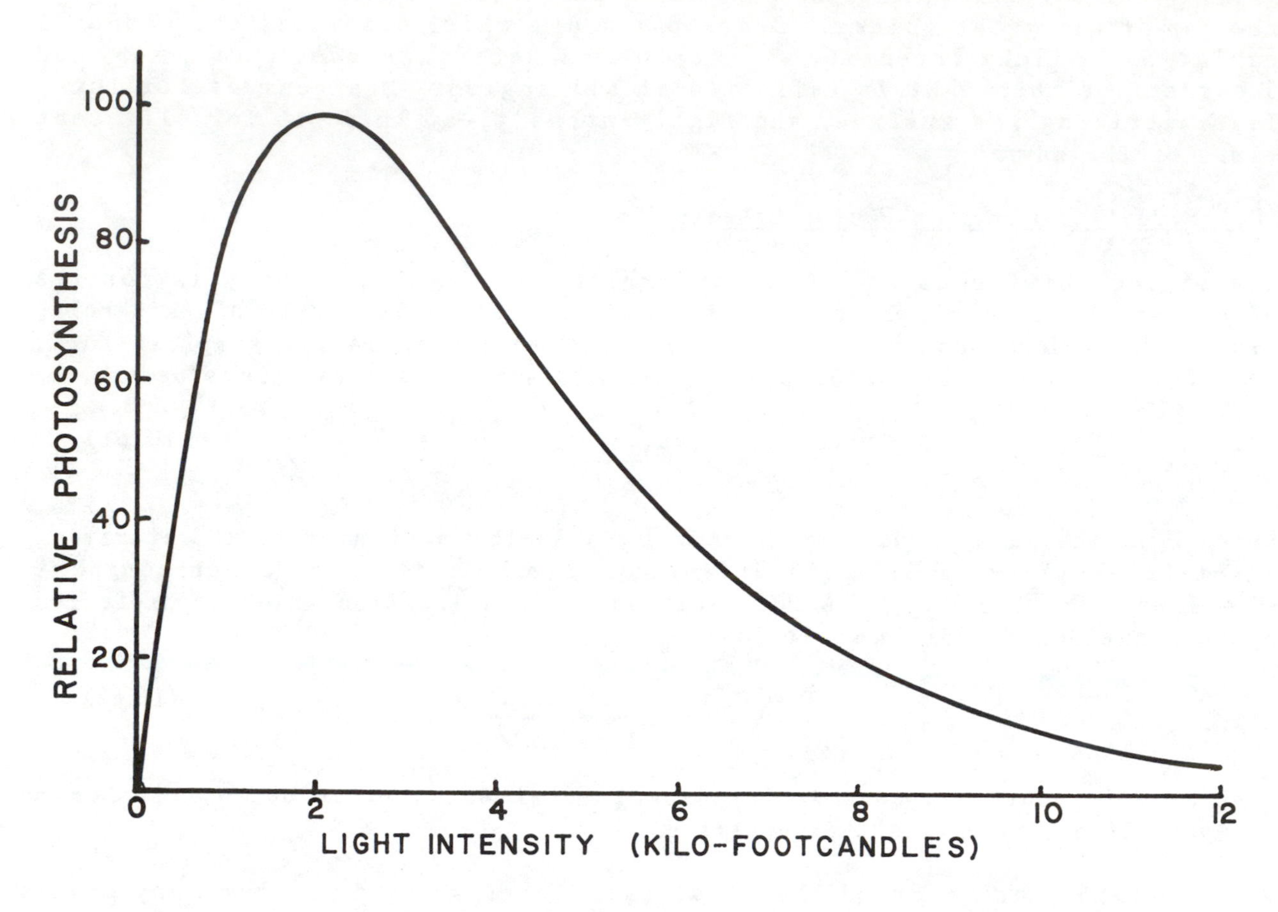

Figure 10.4. Photosynthesis model of Ryther and Steele.

Another simple photosynthesis model which provides for inhibition at high light intensities is a modification of the saturation Equation 10.61, which is analogous to enzyme inhibition equations. This has the following form:

$$P = P_{max} \cdot \left(\frac{1}{1 + I/k + K/I} \right) \qquad (10.65)$$

In this equation K is a constant which affects the inhibitory action and k is the saturation constant. The value of this model lies in its simplicity. However, it is clearly not as flexible as Equation 10.65. To approximate algal phytoplankton data, $K = 4$, $k = 1$, and $P_{max} = 100$ mg $C/m^3 \cdot hr$.

<u>Exercise 10.6</u>**

Compare the model Equations 10.63 or 10.64 with Equation 10.65 by programming each with appropriate parameters and varying light intensity from 0 to 10 kilo-footcandles. Note that all constants in Equations 10.61-10.65 are expressed either as kilo-footcandles or 1/kilo-footcandle.

10.7 Phytoplankton Productivity Curves

An interesting application of the model equations developed in Section 10.5 is the simulation of phytoplankton productivity at different depths in a lake. Basically these models would employ the following simple form of the Beer's Law equation to determine the intensity of light at each depth, assuming light is entering normal to the lake's surface.

$$I_z = I_0 e^{-\eta z} \tag{10.71}$$

where I_0 is surface intensity, and other symbols are as defined in Section 10.5.

The rate of photosynthesis at each depth would be calculated using one of the equations described above (10.63–10.64). The result is the typical photosynthesis-depth curve which exhibits a maximum at the depth of optimum light intensity. See Figure 10.5.

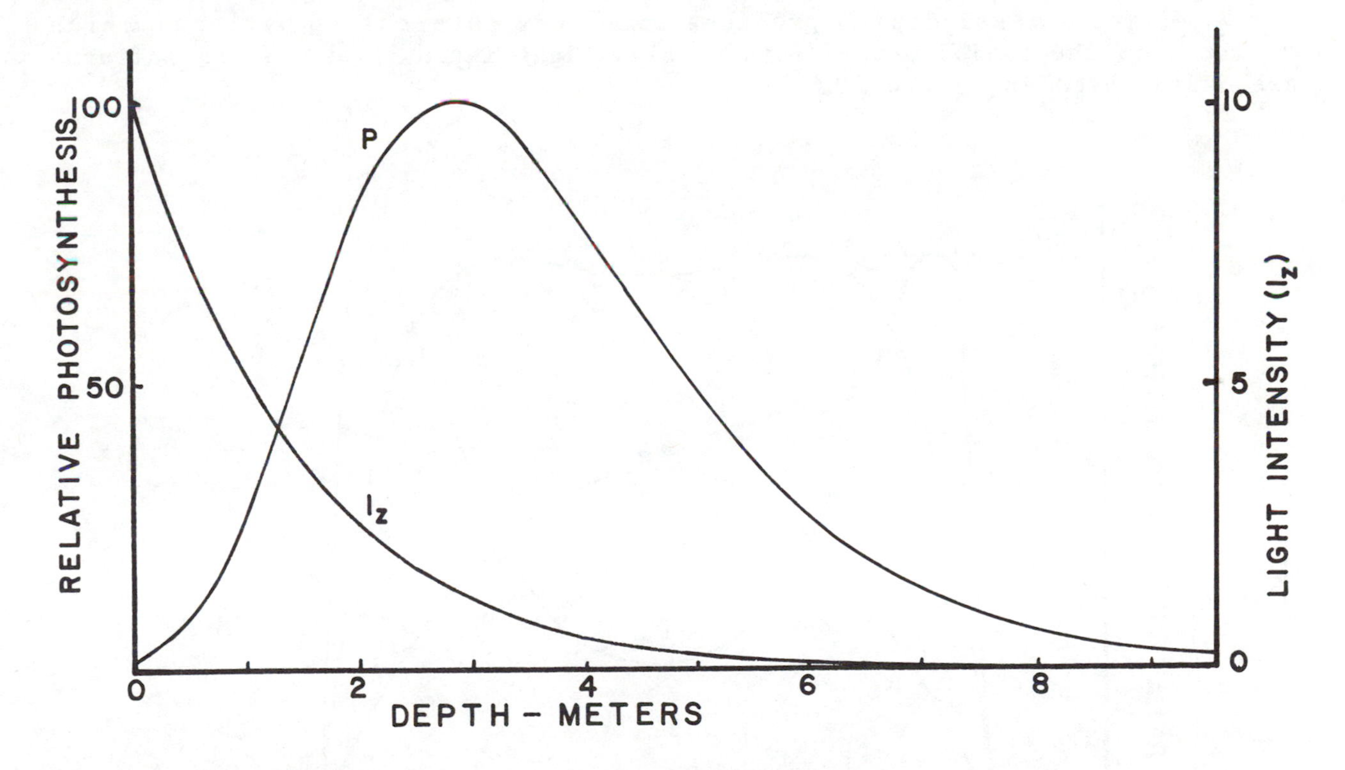

Figure 10.5. Simulation of photosynthesis at different depths in the water column.

Phytoplankton are assumed to be evenly distributed with respect to depth in this model. For further discussion of the assumptions and modeling techniques, refer to Vollenweider (1965) or Fee (1969, 1973).

Exercise 10.7

Model the photosynthesis-depth curve described above. Implement your simulation with the following parameters: surface intensity, I_0 = 10 kilo-footcandles, η = 0.69, and parameters for rate of photosynthesis as given in Section 10.6. Determine the rate of photosynthesis at each half-meter depth down to 10 meters. Also, program the computer to integrate the total photosynthesis below 1 m^2 of lake surface by summing up the productivity at each depth. Plot light intensity, and rate of photosynthesis as a function of depth.

Exercise 10.8

Model diurnal variation in photosynthesis at a given depth by combining one of the photosynthesis-rate equations with one of the sine equations expressing the temporal variation of light intensity. Assume maximum light intensity is 10 kilo-footcandles, and that you are modeling plankton at a depth of 1 meter (η = 0.69). Use the parameters from Section 10.6 in the rate equations, and determine the rate of photosynthesis for each 6-minute (0.1 hr) period during the 12 hours photoperiod. Also program the computer to integrate the total photosynthesis which occurs at 1 meter depth during the 12 hour period. Graph the rate of photosynthesis as a function of time. Repeating this procedure for 2, 4, and 6 meter depths provides some very interesting patterns which result from the inhibitory effects of high light intensities. These patterns are illustrated in Figure 10.6.

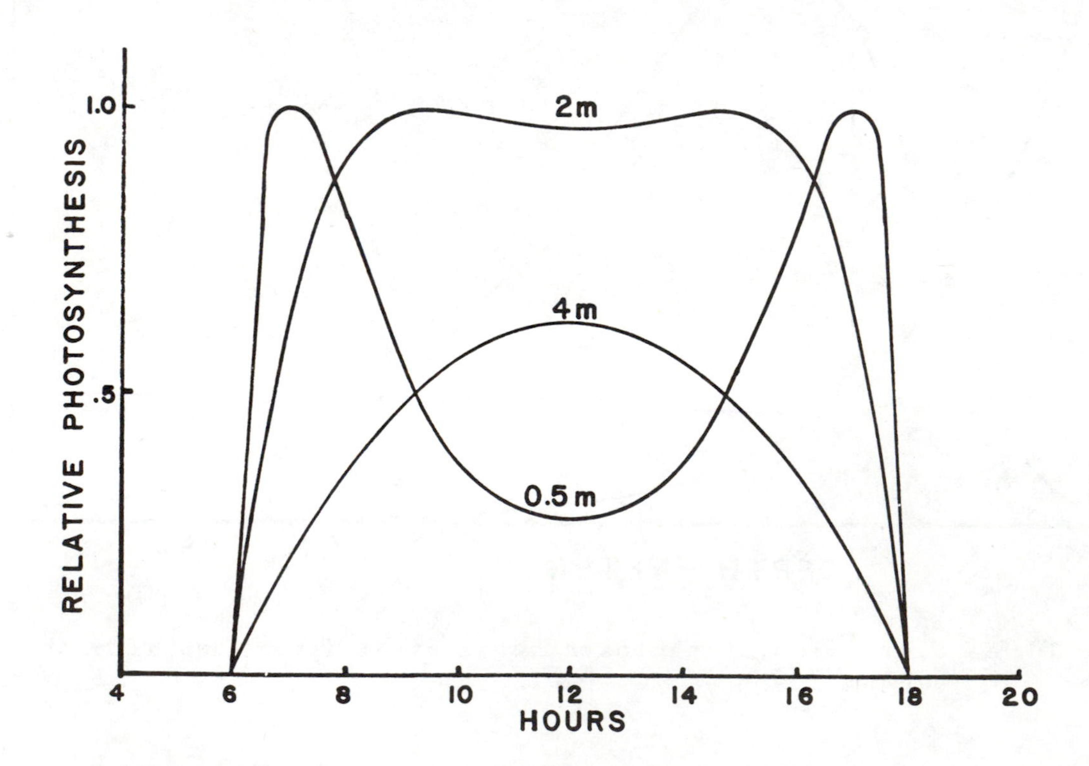

Figure 10.6. Simulation of diurnal variation in photosynthesis at different depths in the water column.

Conclusion

This chapter has been concerned with the temporal and spatial variations in light intensity, and the effect of light on the rate of photosynthesis. The models described above assume a uniform distribution of organisms in the water column. If we combine these models with the phytoplankton growth models described in Chapter 13, we would be able to see an interesting interaction between light intensity, photosynthesis, and phytoplankton growth rate.

For additional information on temporal and spatial variation of light, refer to Hutchinson (1957). For information on models dealing with photosynthesis in terrestrial plants refer to Gates (1962).

The causal relationship between solar light intensity and environmental temperature will be discussed in Chapter 11.

CHAPTER 11
TEMPERATURE AND BIOLOGICAL ACTIVITY

Our interest in temperature as a modeling parameter stems from the marked effect it makes on practically all biological activity. This response results largely from the effect of temperature on chemical reaction rates. At very low temperatures, reaction rates become vanishingly small, and biological activity all but ceases. At high temperatures, competing reactions, such as the denaturation of vital proteins, are favored, and organisms begin to disintegrate. Thus biological activity is largely confined within the limits of 0 and 100 degrees Celsius. This chapter will describe various methods for modeling the temporal variation of temperature, and the general effect of temperature on biological activity.

11.1 Temperature Variation in Terrestrial Systems**

The temperature of the environment is primarily a function of solar radiation. For this reason, temperature tends to follow a sinusoidal pattern not unlike that of solar radiation. However, temperature is a function of the integrated quantity of heat stored in the environment as a result of solar input. Thus, the temperature curve should approximate the integral of the solar sine curve. Analytically, the integral of a sine function is a negative cosine function.

$$\int \sin x \, dx = - \cos x + c$$

The curve for the negative cosine is essentially identical to a sine curve, except that it lags the sine curve by 90° (is out of phase by 1/4 wavelength). This relationship is diagrammed in Figure 11.1. The actual temperature cycle does not lag solar radiation by 1/4 wavelength, because of the different time constants associated with the major integrating elements of the biosphere, namely, air, land, and water. According to air temperature data collected, the actual lag is approximately 3-4 weeks. If we were to adopt the four week

James D. Spain, BASIC Microcomputer Models in Biology

ISBN 0-201-10678-7

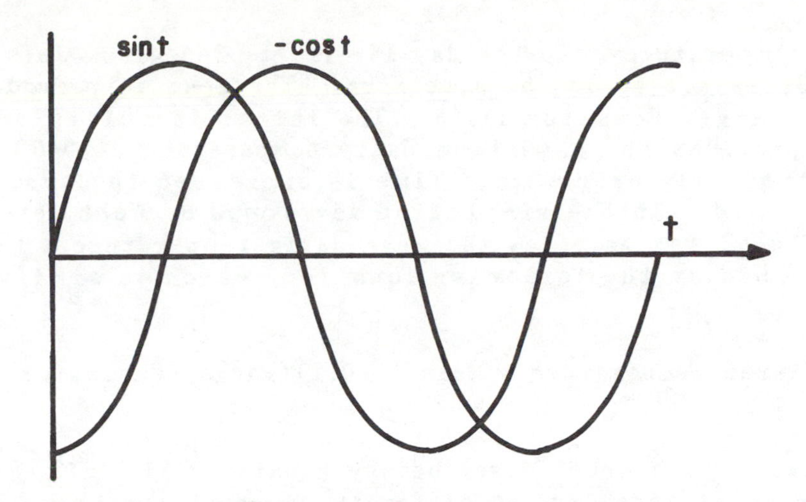

Figure 11.1. Comparison of the sine curve with its integral, the negative cosine curve.

figure, and assume that the annual mean air temperature was 10°C with a daily mean temperature range of 12° on either side of the annual mean, we could describe the daily mean temperature by the following equation:

$$T = 10 + 12\sin\left(2\pi\left(\frac{W - 15}{52}\right)\right) \qquad (11.11)$$

This can be compared with a solar light intensity model which might be

$$I = 400 + 200\sin\left(2\pi\left(\frac{W - 11}{52}\right)\right) \qquad (11.12)$$

In many simulations, temperature extremes are more important than the mean. Webb, et al. (1975) developed an air temperature model for an Oregon forest which was composed of three components: mean daily temperature, daily temperature range, and daily temperature excursion. Air temperature at any time was given by the overall equation

$$\text{Air Temperature} = \text{Mean} + (\text{Range})(\text{Excursion}) \qquad (11.13)$$

The following equations were then used to describe each component:

$$\text{Mean} = 10 + 8\sin\left(2\pi\left(\frac{D - 107}{365}\right)\right) \qquad (11.14)$$

$$\text{Range} = 7 + 2\sin\left(2\pi\left(\frac{D - 107}{365}\right)\right) \qquad (11.15)$$

$$\text{Excursion} = \sin\left(2\pi\left(\frac{H - 10}{24}\right)\right) \qquad (11.16)$$

As presented here, Equation 11.14 states that mean daily temperature has an annual mean of 10 and a range of 8 which is multiplied by the sine-function so that the peak temperature comes on day 198 (107 + 365/4) or July 16. The main equation, 11.13, is set up in the same form. The mean is provided by Equation 11.14, and the range, Equation 11.15. The latter is multiplied by the sine function that provides for a minimum daily temperature at 0400 and a maximum at 1600, i.e. the daily excursion. Time is expressed in days, D, and hours, H, between 0 and 24. In the simulation developed by Webb, et al. (1975) the temperature of soil was taken as the mean daily temperature. Litter temperature was described by the following equation, which is similar to Equation 11.13:

$$\text{Litter Temperature} = \text{Mean} + (0.1)(\text{Range})(\text{Excursion}) \qquad (11.17)$$

Exercise 11.1**
Program the temperature model described by Equations 11.13-11.16 and implement it for four representative days at different times of the year. Increment time in two loops, one which increments days, and another which increments hours. Plot temperature as a function of time in order to demonstrate the behavior of your model.

11.2 Temperature Variation in Aquatic Systems

The major source of heat energy in a lake is the solar radiation that falls on its surface. It is therefore not surprising that the heat content of a lake and its temperature follow a sine function with respect to time. The energy budget equation which describes the heat content of a lake at any time is as follows:

$$Q_{t+\Delta t} = Q_t + I_t + S_t - E_t - O_t \qquad (11.21)$$

where Q_t is the heat content of the lake at any time, I_t is the effective incoming radiation passing through the surface of the lake, S_t is the sensible heat transfer by conduction or convection to or from the atmosphere, E_t is the heat lost or gained through evaporation or condensation, O_t is outgoing radiation to space.

Radiation input by weeks may be described by the following equation:

$$I_t = 2800 + 1400\sin\left(2\pi\left(\frac{w - 11}{52}\right)\right) \qquad (11.22)$$

where all constants have been previously described. This and all other components of the energy balance equation are expressed in calories per cm^2 per week.

Sensible heat transfer, based on Newton's law of cooling is, as follows:

$$S_t = k(T_a - T_w) \qquad (11.23)$$

where k is a transfer coefficient which assumes an average wind velocity, T_w is the temperature of the surface water, and T_a is the temperature of the air, which may be described by Equation 11.11.

Evaporative heat loss is given by

$$E_t = R \cdot H \cdot (P_w - P_a) \tag{11.24}$$

where R is a rate constant which assumes an average wind velocity of 10m/s, and H is the latent heat of evaporation which is a function of temperature. H may be approximated in the range 0 - 50°C by the equation

$$H = 595.8 - 0.54T_w \tag{11.25}$$

P_w is the saturation water vapor pressure that is also a function of surface water temperature. P_w is approximated by the cubic equation

$$P_w = 4.57 + 0.357T_w + 0.0065T_w^2 + 0.0004T_w^3 \tag{11.26}$$

P_a, the partial pressure of water vapor in the air, is, of course, quite variable. For purposes of this simulation we will assume a constant 80 percent relative humidity so that P_a will be a function of air temperature as follows:

$$P_a = 0.8\ 4.57 + 0.357T_a + 0.00065T_a^2 + 0.0004T_a^3 \tag{11.27}$$

According to the Stefan-Boltzmann law, back radiation, O_t, is a function of the absolute temperature of the lake surface and the absolute temperature of the sky, both raised to the fourth power. Since the latter is variable we will employ a simple equation which relates back radiation to the difference between air temperature and water temperature

$$O_t = 77(T_w - T_a) \tag{11.28}$$

where O_t is expressed in calories per square centimeter per week.

The values for other constants given on a weekly basis may be assumed to be as follows:

$$k = 100\ \text{cal/C}^\circ \cdot \text{cm}^2 \cdot \text{wk}$$

$$R = 1.0\ \text{g/mmHg} \cdot \text{cm}^2 \cdot \text{wk}$$

The lake is assumed to have an average depth of 10m and to circulate continuously throughout the year, so that

$$(T_w)_t = Q_t/1000 \tag{11.29}$$

For additional information on models relating to heat transfer in lakes refer to Hutchinson (1957).

Exercise 11.2

Write a flowchart for the heat balance model of a lake. Program the simulation with the suggested transfer coefficients or rate constants, and generate about 100 weeks of simulation data. Assume an initial lake temperature of 6° for week 1. Because of all of the equations involved, this simulation looks more complicated than it is. Please note that this model has been greatly simplified, and as a result, lacks many features that contribute to the temperature regime of a real lake system. For a more sophisticated treatment, at least as far as circulation goes, refer to Exercise 11.9.

Exercise 11.3

When exposed to heat, a partially dehydrated camel will conserve water rather than maintain body temperature, relying on its relatively large size, approximately 1000 lbs, to moderate the changes which occur. During the day the camel absorbs heat, but during the relatively cool nights, it loses heat. Develop a simulation based upon the following assumptions:

1) Metabolic heat production, $\dot{Q}_{in}$, is constant at 250 kcal/hr.

2) Effective air temperature varies sinusoidally on a 24 hr. cycle with a maximum of 46°C at 2:00 p.m. and a mean of 33°C.

3) The dehydrated camel exhibits no physiological or behavioral temperature regulation mechanisms.

4) Thermal conductance, k, is 62.5 kcal/hr.C°.

Use a sinusoidal forced temperature input, and metabolic heat production to determine body temperature, and heat loss at hourly intervals throughout a 48 hour period. Graph the data as a function of time. The following equations are needed:

$$\text{Air temp} = T_A = 13\sin\left(2\pi\cdot\frac{(t - 8\pi)}{24}\right) + 33$$

$$\text{Heat loss} = \dot{Q}_{out} = k(T_B - T_A)$$

$$\text{Body temp} = T_B = \frac{Q}{W}$$

$$Q_{t+1} = Q_t - \dot{Q}_{out} + \dot{Q}_{in}$$

where W is weight in kilograms, Q_t is total heat content, and $\dot{Q}_{in}$ is metabolic heat production. Assume initial heat quantity (above 0°C) is 17300 kcal and W is 455 kg. For additional information about this model, refer to Riggs (1970).

11.3 Effect of Temperature on Chemical Reaction Rates**

Biological activities such as feeding, digestion, respiration, photosynthesis, heart rate, and movement are all dependent upon chemical reactions which are catalyzed by enzyme proteins. Therefore, temperature affects biological activity in two ways. First, a temperature rise of 10 degrees Celsius causes an approximate doubling or tripling of reaction rate by increasing the proportion of reactant molecules having an energy equal to or greater than the activation energy for the reaction. Second, high temperature denatures enzyme molecules, causing them to lose their catalytic activity. These two competing temperature effects result in the typical optimum temperature curve associated with essentially all biological activity. The objective of this section is to explore the equations which have been devised to model temperature dependent biological systems.

One of the earliest models dealing with the effect of temperature on chemical reaction rates was developed by Arrhenius in 1889, based on the activation energy concept. This equation defines the change in reaction rate with temperature as follows:

$$\frac{d\ln k}{dT} = \frac{E_a}{RT^2} \tag{11.31}$$

where k is the reaction rate constant, T is absolute temperature in K°, R is the gas law constant, and E_a is the activation energy for the reaction. When integrated analytically, this equation can take the following form:

$$\ln k = \frac{-E_a}{R}\cdot\frac{1}{T} + \ln k_\infty \tag{11.32}$$

where $\ln k_\infty$ is an integration constant, here representing the limiting value of k.

Notice that this is an equation for a straight line which allows one to graph lnk as a function of 1/T to obtain $-E_a/R$ as the slope. These Arrhenius plots are often used and obtain a measure of activation energy for various biological reactions.

By integrating between limits we obtain the following equation which is more useful for modeling:

$$\ln\frac{k_2}{k_1} = \frac{E_a}{R}\cdot\frac{(T_2 - T_1)}{T_2 \cdot T_1} \tag{11.33}$$

If the rate at T_0 is used as the reference rate and antilogs of each side are taken, then the equation becomes

$$k_T = k_{T_0} e^{\left[\frac{E_a}{R}\cdot\frac{(T - T_0)}{T \cdot T_0}\right]} \tag{11.34}$$

where k_T is the rate at some temperature, T, and k_{T_0} is the reference rate at temperature T_0.

Expressing temperatures in Celsius rather than Kelvin degrees gives

$$k_C = k_0 e^{\left[\frac{E_a}{R \cdot 273}\cdot\frac{C}{(C + 273)}\right]} \tag{11.35}$$

where k_C is the rate at temperature C, and k_0 is the rate at 0°.

If $E_a/R\cdot 273$ is replaced by a single constant, A, then we have a useful model describing the effect of temperature on simple chemical reactions.

$$k_C = k_0 e^{A \cdot C/(C + 273)} \qquad (11.36)$$

Constant A, as used here, may be obtained from the slope of the Arrhenius plot, where A = (slope)·(−1/273), or it may be obtained from the Q_{10} value (discussed below) and Equation 11.36. For further discussion on the origin of the Arrhenius equation, see Hamil, et al. (1966).

A simple, widely used method for modeling the effect of temperature on chemical reactions is called the Q_{10} approximation. Q_{10} is the factor by which reaction velocity is increased for a temperature rise of 10°C. The equation, which is analogous to Equation 11.36, is as follows:

$$k_C = k_0 \cdot Q_{10}{}^{C/10} \qquad (11.37)$$

or:

$$k_2 = k_1 \cdot Q_{10}{}^{(C_2 - C_1)/10}$$

The difficulty with this equation is that it is useful only for a narrow range of temperatures. For further discussion see Prosser and Brown (1962).

11.4 Simulation of Temperature Fffect on Enzyme Activity**

The amount of product produced by an enzyme catalyzed reaction is a function of at least two reactions which have different temperature dependencies. One is the reaction involved in product formation. The other is the reaction involved in denaturation of the enzyme. These reactions are written together as follows:

$$\begin{array}{ccccc} & K & & k_1 & \\ S + E & \dashrightarrow & ES & \dashrightarrow & P + E \\ k_d \downarrow & & & & \\ D & & & & \end{array} \qquad (11.41)$$

where E is the active form of the enzyme, and D is the denatured form, P is the product, S is the substrate, and ES is the enzyme-substrate compound. The rate of product formation is

$$\frac{dP}{dt} = k_1 [ES] \qquad (11.42)$$

Assuming substrate is in excess and k_1 is the rate constant for the limiting reaction, we may assume that ES is in equilibrium with E and S so that K is [ES]/[E][S], and [ES] is K[S][E]. Then by substitution into Equation 11.42:

$$\frac{dP}{dt} = (k_1 K[S])[E] = k_r [E] \qquad (11.43)$$

where k_r is an overall constant involving k_1, K, and the substrate concentration, [S], which will be essentially constant if in great excess.

The reaction now simplifies to the following:

$$D \overset{k_d}{\longleftarrow} E \overset{k_r}{\longrightarrow} P + E \qquad (11.44)$$

The effect of temperature on k_r is modeled using a modification of Equation 11.36.

$$(k_r)_T = (k_r)_0 \cdot e^{A_r T/(T + 273)} \qquad (11.45)$$

where T refers to temperature in degrees Celsius.

The denaturation reaction rate, k_d, is also subject to temperature, but because of the high activation energy involved, this reaction has a very high value for A_d, equivalent to a very high Q_{10} value. Here we are assuming that denaturation is controlled totally by kinetic effects. The equation would be much the same if instead, we were to assume a temperature effect on the equilibrium constant involved in the denaturation reaction.

This simulation must also take into account that to produce a measurable product, a reaction must proceed for a finite period of time. The activity of the enzyme at different temperatures is based upon the amount of product produced in a specified period of time. Since denaturation of the enzyme is also a function of time, the resulting optimum is a function of time. A flowchart describing the effect of temperature on enzyme activity is presented in Figure 11.2.

Exercise 11.4**
Program the enzyme optimum-temperature simulation described in Section 11.4. Use the following parameters to implement your simulation:

$k_r = 1$, $k_d = 0.0001$, $A_r = 18$, $A_d = 55$, $E_0 = 2$, $t_{max} = 20$, $T_0 = 0$, $T_{max} = 80$

Plot the amount of product produced as a function of temperature to obtain a typical optimum temperature curve. Compare the form of the curves shown in Figure 11.3 with the curve you obtain.

11.5 O'Neill Model for the Effect of Temperature on Biological Activity**

Because of the denaturation effect, Equation 11.36 and 11.37 are less and less accurate as the optimum temperature is approached. Since many simulation models are particularly concerned with conditions in the optimum and/or the lethal temperature range, these equations have limited value. The general form of optimum-temperature curves is that which was obtained in the enzyme simulation, Figure 11.3. The typical curve is characterized by an exponentially ascending limb at temperatures below optimum, and an abruptly descending limb at temperatures above optimum. A variety of empirical and semi-theoretical equations have this form. Watt (1968) examined three types of equations and compared their curve-fitting ability to some real data. His approach to the problem has been adopted by others attempting to develop adequate temperature dependent models.

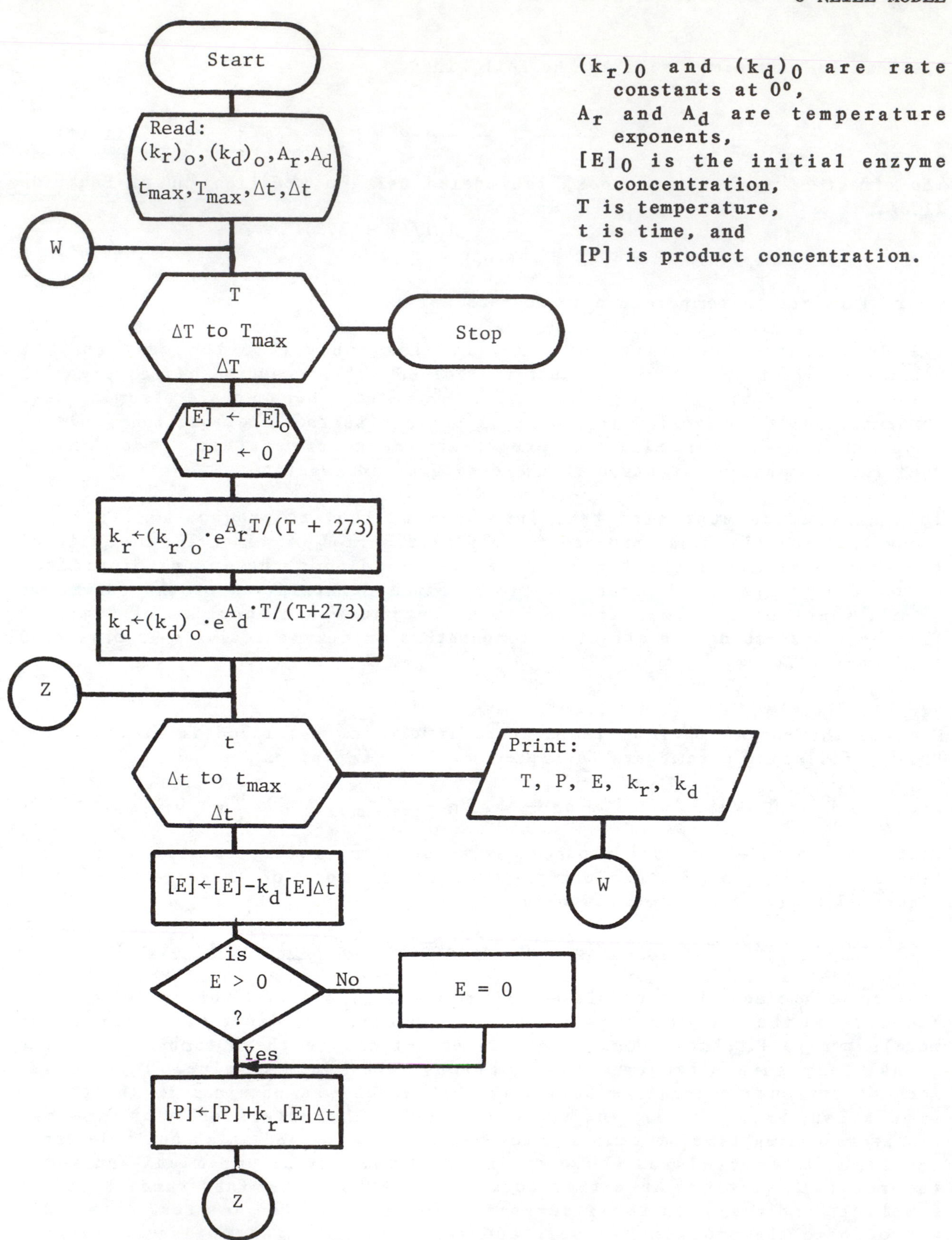

Figure 11.2. Flowchart of the effect of temperature on enzyme activity

O'Neill (1968, 1972) has developed an equation which results in a curve with the necessary form, and at the same time is based upon constants which may be readily derived from most experimental data. The O'Neill equation has the following form:

$$k_T = k_{max} \cdot \left(\frac{T_{max} - T}{T_{max} - T_{opt}} \right)^x \cdot e^{\left[\frac{x(T - T_{opt})}{(T_{max} - T_{opt})} \right]} \qquad (11.51)$$

where k_T is the rate of a temperature-dependent biological function such as feeding of poikilothermic organisms at temperature, T, k_{max} is the rate at optimum temperature, T_{opt}, T_{max} is the lethal temperature, and x is defined by the following equation:

$$x = \frac{W^2 \left[1 + \sqrt{1 + \frac{40}{W}} \right]^2}{400} \qquad (11.52)$$

W is defined by Q_{10} and the difference between T_{max} and T_{opt} as follows:

$$W = (Q_{10} - 1)(T_{max} - T_{opt}) \qquad (11.53)$$

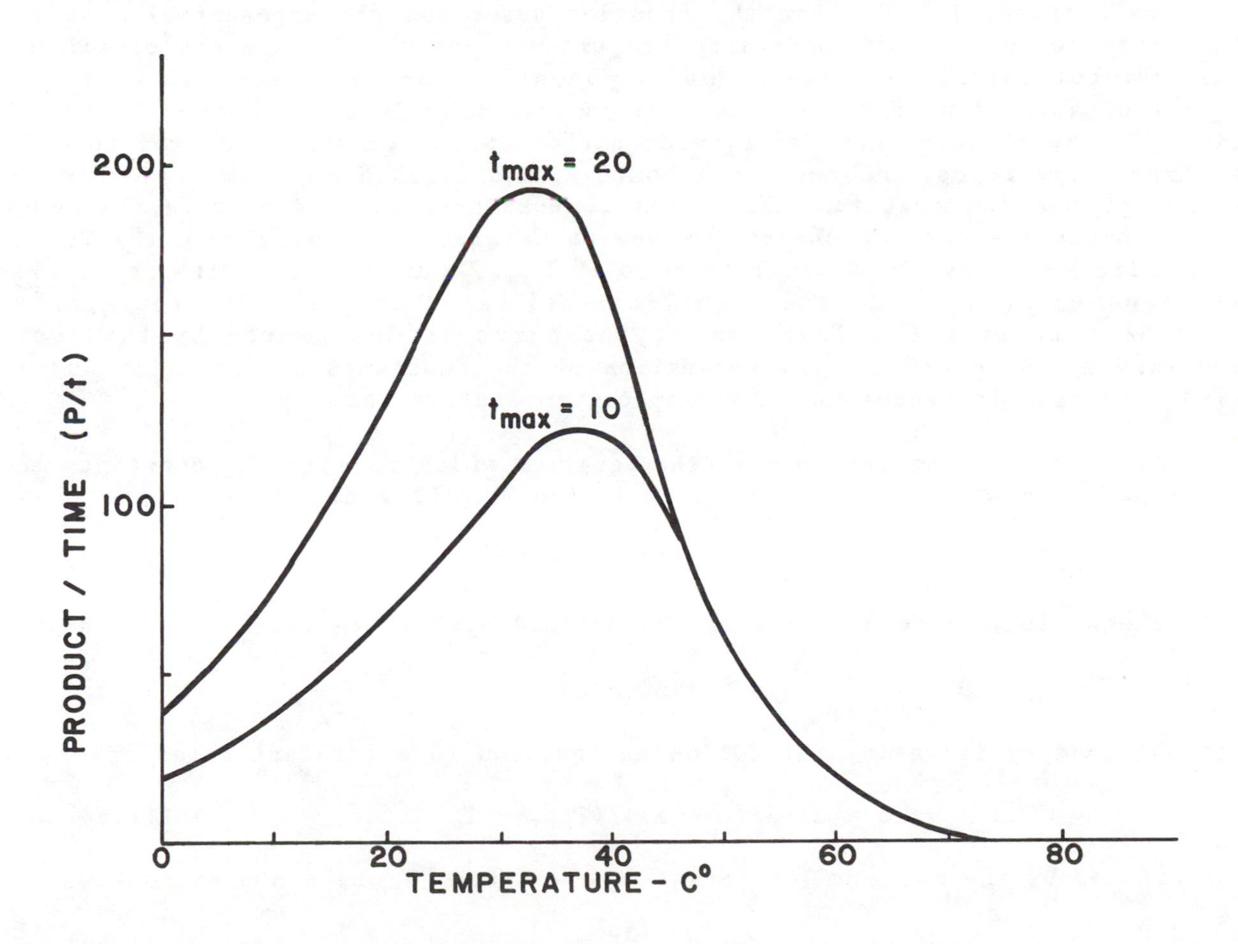

Figure 11.3. Simulated optimum temperature curves based on the enzyme activity model.

This equation has the advantage of being defined by key points readily discernible from most real data. The curve is apparently empirical. Note that, for temperatures above T_{max}, it is necessary to specify $k_T = 0$, since the equation is not designed to predict values beyond that point.

Exercise 11.5**
Write a program which determines k_T for temperatures between 0 and 45°C. Assume

$$T_{opt} = 37, \quad T_{max} = 42, \quad Q_{10} = 2.5, \text{ and } k_{max} = 100.$$

Implement the program, and draw a curve of k_T as a function of T.

11.6 Exponential Temperature-Activity Model

Another model for describing the effect of temperature on biological activity is based on the following product equation:

$$k_T = k_0\left[e^{AT/(T + 273)}\right]\cdot\left[1 - be^{CT/(T + 273)}\right] \qquad (11.61)$$

The theory behind this equation is similar to that described in Section 11.4. The first exponential term in the equation describes the exponential rise in biological activity with increasing temperature due to the general effect on chemical reaction rate. The second exponential term is concerned with the rate of denaturation of a rate-limiting enzyme. The latter term may be visualized as the steady-state fraction of native enzyme present compared to 0°C. As temperature rises, the two terms compete, and reach a maximum value for k_T at the optimum temperature. Above this temperature, the second term predominates because its rate of change per degree Celsius is greater than the first term. Its value becomes equal to zero at T_{max}, and therefore the rate, k_T, also becomes zero. In this equation, all rates are relative to k_0, the apparent rate at 0°C. The form of the curve is determined by the four constants k_0, A, b and C. The evaluation of the constants in the equation as given is difficult because of the complex temperature term.

The following is a simpler form of the equation which permits the constants to be defined from Q_{10}, T_{max}, and T_{opt} as in the O'Neill model.

$$k_T = k_0 \cdot e^{aT}(1 - be^{cT}) \qquad (11.62)$$

In this equation, the constant a may be defined by the Q_{10} value as follows:

$$a = (\ln Q_{10})/10 \qquad (11.63)$$

c is obtained by iterating the following equation to a constant value of c.

$$c = \ln(a/(c + a))/(T_{opt} - T_{max}) \qquad (11.64)$$

b is defined by the maximum (or lethal) temperature, T_{max}, and c as follows:

$$b = e^{(-cT_{max})} \qquad (11.65)$$

Approximate values for A and C in Equation 11.61 would be obtained by multiplying a and c by 300, the approximate value for (T + 273). The value for b is the same in Equations 11.61 and 11.62. A may also be obtained from the slope of the Arrhenius plot for temperature data below T_{max}, as described in Section 11.3.

Exercise 11.6
Repeat Exercise 11.4 using Equations 11.61 or 11.62. Compare the curve obtained with that obtained using Equation 11.51.

Exercise 11.7
Use any of the equations from this chapter to model the following temperature dependent data derived from Messenger and Flitters (1958).

Effect of Temperature on Oriental Fruit Fly Egg Development

Temp(C°)	13	14	15	17	18	20	21	24	27
Rate(/hr)	3.4	4.4	5.9	7.9	10.2	12.4	15	20	25

29	31	32	34	35	35.5	36	36.4	36.7
28.5	30	31	32	30	27	24	18.5	.1

Use the Arrhenius plot or any other appropriate curve fitting techniques to obtain the constants necessary for the model equation you use.

11.7 Use of the Polynomial to Model Temperature Effects

One may question the legitimacy of modeling the effect of temperature on a whole organism using an equation which is based on the response of a single enzyme to temperature. The assumption is that there will be a single limiting enzyme in any system, and response of the organism to different temperatures will be similar to that of the limiting enzyme. This assumption is probably incorrect in that different enzymes could be limiting at different temperatures. If such is indeed the case then the overall response of an organism to temperature would be the product of the response curves of all enzymes which have the potential of being limiting anywhere within the temperature range for that organism. Thus, the overall response curve could be very different from that which is produced by a single enzyme.

Based on these considerations, and the relatively poor fit of theoretical equations to real data, perhaps another approach is called for. If one is only interested in providing a function generator capable of predicting biological activity at different temperatures, the most direct approach would be to fit one of the polynomial equations to the data, and then incorporate this into the model where needed. This empirical procedure is perfectly justified if the theoretical basis for the temperature effect is not vital to the model.

Exercise 11.8
Fit one or more polynomial equations to the temperature data provided in Exercise 11.7. Compare the resulting fit with that obtained using the previous equations.

Exercise 11.9

If you are interested in a more realistic model of heat transport in a lake, try the following one-dimensional circulation model. In this model, assume that transport occurs only in the vertical dimension between 20 compartments, each representing 1 meter of depth below 1 cm^2 of surface for a total of 20 meters. The volume of each compartment is 100 cm^3, so that the temperature of each is obtained by dividing the heat content in calories by 100 calories/C°. As with the simpler model (Section 11.2), solar radiation provides the primary energy input. Since only absorbed radiant energy contributes to the heat content, the following modification of Beer's law would be employed:

$$I_a = I_z - I_{z+1} = I_0(e^{-\eta z} - e^{-\eta(z + 1)}) \qquad (11.71)$$

where I_a is energy absorbed, I_z is light intensity at depth z, I_{z+1} is the intensity 1 meter lower, η is the absorption coefficient,and I_0 is surface intensity. Since the temperature increase is a function of energy absorbed, I_a, the new temperature after one time interval will be given by:

$$T_z \longleftarrow T_z + I_a/100 \qquad (11.72)$$

Other modes of heat transfer are assumed to occur only at the surface, so that only the top meter is affected.

$$Q_0 = S - O - E \qquad (11.73)$$

where Q_0 is the heat transport across the surface, O is heat radiation to space, E is evaporative heat loss, and S is sensible (convective or conductive) heat exchange.

The amount of heat, Q_z, transported from compartment z to compartment z + 1 is assumed to be a function of the temperature difference and the eddy diffusivity coefficient, A_z, as follows:

$$Q_z = A_z(T_z - T_{z+1})\cdot F \qquad (11.74)$$

where F is a proportionality constant. The eddy diffusivity coefficient is a function of both the force acting to produce mixing, i.e., wind, and the force tending to resist mixing, i.e. the density gradient associated with the thermocline. We will assume A_z is described by the following semi-empirical equation:

$$A_z = 0.9A_{z-1}\cdot e^{[-G(\rho_{z+1} - \rho_z)]} + A_{min} \qquad (11.75)$$

where A_0 is the wind-induced surface value of eddy diffusivity. The constant, 0.9, describes how this is diminished exponentially with increasing depth. G is a proportionality constant, A_{min} is a minimum value of approximately 10^{-2} cal/cm^2, and ρ_{z+1} and ρ_z are the densities in compartment z+1 and z respectively. Water density is a function of temperature, and is obtained from the following empirical equation:

$$\rho = 0.99987 + 0.69 \times 10^{-5}T - 8.89 \times 10^{-6}T^2 + 7.4 \times 10^{-8}T^3 \qquad (11.76)$$

Set up the model so that eddy diffusion occurs between compartments in much the same way as molecular diffusion was modeled in Exercise 6.6. Initial temperature values for each compartment should be stored in an array, and changes in temperature described above should be used to calculate the various energy inputs and outputs to each compartment. As before, the numerical integration process is best accomplished in two stages. Calculate the changes for each compartment in the first stage, and then apply these results to each compartment in the second stage by using the following temperature update equation:

$$T_z \longleftarrow T_z + ((I_a)_z + Q_{z-1} - Q_z)/100 \qquad (11.77)$$

Using this approach, heat transfer across the surface is best handled by making it equal to Q_0 as in Equation 11.73. Note that Equation 11.72 is included in Equation 11.77. This simulation requires the use of subscripted variable, and should not be attempted until you have become familiar with this programming technique.

Implement your model with the following parameters:

$I_0 = 600$, $A_0 = 0.5$, $O = 150$, $E = 500$, $S = 30$, $\eta = 0.693$, $F = 100$, $G = 2000$.

Assume an initial temperature value of 4°C at all depths and then use the constant values for heat input and output to simulate the onset of summer stratification conditions. Values provided above represent daily inputs and outputs.

You may simulate the effects of convective mixing by letting the temperature of two compartments equal the average of the two when the density of the lower compartment is less than the one above it. If you wish to combine this model with that of Exercise 11.2, it will show the annual cycle of the lake. Good luck! Putting together a model of this magnitude is a real accomplishment.

Conclusion

This chapter has introduced you to some representative models dealing with the temporal and spatial variation of temperature, and the effect of temperature on biological activity. Models of this type usually play an auxiliary role in multicomponent system models. For example, one of the lake temperature models might be employed in conjuction with a temperature-activity model to provide the necessary input forcing functions for a phytoplankton productivity simulation. Temperature-activity models are also essential to physiological simulations in which temperature is a variable as would be the case with poikilothermic organisms. It is through such model-clustering that one builds up the large multi-component simulations.

CHAPTER 12
MATERIAL AND ENERGY FLOW IN ECOSYSTEMS

The techniques of mathematical modeling and simulation have been extremely useful for understanding the complex interactions that occur in an ecosystem. Energy or material flow simulations are especially valuable for providing a basis for comparison between ecosystems, as well as a means of assessing the importance of various components within an ecosystem. This type of model treats the ecosystem as an integrated unit having a metabolism not unlike that of an individual organism.

12.1 The Energy Flow Concept**

E. P. Odum (1968) presented an interesting review of the development of modern material and energy flow models starting with the early conceptual models of food webs and ecological pyramids. The first detailed study of nutrient cycling in an ecosystem was carried out by Elton in 1923. A fascinating review of this and other historical developments in ecological modeling is provided by Hutchinson (1979). Subsequently, biologists began to be more rigorous in their application of thermodynamic principles, and the concept of material and energy balance emerged. A landmark paper, which led to the quantitative description of energy flow, was written by Lindeman (1942).

H. T. Odum (1956) was the first to assign numbers to the different energy conversions involved in a typical energy-flow diagram. This technique, illustrated in Figure 12.1, employs methods which had been used successfully in chemical engineering for a number of years (Hougen and Watson, 1943). More recently, these conceptual and mathematical models have been used as the basis for computer simulations by Watt (1966), Patten (1971), and many others.

An energy or material flow model describes the amount and direction of flow within an ecological community, and between the community and its environment.

James D. Spain, BASIC Microcomputer Models in Biology

ISBN 0-201-10678-7

To develop such a model, one must have quantitative data about the standing crop, the energy or material inputs, and the energy or material outputs. The organisms in the system may be grouped in a variety of ways. A common division is by trophic level. Moreover, a given trophic level may be sub-divided in a variety of ways which depend upon the interests of the modeler, and the objectives of the model.

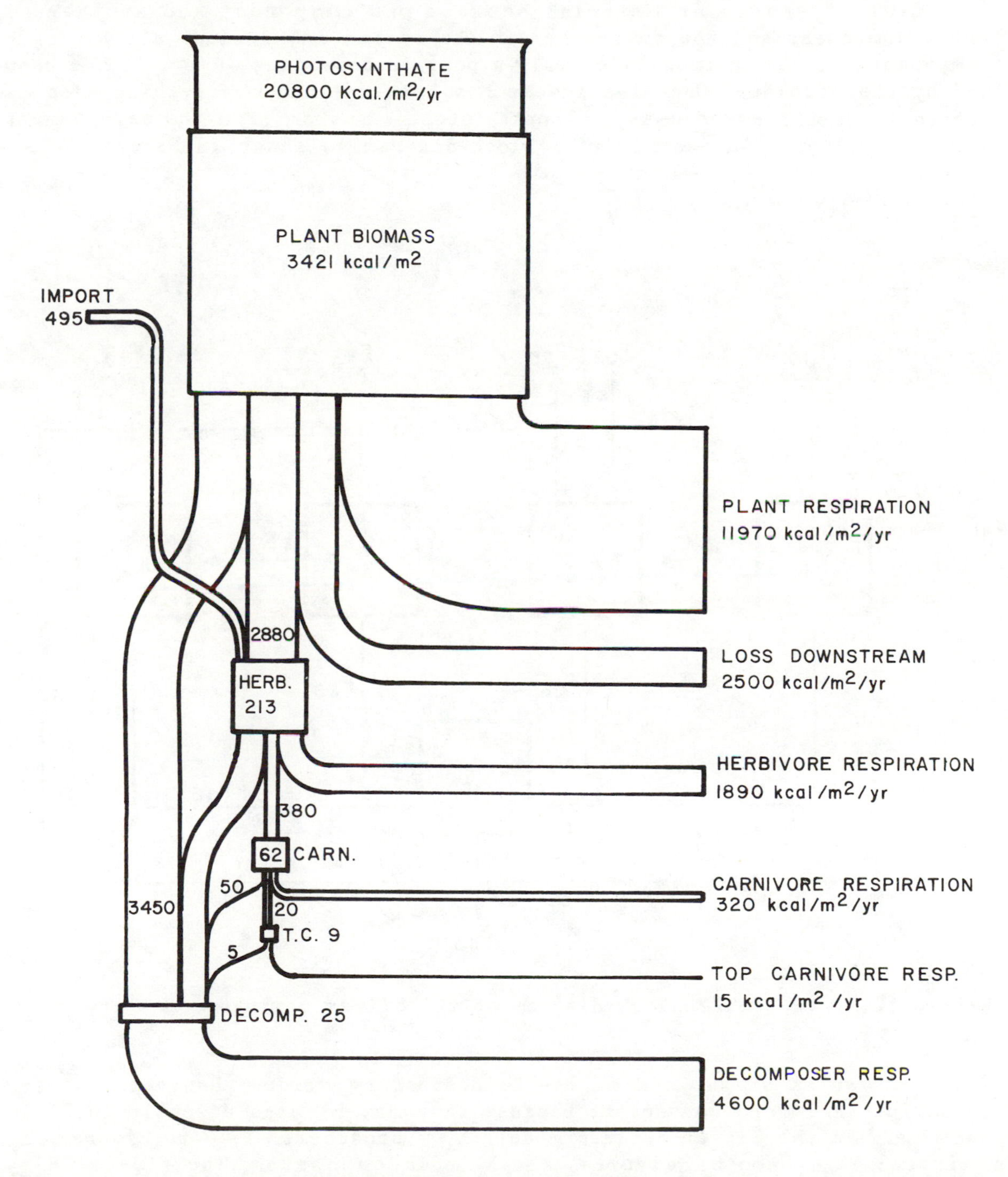

Figure 12.1. An energy flow diagram of the Silver Springs community. Based on diagrams by Odum (1956) and Patten (1971).

12.2 Block Diagrams and Compartment Models**

The conceptual model which is most useful for the development of energy or material flow simulations is the block diagram. This is a pictorial representation of a system in which each rectangular block represents some system component or state variable. The blocks are connected by arrows representing flow or flux of energy or material between one component and another or between a component and the environment. The system may include all biological components of the community or only a portion depending on how it has been defined by the modeler. Once the system has been defined, everything else is considered to be the environment. Inputs into the system from the environment are called forcings. An example of a block diagram is shown in Figure 12.2.

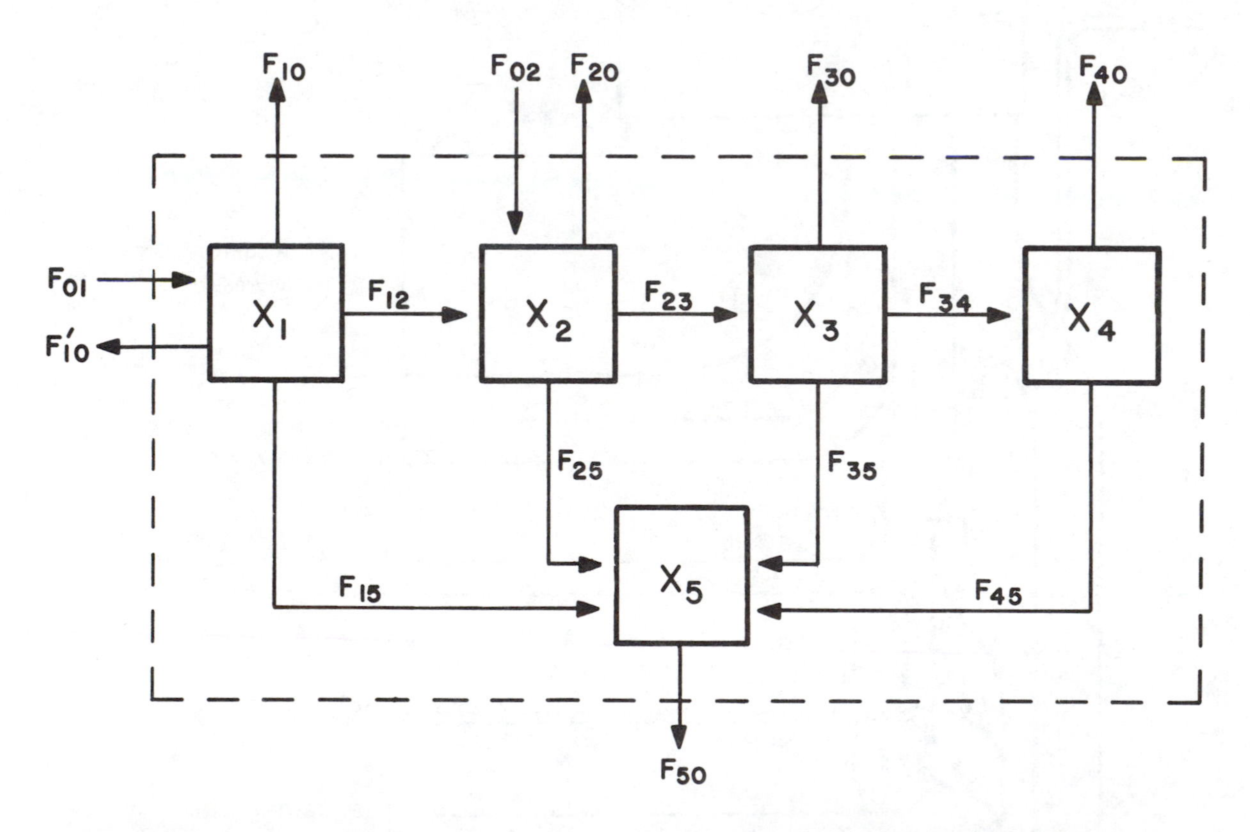

Figure 12.2. A simple block diagram of the Silver Springs community.

The state variables of the system may be defined by words. However, letters and subscripts are more convenient because they may be used directly in BASIC programming. In the Silver Springs model, X_1 = producers, X_2 = herbivores, X_3 = carnivores, X_4 = top carnivores, and X_5 = decomposers. The fluxes, F_{12}, F_{13}, etc., have subscripts which describe the donor and the recipient blocks involved. F_{01} and F_{02} are the two forcings to the system. F_{10}, F_{20}, F_{30}, F_{40}, and F_{50} represent losses from the system to the environment.

Because of the laws of conservation of mass and energy, under steady-state conditions, the sum of the inputs to the system, F_{01} and F_{02}, in either mass or energy units must equal the sum of the outputs from the system, F_{10}, F_{20}, F_{30}, F_{40}, and F_{50}. Also, under steady-state conditions, the sum of the inputs to any block or system component must equal the sum of the outputs from that block.

The state variables, X_1, X_2, etc., express the biomass of each system component on either a volumetric or areal basis. These variables may be expressed in such units as grams of dry weight per square meter, or grams of organic carbon per cubic meter, or they may be a biomass energy equivalent such as kilocalories per square meter. Fluxes employ these same units in conjunction with units of time, such as kilocalories per square meter per year, or grams per cubic meter per day.

The compartment model utilizes the concepts embodied in the block diagram, but, in addition, assigns appropriate mathematical functions to each of the fluxes so that the model may be implemented in the computer to produce simulation data. Fluxes are used in the system equations to provide an energy or material balance for each state variable. The rate of change of each state variable would be described as follows, using Newton's notation for the differential with respect to time:

$$\dot{X}_1 = F_{01} - F_{12} - F_{15} - F_{10} \tag{12.21}$$

$$\dot{X}_2 = F_{12} + F_{02} - F_{20} - F_{23} - F_{25} \tag{12.22}$$

Exercise 12.1**
Write the three other differential equations required for the Silver Springs model (Figure 12.2).

$$\dot{X}_3 =$$

$$\dot{X}_4 =$$

$$\dot{X}_5 =$$

12.2 (Continued)

Each flux must be defined by an equation which usually involves a rate coefficient often designated by a Greek letter (in this case, Ø) to distinguish it from the flux. These equations may take one of the following forms:

a) $F_{ij} = k$ — Flow from compartment i to j is independent of time or system state.

b) $F_{ij} = \phi_{ij} X_i$ — Flow to j is proportional to the content of i. This is a linear equation with control by the donor compartment only. ϕ_{ij} is the rate constant.

c) $F_{ij} = \phi_{ij} \cdot X_j$ — Flow to j is proportional to the content of j. This is a linear equation with receptor control.

d) $F_{ij} = \phi_{ij} \cdot X_i \cdot X_j$ — Flow is regulated by both donor and receptor in a cross product manner (mass action approach). This is a non-linear equation.

e) $F_{ij} = \phi_{ij} \cdot X_j(1 - \beta_{ij} X_j)$ — Flow is regulated by a positive linear term and a negative non-linear term similar to the logistic equation.

f) $F_{ij} = \phi_{ij} X_j(1 - \alpha_{ij} X_i - \beta_{ij} X_j)$ — Flow is regulated in a manner similar to the competition equations.

g) $F_{ij} = \phi_{ij} \cdot f(\text{time})$ — Flow is a forcing function of time, such as the sine function equation in Section 10.11.

h) $F_{ij} = \phi_{ij} \cdot X_i \cdot X_j / (K + X_i)$ — Flow is regulated in a manner similar to the Holling-Tanner predation model.

For additional examples, see Patten (1971).

Compartment models of the type described above are often implemented on the computer by solving the system equations with matrix algebra, a method which will be discussed in Chapter 16. At present, we will solve the equations using the direct approach. Since most of the flux equations are functions of one or more state variables, it is necessary to calculate the fluxes first, then utilize them and the Euler integration method to calculate the up-dated values of the state variables. You will recognize this procedure as the standard two stage approach.

12.3 The Silver Springs Model**

The flux equations for the linear form of the Silver Springs model have been adapted from Patten (1971). The following equations define the fluxes involved:

A. Forcing:

$$F_{01} = [M + R \cdot \sin(2\pi(T-11)/52)]\Delta t$$

$$F_{02} = k\Delta t$$

F_{01} is the energy input from photosynthesis, assumed to be proportional to light intensity. M is the annual mean, and R is the range of the sinusoidal forcing.

B. Feeding: $F_{12} = (\tau_{12}X_1)\Delta t$

$F_{23} = (\tau_{23}X_2)\Delta t$

$F_{34} = (\tau_{34}X_3)\Delta t$

Here τ_{ij} is a donor dependent linear feeding coefficient having units of inverse weeks (wk^{-1})).

C. Mortality: $F_{15} = (\mu_{15}X_1)\Delta t$

$F_{25} = (\mu_{25}X_2)\Delta t$

$F_{35} = (\mu_{35}X_3)\Delta t$

$F_{45} = (\mu_{45}X_4)\Delta t$

Here μ_{ij} is a donor dependent mortality coefficient (wk^{-1}).

D. Respiration: $F_{10} = (\rho_{10}X_1)\Delta t$

$F_{20} = (\rho_{20}X_2)\Delta t$

$F_{30} = (\rho_{30}X_3)\Delta t$

$F_{40} = (\rho_{40}X_4)\Delta t$

$F_{50} = (\rho_{50}X_5)\Delta t$

Here ρ_{ij} is a donor dependent respiration coefficient (wk^{-1}).

E. Export: $F'_{10} = (\lambda_{10}X_1)\Delta t$

Here λ_{10} is the export coefficient expressing loss downstream per week.

The updated values for the state variables are obtained by Euler integration as follows:

$$X_1 = X_1 + F_{01} - F_{12} - F_{15} - F_{10} - F'_{10} \quad (12.31)$$

$$X_2 = X_2 + F_{02} + F_{12} - F_{23} - F_{25} - F_{20} \quad (12.32)$$

$$X_3 = X_3 + F_{23} - F_{34} - F_{35} - F_{30} \quad (12.33)$$

$$X_4 = X_4 + F_{34} - F_{45} - F_{40} \quad (12.34)$$

$$X_5 = X_5 + F_{15} + F_{25} + F_{35} + F_{45} - F_{50} \quad (12.35)$$

An alternate two stage approach would be to write equations for the changes in each compartment, $X_1, X_2, X_3, \ldots, X_n$, then add the change to the old value of $X_1, X_2, X_3, \ldots, X_n$ to obtain the updated state variables.

Exercise 12.2
Draw a flow chart of the linear form of the Silver Springs simulation model described above. Then program the simulation, and implement it with time incremented weekly through the year, using the following parameters from Patten (1971).

X_1 = 3421 kcal per m^2	μ_{15} = 1.01 per year	ρ_{10} = 3.50 per year
X_2 = 213 ''	μ_{25} = 5.13 ''	ρ_{20} = 8.86 ''
X_3 = 62 ''	μ_{35} = 0.74 ''	ρ_{30} = 5.10 ''
X_4 = 9 ''	μ_{45} = 0.676 ''	ρ_{40} = 1.47 ''
X_5 = 24 ''	λ_{10} = 0.73 ''	ρ_{50} = 188.6 ''

τ_{12} = 0.84 per year	k = 486 per year
τ_{23} = 1.79 ''	R = 175 kcal per wk
τ_{34} = 0.339 ''	M = 400 kcal per wk

Note that these rate coefficients are based on an annual flux through the system. You must divide by 52 in order to put them on a weekly basis.

For an initial run, enter all parameters as suggested, except let R = 0. This procedure will keep F_{01} constant. The system should then tend to a steady-state condition. If it does not, check inputs for errors.

After this initial check, carry out a three year run using R = 175 kcal/wk. Graph the contents of each compartment in each time interval. How stable is the system? Which compartments exhibit the most instability? If the system is unstable, reduce the time increment to 0.1 week.

To provide a rather crude simulation of succession, set the solar input at a constant value and reduce the size of all state variables to some minimum level such as 5.0 kcal/m^2), and then allow the system to run until a moderately steady-state exists (about 150 weeks).

Exercise 12.3
Modify the above simulation program so as to employ the non-linear form of the equations for the three feeding fluxes as follows:

$$F_{12} = \tau'_{12}X_1X_2 \cdot \Delta t$$

$$F_{23} = \tau'_{23}X_2X_3 \cdot \Delta t$$

$$F_{34} = \tau'_{34}X_3X_4 \cdot \Delta t$$

Patten (1971) gives the values for the non-linear feeding coefficients as follows:

$$\tau'_{12} = 0.0039/\text{yr}, \ \tau'_{23} = 0.0272/\text{yr}, \text{ and } \tau'_{34} = 0.0382/\text{yr}$$

Graph the data in the same fashion as in Exercise 12.2. Compare the output of this non-linear system with the linear one in Exercise 12.2. How does this change affect the stability of the system?

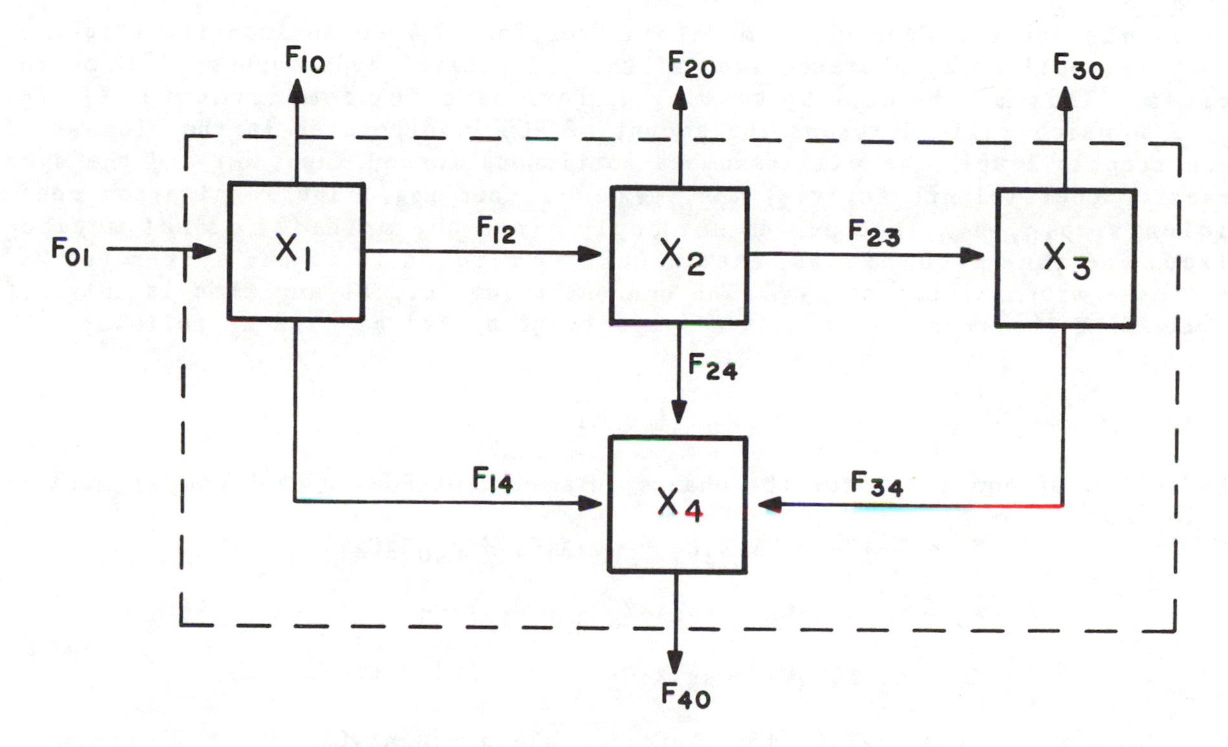

Figure 12.3. Block diagram of a four compartment aquatic food chain.

<u>Exercise 12.4</u>**

Develop an energy-flow simulation for a four compartment aquatic food chain composed of a primary producer (X_1), an herbivore (X_2), a carnivore (X_3), and a decomposer (X_4). Use a single sinusoidally-varying forced input (F_{01}) based on the annual light cycle. Use non-linear energy transfers between compartments X_1 and X_2 ($F_{12} = \tau'_{12}X_1X_2$), and between X_2 and X_3 ($F_{23} = \tau'_{23}X_2X_3$). This hypothetical system has been simplified to make it easily adaptable to the subsequent exercises involving the flux of non-metabolizable substances, and the effect of varying temperature. Figure 12.3 is a block diagram for this system.

Write the four differential equations required for the model and program these in the simulation as difference equations. Test the program with the following input parameters, incrementing on a weekly basis or less.

$X_1 = 3000\ kcal/m^2$	$\rho_{10} = 0.07/wk$	$\mu_{14} = 0.03/wk$	$\tau'_{12} = 0.00008/wk$
$X_2 = 200\ kcal/m^2$	$\rho_{20} = 0.15/wk$	$\mu_{24} = 0.06/wk$	$\tau'_{23} = 0.0006/wk$
$X_3 = 50\ kcal/m^2$	$\rho_{30} = 0.10/wk$	$\mu_{34} = 0.02/wk$	
$X_4 = 34\ kcal/m^2$	$\rho_{40} = 3.0/wk$	$F_{01} = 400 \pm 150\ kcal/m^2/wk$	

Program this model so that you can graph F_{01}, X_1, X_2, X_3, X_4 for about 104 weeks.

Exercise 12.5

Modify the aquatic food chain model in Exercise 12.4 to include the effect of a non-metabolizable substance such as the chlorinated hydrocarbon, PCB, on the system. This may be done by setting up four parallel compartments, Z_1, Z_2, Z_3, Z_4, which will represent the amount of PCB incorporated in the biomass of each trophic level. We will assume a continuous forced input F'_{01} and the same transfer coefficients for τ_{12}, τ_{23}, μ_{14}, μ_{24}, and μ_{34}. The respiration coefficients, ρ_{10}, ρ_{20}, and ρ_{30} do not apply since the material is not metabolized, and thus accumulates, except that X_4 returns it to the system (at Z_1) in a rate proportional to ρ_{40}. The concentration, C_i, at any time is obtained by dividing the amount of PCB in a compartment by its biomass as follows:

$$C_i = \frac{Z_i}{X_i}$$

The system of equations for the change in amount of PCB in each compartment is as follows:

$$\dot{Z}_1 = F'_{01} - \tau_{12}X_1X_2C_1 - \mu_{14}X_1C_1 + \rho_{40}X_4C_4$$

$$\dot{Z}_2 = \tau_{12}X_1X_2C_1 - \tau_{23}X_2X_3C_2 - \mu_{24}X_2C_2$$

$$\dot{Z}_3 = \tau_{23}X_2X_3C_2 - \mu_{34}X_3C_3$$

$$\dot{Z}_4 = \mu_{14}X_1C_1 + \tau_{24}X_2C_2 + \mu_{34}X_3C_3 - \rho_{40}X_4C_4$$

Notice that each equation involves the same fluxes as employed in the main equation, except that they have been modified by using the concentration term to convert biomass flow into PCB flow. Hence, the difference equations may be simplified to the following form:

$$Z_1 = Z_1 + F'_{01} - F_{12}C_1 - F_{14}C_1 + F_{40}C_4$$

Set up the other three difference equations and program this model so that you will obtain simulation data on the concentration of PCB in each trophic level at each time interval over a two year period. Prepare a graph of the resulting data. Repeat the simulation using a single pulsed input instead of a continuous input.

Exercise 12.6

Modify the aquatic food chain simulation in Exercise 12.4 to include the effect of temperature on respiration and feeding rates. Assume that the water temperature varies sinusoidally with a mean of 12°C, a maximum of 22°, and a minimum of 2°. Also assume that the temperature curve lags the light curve by five weeks. Assume that the ambient temperature stays below the optimum at all times, so that Equation 11.36 accurately describes the response of all organisms to different temperatures. Assume the same temperature coefficient, A = 20, for both feeding and respiration on all organisms. This is obviously a tremendous over-simplification since we would expect each process in each organism to have its own unique temperature dependency. Therefore, this model must be considered as only a first approximation of the effect of temperature on the system. Assume the following values for the temperature-dependent rate constants at 0°C.

$(\tau_{12})_0 = 0.00004 \qquad (\rho_{20})_0 = 0.07$

$(\tau_{23})_0 = 0.0003 \qquad (\rho_{30})_0 = 0.05$

$(\tau_{10})_0 = 0.3 \qquad (\rho_{40})_0 = 1.4$

Generate two years of simulation data and graph biomass as a function of time. Compare the results with those obtained in Exercise 12.4.

12.4 The English Channel Marine Community

Figure 12.4, adapted from Brylinsky (1972), summarizes the energy flow in the English Channel marine community.

Exercise 12.7

Develop a complete energy-flow model based on the block diagram describing the marine community of the English Channel. Successful completion of this exercise will involve the following tasks:

1) Select an appropriate time increment.
2) Set up the equation for sinusoidal forced input based upon F_{01}, the time increment, and the range for that latitude (50°N).
3) Calculate linear transfer coefficients based upon annual flux, standing crops, and time increment.
4) Write flux equations and system equations.
5) Program the model and generate two years of simulation data.
6) Graph the data as a function of time.

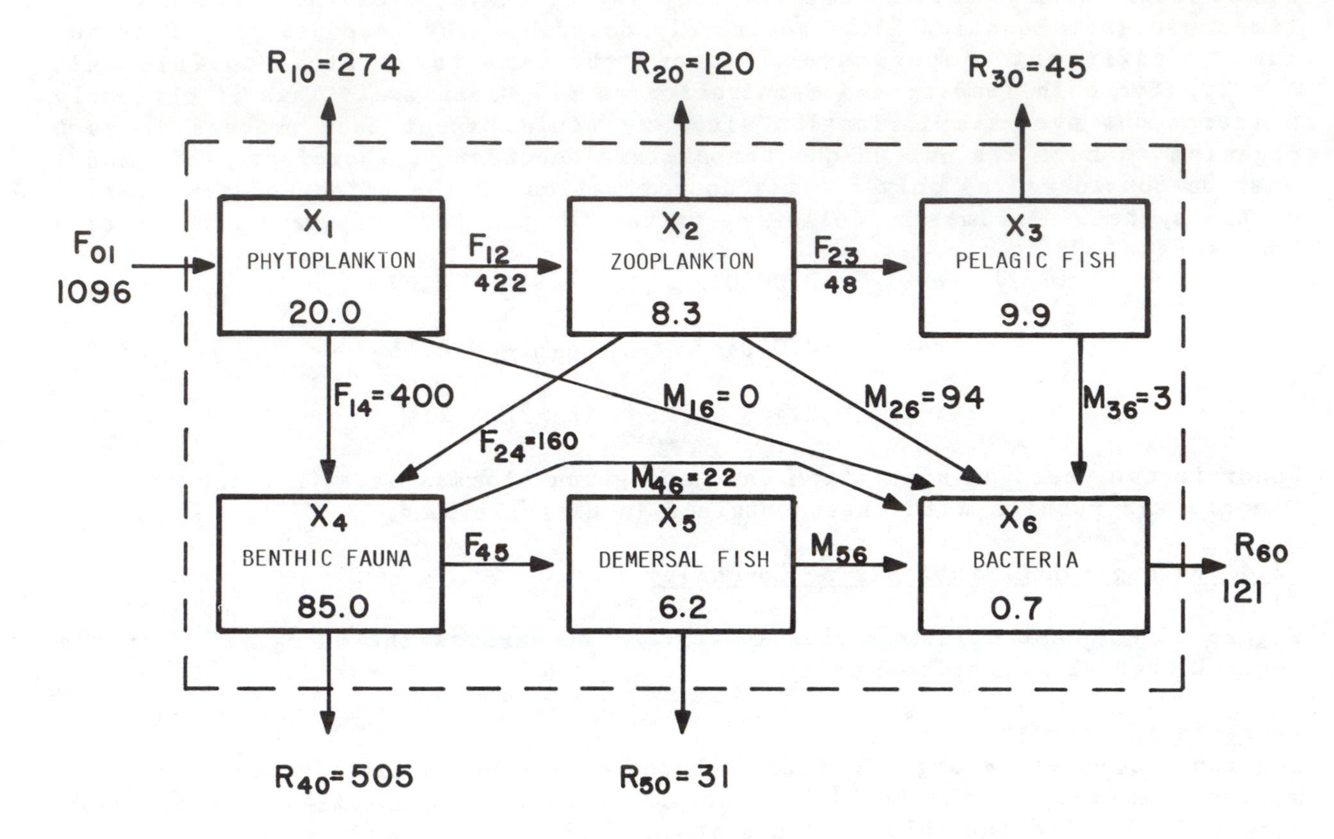

Figure 12.4. Energy flow in the English Channel. Units are expressed in either kcal/m^2 or kcal/m^2·yr. Adapted from Brylinsky (1972).
Adapted from Figure 1, p. 85 in *Systems Analysis and Simulation in Ecology*, Vol. 2 by B. C. Patten (ed.), Academic Press, Inc., 1972. Reprinted by permission of Academic Press, Inc.

<u>Exercise 12.8</u>

Use the energy-flow diagram from Figure 12.5 based upon the model data provided by Golley (1965) to develop a simulation of an old-field broomsedge community. Energy fluxes are expressed in kcal/m^2/yr. and energy equivalents of standing crops are expressed in kcal/m^2.

The location is in South Carolina so that the latitude is approximately 32°N. Thus, the temperature regime for this area would have a mean of approximately 18° with a range of about 10°.

There are several approaches that may be taken with this model. It may be totally light dependent, or it might also include temperature dependence in plant growth and decline. On the other hand, green vegetation might be forced to conform to the biomass data provided by Golley.

Month	J	F	M	A	M	J	J	A	S	O	N	D
Biomass	93	26	201	254	246	380	473	536	852	813	127	132

Alternatively, growth of green vegetation might be a function of both light and root biomass. The conversion of green vegetation to standing dead vegetation is clearly time or temperature dependent. This dependency might be programmed into the model. Decide on one of these strategies for modeling plant growth and perform the exercise in much the same fashion as 12.7.

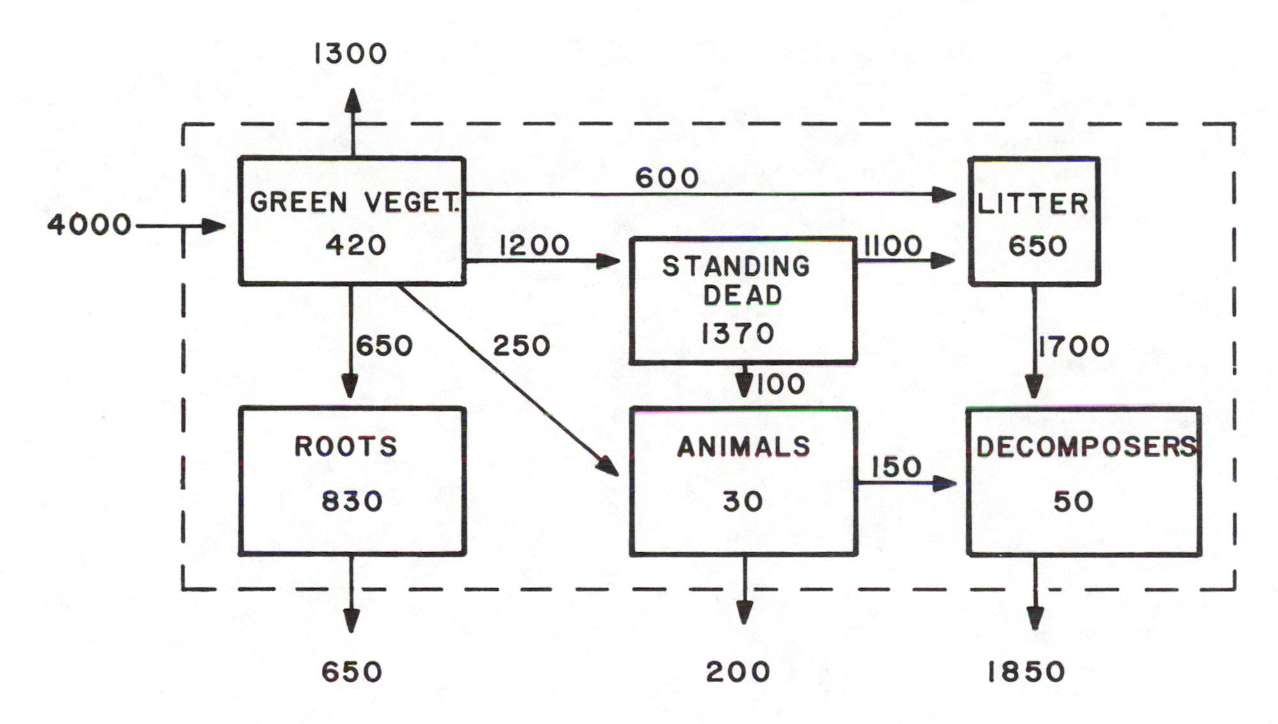

Figure 12.5. Energy flow in an old-field broomsedge community described by Golley (1965). Units are expressed in $kcal/m^2$ or $kcal/m^2/yr$.

Conclusion

This chapter has introduced you to some of the concepts and techniques of energy and material flow models. This type of model deals with the ecosystem in a holistic manner without concern for the multitude of mechanistic details which are undoubtedly important to its operation. In this chapter, we employed the direct two-stage programming approach to implement the models. The more elegant matrix approach will be discussed in Chapter 16. The examples and exercises discussed were of simple, low diversity systems to promote better understanding. However, the same principles could be employed in more complex, high diversity systems in which essentially all major organisms are represented by state variables. The shortgrass prairie ecosystem model of

Patten (1972) is an impressive example of this type of simulation. However, Patten, et al., (1975) make the following comment about models of this nature:

> "The model reduces the intricate beauty and awesome complexity of a piece of living nature to what is by comparison a flat, pallid image of the reality. It is in the homomorphic character of models to do this, and ecologists would make a grave error ever to begin confusing image with reality. An ecosystem model, no matter how sophisticated or difficult to produce, is but a shadow of its prototype, and modeling, simulation, and systems analysis are means to understanding the latter, not ends."

CHAPTER 13

SIMPLE MODELS OF MICROBIAL GROWTH

For a very long time, microbiologists have recognized a similarity between enzymes and microbes in their ability to act upon a substrate or nutrient. This resemblance is most striking in the hyperbolic response of microbial population growth rate to increasing concentrations of a limiting nutrient. The saturation behavior observed suggests that equations of the Michaelis-Menten type can serve as models for certain types of microbial growth. Thus, there is an extensive body of knowledge, including many well verified models, which by analogy have high potential for modeling growth in microorganisms, particularly bacteria and algae.

This is the first of two chapters devoted to microbial growth models. A subsequent chapter will examine models which consider growth as a multistage process. The present chapter describes simple models based on single equations of the Michaelis-Menten type.

13.1 The Monod Model**

The hyperbolic relationship between microbial growth rate and limiting nutrient concentration was first formalized as an equation of the Michaelis-Menten type by Monod (1942). This model describes the growth rate of cell biomass as follows:

$$\frac{dX}{dt} = \mu X = \bar{\mu} \cdot \left(\frac{S}{K_s + S}\right) X \qquad (13.11)$$

where μ is the specific growth rate for a free nutrient concentration, S, $\bar{\mu}$ is the maximum growth rate obtained under saturating concentrations of nutrient, K_s is the half-saturation constant (equal to substrate concentration when $\bar{\mu} = 0.5$), and X may be either cell biomass in g/liter, or cell population density, assuming uniform cell size. Some typical growth data are plotted in Figure 13.1.

James D. Spain, BASIC Microcomputer Models in Biology ISBN 0-201-10678-7

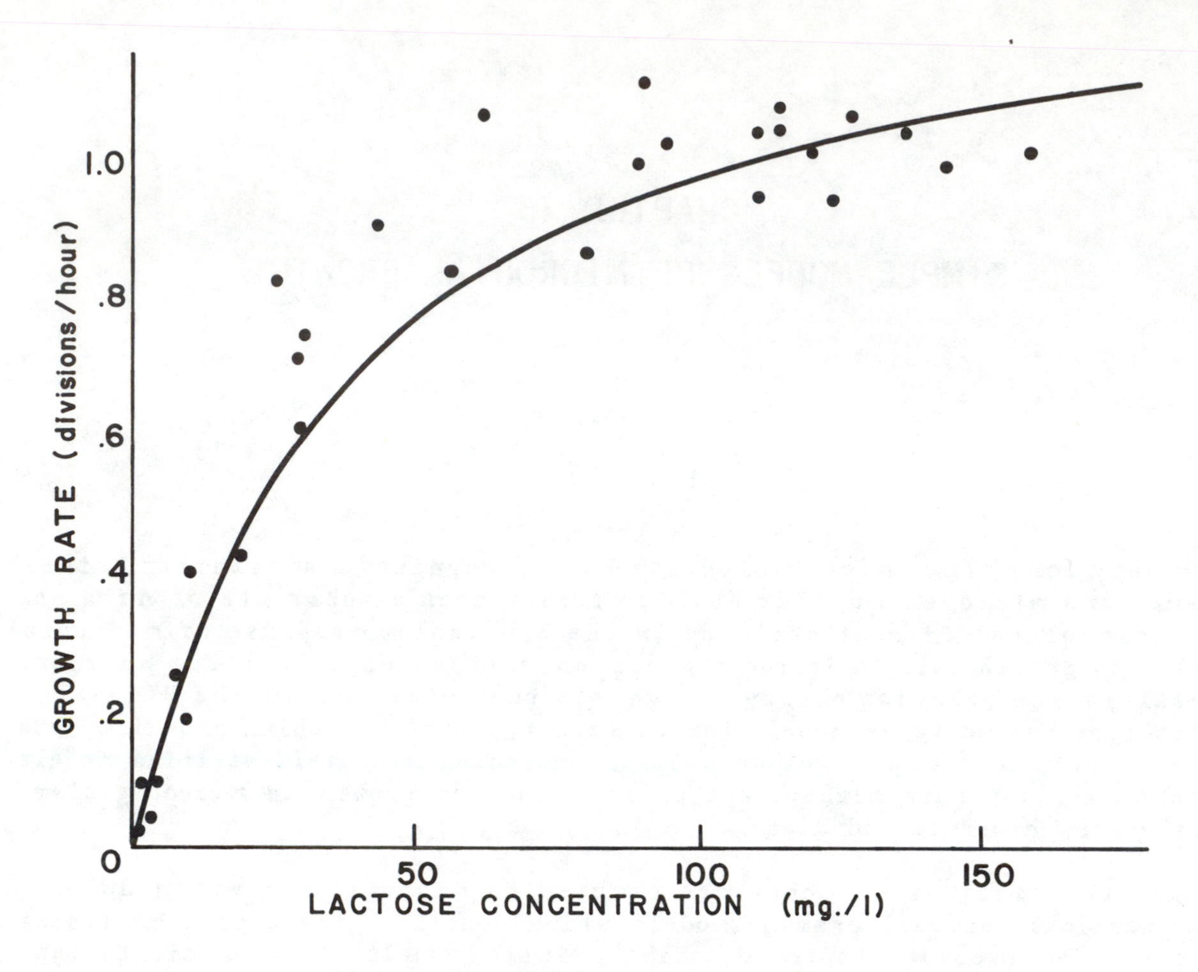

Figure 13.1. Growth rate of E. coli at different concentrations of lactose. Based on data from Monod (1942).

A formal derivation of this equation is probably not possible. However, there is an analogy to the Michaelis-Menten equation (See Section 2.2). In Chapter 2 we discussed how certain assumptions about the following chemical equilibrium led to the Michaelis-Menten model:

$$E + S \underset{k_2}{\overset{k_1}{\rightleftharpoons}} ES \overset{k_3}{\longrightarrow} P + E$$

where E is the free enzyme, S is the substrate, ES is the enzyme bound to substrate, P is the product, and k_1, k_2 and k_3 are rate constants for the reactions shown. Then

$$v = V_{max} \cdot \frac{S}{K_m + S} \qquad (13.12)$$

The derivation of this equation came from assuming a steady-state condition for ES and assuming that $v \rightarrow V_{max}$ as $ES \rightarrow E_{total}$. In a similar manner, the growth of cells might be described by a hypothetical reaction of the following type:

$$X + S \underset{k_2}{\overset{k_1}{\rightleftarrows}} XS \overset{k_3}{\longrightarrow} X + P$$

where X refers to living cell biomass, S is a limiting nutrient, XS is nutrient absorbed into the cellular material, P is new cell protoplasm (assimilated nutrient), k_1 is the rate of feeding, k_2 is the rate of nutrient loss through excretion and k_3 is the rate of nutrient assimilation into protoplasm.

As with enzymes, maximum formation of protoplasm may be assumed to occur when cell biomass and the associated synthetic systems are saturated with the limiting nutrient in the form of XS. Here P is identified as new protoplasm, however, it is really indistinguishable from cell biomass, X. Thus, the process exhibits the positive feedback characteristics seen in the autocatalytic enzyme reaction (Section 6.5). Considerable evidence for the validity of the Monod model has accumulated since it was first proposed (Dugdale, 1967, Powers and Canale, 1975). K_s and $\bar{\mu}$, the parameters in the equation, are usually obtained by plotting various linear transformations of growth rate and nutrient concentration data. This may be accomplished by plotting $1/\mu$ as a function of $1/S$, or more frequently, S/μ as a function of S (the transformation used in Table 3.1, Equation No. 4).

The growth rate equation may be numerically integrated to model the change of total cell biomass with time, as follows:

$$\Delta X_t = (\bar{\mu} \cdot \frac{S}{K_s + S} \cdot X)\Delta t \qquad (13.13)$$

$$X_{t+\Delta t} = X_t + \Delta X_t \qquad (13.14)$$

Since cells are normally undergoing some autolysis and loss of biomass from respiration and mortality, Equation 13.13 usually includes an additional term as follows:

$$\Delta X_t = (\bar{\mu} \cdot \frac{S}{K_s + S} \cdot X_t - R \cdot X_t)\Delta t \qquad (13.15)$$

where R is the first order rate constant describing loss of biomass.

<u>Exercise 13.1</u>**

Use Equation 13.15 to develop a simulation for the growth of bacteria in the presence of limiting nutrient. Assume batch culture conditions in which there is a fixed amount of total nutrient, S_T. The relationship between total nutrient, free nutrient and bound nutrient would be described by the following equation:

$$S_T = S + f \cdot X_t \qquad (13.16)$$

where f is the fraction of bacterial biomass, X_t, which represents bound nutrient.

Use the following parameters to implement your simulation:

$\bar{\mu}$ = .3/hour, S_T = 100mg/1, K_s = 25mg/1, R = .03/hour, and f = .03.

Assume an initial biomass of 1 mg/liter and allow the program to simulate about 60 time intervals (hours). Plot X_t and S as a function of time.

The numerical integrations in this chapter can be rather unstable unless the time increments are made sufficiently small. A suggested value for Δt in this exercise is 0.1 hr.

13.2 The Continuous Culture or Chemostat Simulation**

Many systems, especially aquatic ecosystems such as plankton populations in streams or small lakes, behave more like a continuous culture than a batch culture. For this reason the kinetics of the continuous culture system are very interesting. The approach is that used by Novick and Szilard (1950).

The following is a diagram of the basic continuous culture system:

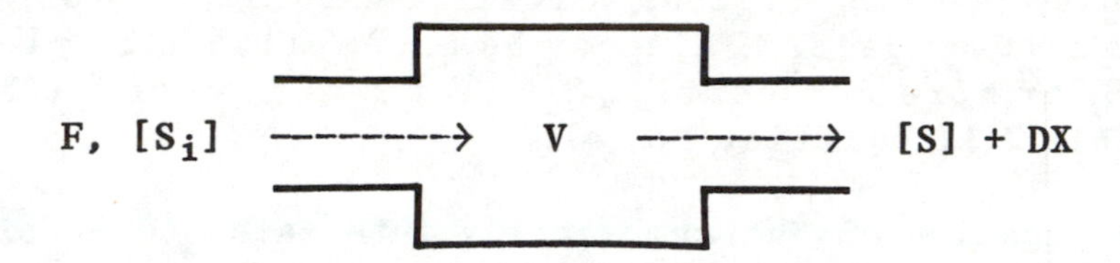

V is the volume of the culture vessel or chemostat in cm^3, F is the flow rate in cm^3/min, X is the cell density in $cells/cm^3$, $[S_i]$ is the concentration of substrate coming into the chemostat expressed in g/cm^3, [S] is the substrate concentration in the chemostat and leaving the chemostat and D is the dilution rate, equivalent to F/V in min^{-1}.

The net increase of cells in the chemostat during one time interval is given by the expression

Net Increase = Growth Rate − Washout Rate

$$\Delta X_t = \mu X_t \Delta t - D X_t \Delta t \qquad (13.21)$$

where μ is the growth rate constant and μX is the growth rate in cells per min per cm^3. At steady-state $\Delta X/\Delta t$ equals 0 and μ equals D. Cell growth may be characterized by the growth equations, 13.13 and 13.14.

The change in substrate is defined by

$$\frac{dS}{dt} = [S_i]D - [S]D - \mu X f \qquad (13.23)$$

where $[S_i]\cdot D$ is the rate coming in, μXf is the rate used in the formation of new cells, $f = dS/dX \cong \Delta S/\Delta X$, and [S]D is the rate at which substrate is lost from the chemostat.

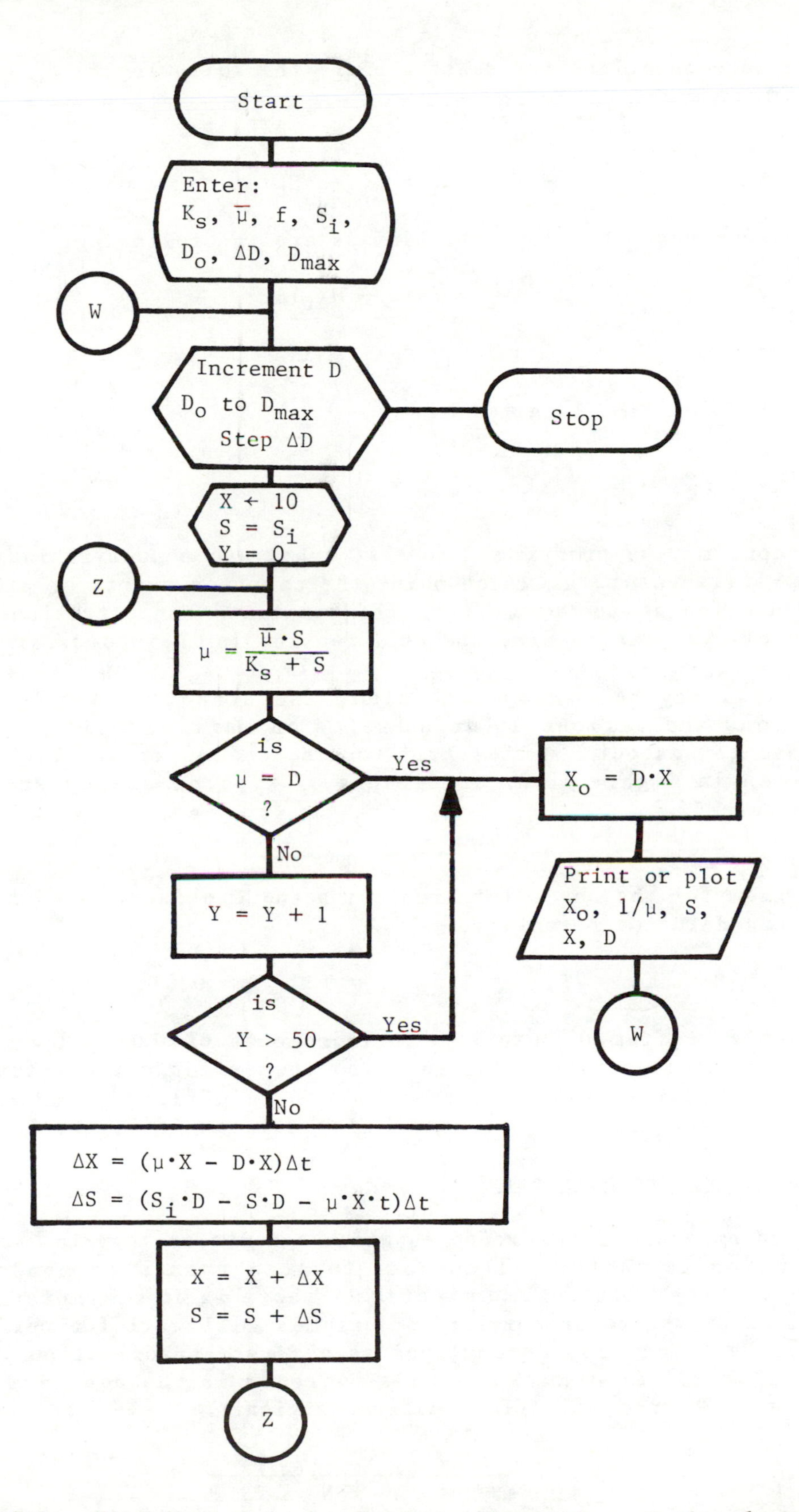

Figure 13.2. Flowchart for steady-state simulation of the chemostat.

The mass balance equations for substrate are the following:

$$\Delta S_t = ([S_i]D - [S]_t - \mu Xf)\Delta t \qquad (13.24)$$

$$S_{t+\Delta t} = S_t + \Delta S_t \qquad (13.25)$$

The mass balance equations for cell biomass are the following:

$$\Delta X_t = (\mu X_t - DX_t)\Delta t \qquad (13.26)$$

$$X_{t+\Delta t} = X_t + \Delta X_t \qquad (13.27)$$

The cell generation time is given by

$$G = \frac{1}{\mu} \qquad (13.28)$$

These equations may be programmed so that substrate concentration is iterated to a steady-state value, at which point the computer prints or plots the rate of cell production by the chemostat, the substrate concentration leaving the chemostat, cell generation time and cell density in the chemostat.

The simulation may be employed to study the effect of various substrate concentrations and various dilution rates on maximum cell production and dilution rate at washout. A flowchart for the chemostat simulation described above is seen in Figure 13.2, and graphs of typical steady-state data are shown in Figure 13.3.

<u>Exercise 13.2</u>**
Write a program for the chemostat simulation and implement it with the following growth and dilution parameters:

$$K_s = 75, \ \bar{\mu} = 1.5, \ S_i = 100, \ f = 0.01.$$

Vary dilution rate from 0.02 to 1.20 in increments of 0.02. Have the program iterate to the steady-state condition, and then print out or plot generation time, cell density, substrate concentration in effluent, and rate of cell production or total cells in effluent. Plot data as in Figure 13.3.

13.3 Multiple Limiting Nutrients**

One of the advantages of the Monod equation and its analogy to the Michaelis-Menten equation is that it allows one to draw upon a tremendous body of literature dealing with enzyme kinetics. There is no guarantee, of course, that equations proven to be correct for enzymes will work for cell population growth. However, they may be employed as a first approximation when nothing else is available. An example of this approach is the equation adapted by Chen (1970) to describe multiple limiting nutrients.

$$\mu = \bar{\mu}\left(\frac{S}{K_s + S}\right)\left(\frac{N}{K_n + N}\right)\left(\frac{P}{K_p + P}\right) \qquad (13.31)$$

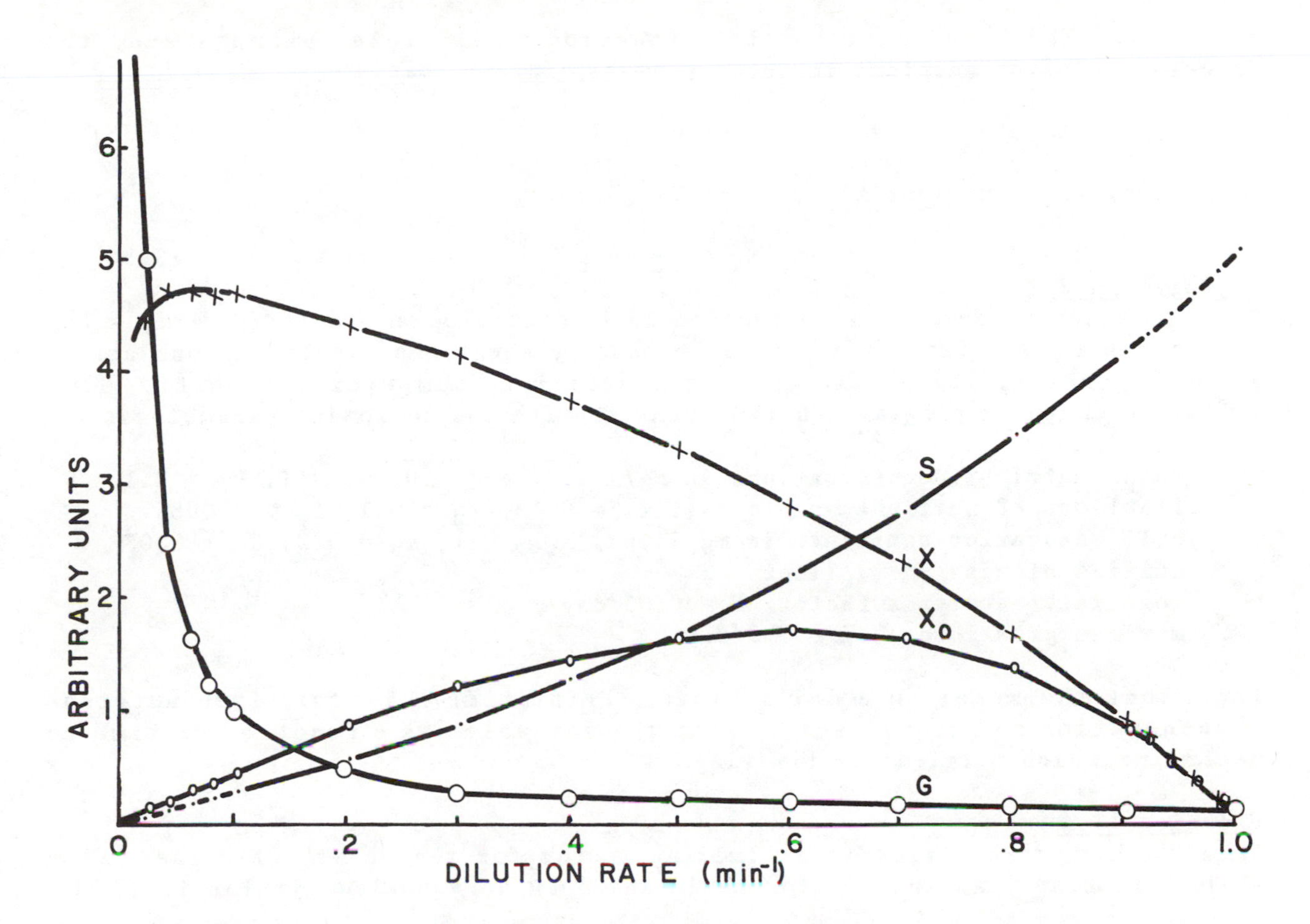

Figure 13.3. The simulated effect of dilution rate on steady-state conditions in a chemostat. K_s = 50 g/1, $\bar{\mu}$ = 2 min^{-1}, f = $\Delta S/\Delta X$ = 0.01.

In this equation, which could be employed to model diatom growth rate, S, N and P are three potentially limiting nutrients, silicon, nitrogen, and phosphorus. The half saturation constant for each nutrient is symbolized by K_s, K_n, and K_p. The origin of equations of this type is presumably the equation for enzyme reaction kinetics involving two or more substrates developed by Cleland (1970).

The Chen equation (13.31) may be employed to simulate the principle called Liebig's law of the minimum. This law states that growth of an organism or population of organisms will be limited by the factor which exists in a minimum concentration relative to its need. In terms of this equation, it means that growth will be limited by the free nutrient having the lowest concentration relative to its requirement by the organism. Free nutrient at any time is a function of the total nutrient and the bound nutrient according to the equation

$$N_T = N_s + N_b \tag{13.32}$$

where N_T refers to total nutrient, N_s to soluble nutrient, and N_b to bound nutrient. The bound nutrient is a function of the total biomass and, the fraction relating nutrient to total biomass, f_n, so that

$$N_b = f_n \cdot X \tag{13.33}$$

Therefore, free nutrient at any time is given by

$$N_s = N_T - f_n \cdot X \tag{13.34}$$

Exercise 13.3**
Use the equation described in Section 13.3 to develop a model of Liebig's law of the minimum. First write up all necessary equations, including one analogous to Equation 13.15. Diagram an appropriate computer flowchart. Then write the computer program and implement it with the following parameters:

total nutrient concentrations in mg/liter, $S_T = 10$, $N_T = 1$, $P_T = 0.1$
fractions of nutrient in biomass, $f_s = 0.15$, $f_n = 0.1$, $f_p = 0.008$
half saturation constants in mg/liter, $K_s = 0.1$, $K_n = 0.2$, $K_p = 0.03$
initial biomass in mg/liter = 1
mortality/autolysis factor, R = 0.05/day
maximum growth rate, $\bar{\mu} = 1.2$/day.

Increment the model on a daily basis. Print time, biomass, free nutrient concentration and growth rate. Plot the variables as a function of time to determine which nutrient is limiting.

Exercise 13.4
In most instances, light is a limiting factor for the growth of algae. Chen (1970) proposed that this factor be included in an equation similar to 13.31.

$$\mu = \bar{\mu}\left(\frac{I}{K_i + I}\right)\left(\frac{N}{K_n + N}\right)\left(\frac{P}{K_p + P}\right) \tag{13.35}$$

Here we are assuming a hyperbolic relationship between algal growth and light intensity as in Section 10.51. This equation is assumed to apply to algae which do not have a requirement for silicon.

Use Equation 13.35 to model the response of an algal population to constant total nutrient, but varying light intensities. Include other equations such as 13.15, and 13.32. Assume the same input parameters as in Exercise 13.3, except assume that average daily light intensity varies in an annual cycle with a maximum of 600 cal/cm^2/day and a minimum of 100 cal/cm^2/day. Let K_i = 400 cal/cm^2/day. Increment on a daily basis and obtain about two years of simulated data. Graph biomass and free nutrient concentrations as a function of time to observe the effect of varying light intensity on such a system.

13.4. Competition of Microbes for a Single Limiting Nutrient

It is a general principle of ecology that in order to survive in an ecosystem, two organisms requiring the same limiting nutrient must each adopt a strategy which give themselves a competitive advantage for that nutrient at least some portion of the time. Usually, nutrient concentration will vary according to some annual or daily pattern. Part of this variation results from differential utilization and release of the nutrient by the organisms themselves.

The Monod equations may be employed to model this type of competition between two microbial species for a single limiting nutrient. Let us assume two species, X_1 and X_2, which have growth rates described by the following equations:

$$\mu_1 = \frac{\bar{\mu}_1 \cdot N}{(K_1 + N)} \tag{13.41}$$

$$\mu_2 = \frac{\bar{\mu}_2 \cdot N}{(K_2 + N)} \tag{13.42}$$

where μ_1 and μ_2 are the growth rates for X_1 and X_2, $\bar{\mu}_1$ and $\bar{\mu}_2$ are maximum growth rates, and K_1 and K_2 are the half-saturation constants of the two organisms for the nutrient, N.

The difference equations for describing change in biomass for each organism are as follows:

$$\Delta X_1 = (\mu_1 X_1 - R_1 X_1)\Delta t \tag{13.43}$$

$$\Delta X_2 = (\mu_2 X_2 - R_2 X_2)\Delta t \tag{13.44}$$

where R_1 and R_2 are respiration/mortality rate constants.

If there are no additional sources and sinks for free nutrient, and total nutrient is assumed to be constant, then free nutrient may be described by the equation

$$N = N_T - f_1 X_1 - f_2 X_2 \tag{13.45}$$

where N_T is total nutrient, and f_1 and f_2 are the fractions of nutrient in the biomass of X_1 and X_2. If N_T is not constant, as a result of additional nutrient sources, then a difference expression of the following type is required:

$$N \longleftarrow N + (R_1 X_1 f_1 + R_2 X_2 f_2 - \mu_1 X_1 f_1 - \mu_2 X_2 f_2 + S_1 - S_2)\Delta t \tag{13.46}$$

where S_1 and S_2 are additional sources and sinks such as influents, effluents, mixing, sedimentation, gaseous exchange, predation, and decomposition.

Dugdale (1967) suggested that one way in which two organisms might successfully compete for a single nutrient is by matching their half-saturation constants and maximum growth rates in such a way that both organisms have a range of nutrient concentrations in which their specific growth rates are greater than their competitor. Figure 13.4 shows the growth rate curves of two hypothetical organisms which demonstrate this relationship. Organism X_2 has the higher specific growth rate, and hence the competitive advantage, at concentrations above the critical nutrient concentration, N_c, while organism X_1 has a half-saturation constant which gives it an advantage at nutrient concentrations below N_c.

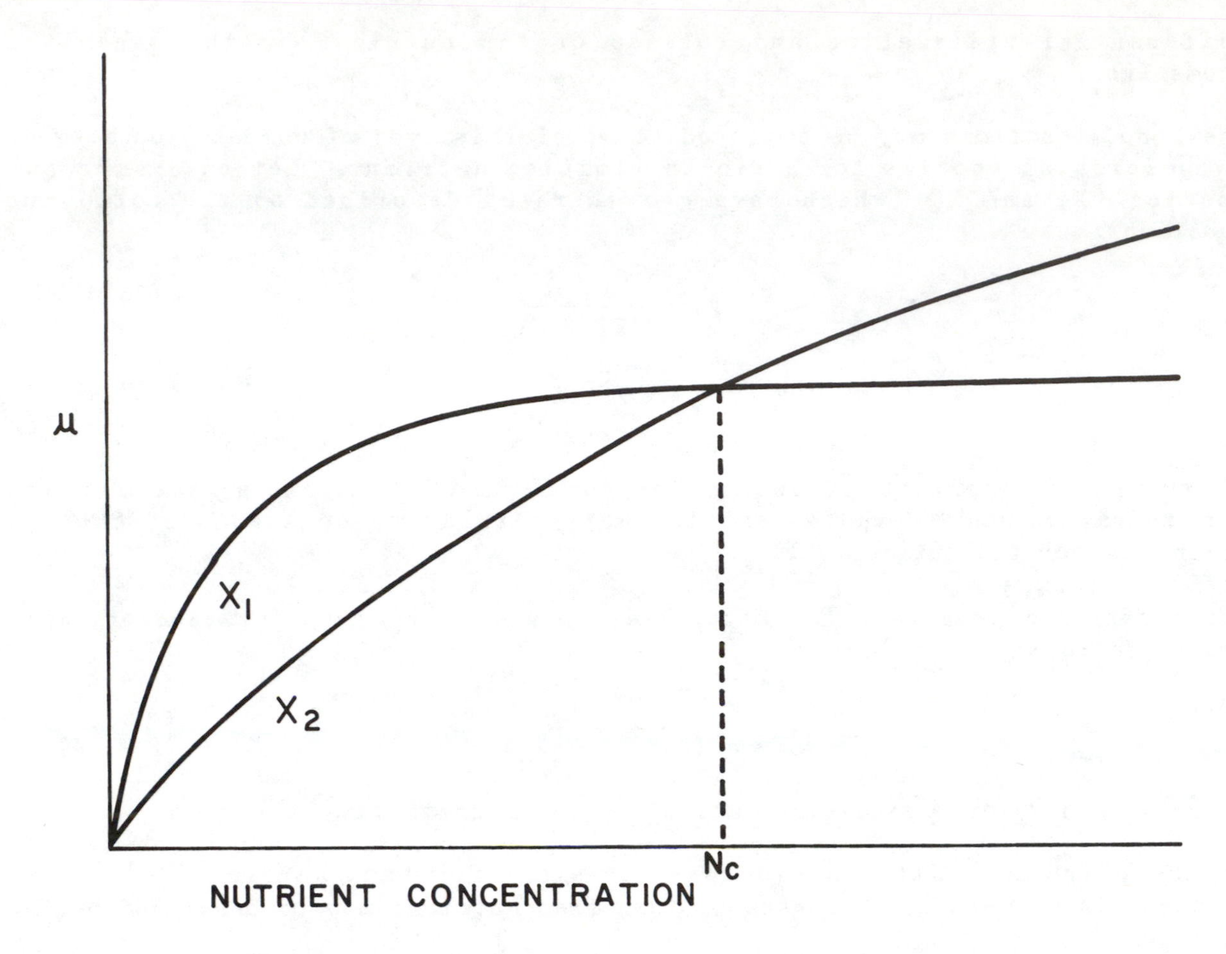

Figure 13.4. Curves for two hypothetical organisms having different nutrient dependent growth rate expressions. Based on Dugdale (1967).

<u>Exercise 13.5</u>
Use the equations in Section 13.4, including Equation 13.45, which assumes a constant total nutrient concentration, to develop a competition model for two microbes having different requirements for a limiting nutrient. Implement the model with the following input parameters:

$X_1 = 0.01$ mg/liter	$R_1 = 0.1$/day
$X_2 = 0.01$ mg/liter	$R_2 = 0.1$/day
$\bar{\mu}_1 = 1$/day	$f_1 = 0.01$
$\bar{\mu}_2 = 2$/day	$f_2 = 0.01$
$K_1 = 1$ mg/liter	$N = 10$ mg/liter
$K_2 = 5$ mg/liter	

Graph species biomass and free nutrient concentration as a function of time for about 100 days. Iterate 10 times per time interval, and check to make sure that N does not go below zero. Can you see what would happen if total nutrient varied in some orderly fashion between a minimum of 0.5 and a maximum of 5.5? To answer this question, program a forced step cycle of 20 days at each of these total nutrient concentrations, giving both X_1 and X_2 an initial biomass of 500 mg/liter.

13.5 Competition for Multiple Limiting Nutrients

Equations of the Monod type may be employed to model phytoplankton succession in lakes involving several species of algae and two or more limiting nutrients. A model of this type was developed by Powers and Canale (1975), and what follows is largely derived from that source. They assumed a warm eutrophic lake with dominant phytoplankton assemblages limited to green and blue-green forms. If algal growth is limited by the supply of nitrogen and phosphorus, the growth rate for each type of algae may be approximated by the following equation of the Chen type:

$$\mu_g = \bar{\mu}_g \left(\frac{P}{K_{pg} + P}\right)\left(\frac{N}{K_{ng} + N}\right) \tag{13.51}$$

$$\mu_b = \bar{\mu}_b \left(\frac{P}{K_{pb} + P}\right)\left(\frac{N + N_2}{K_{nb} + N + N_2}\right) \tag{13.52}$$

In these equations,

- μ_g is the growth rate of green algae,
- $\bar{\mu}_g$ is the maximum growth rate of green algae,
- μ_b is the growth rate of blue-green algae,
- $\bar{\mu}_b$ is the maximum growth rate of blue-green algae,
- P is the concentration of dissolved inorganic phosphorus,
- N is the concentration of dissolved inorganic nitrogen,
- N_2 is the concentration of dissolved nitrogen gas,
- K_{pg} is the half-saturation constant for phosphorus in green algae,
- K_{pb} is the half-saturation constant for phosphorus in blue-green algae,
- K_{ng} is the half-saturation constant for nitrogen in green algae,
- K_{nb} is the half-saturation constant for nitrogen in blue-green algae.

Equation 13.52 includes a term which accounts for the ability of some blue-green species to fix dissolved nitrogen gas.

The growth rate constants, $\bar{\mu}_g$ and $\bar{\mu}_b$, are a function of temperature and light intensity. However, they are assumed to be constant for this model. Powers and Canale (1975) assumed that they were dealing with a chemostat-like lake having a significant input and output. However, in order to simplify this model, we shall assume that nutrient input is equal to output and that dilution rate is slow so that the system may be treated as a batch culture.

The difference equations describing the biomass of each type of algae are as follows:

$$\Delta G = (\mu_g G - R_g G)\Delta t \tag{13.53}$$

$$\Delta B = (\mu_b B - R_b B)\Delta t \tag{13.54}$$

where G and B refer to biomass of green and blue-green algae, and R_g and R_b

are first order respiration/mortality constants which describe the rate of biomass loss.

Free phosphorus concentration is determined by the following equation:

$$P = P_t - f_{pg}G - f_{pb}B \qquad (13.55)$$

where P_t = total phosphorus, f_{pg} is the fraction of phosphorus in green algal biomass, f_{pb} is the fraction of phosphorus in blue-green algal biomass. Because of the additional source involved, the free nitrogen concentration must be obtained by numerical integration using an expression of the type

$$N \longleftarrow N + [R_gGf_{ng} + R_bBf_{nb} - \mu_gGf_{ng} - \mu_bBf_{nb}(N/(N+N_2)) - DN]\Delta t \qquad (13.56)$$

where dissolved nitrogen in one time increment is increased by respiration/mortality of greens and blue-greens and decreased by growth and denitrification. D is a first order denitrification constant. f_{ng} and f_{nb} are constants relating dissolved nitrogen to biomass for greens and blue-greens, i.e., the fraction of nitrogen in algal biomass. Powers and Canale (1975) assumed that there was inhibition of blue-green algae by green algae. However, this feature of their model will be ignored here so that the major attention can be focused upon the role of half-saturation constants in interspecific competition.

Exercise 13.6

Use the information in Section 13.5 to develop a simulation model for competition between blue-green algae and green algae under limiting nutrient conditions. This model is limited in that it lacks some rather important nutrient sources and sinks, such as runoff, mixing, sinking of algae and predation. It also assumes constant light and temperature. For this reason, it will not show the dramatic bloom and decline normally seen in algal populations, although it does demonstrate competition of algae for limiting nutrients. After you have written the computer program, implement it with the following data, based on that provided by Chen (1970).

parameter	units	parameter	units
$\bar{\mu}_g$ = 2	/day	R_g = 0.06	/day
$\bar{\mu}_b$ = 1	/day	R_b = 0.04	/day
K_{pg} = 0.05	mg/liter	f_{pg} = 0.01	mg P/mg biomass
K_{pb} = 0.03	mg/liter	f_{pb} = 0.01	mg P/mg biomass
K_{ng} = 0.3	mg/liter	f_{ng} = 0.08	mg N/mg biomass
K_{nb} = 0.2	mg/liter	f_{nb} = 0.08	mg N/mg biomass
G_0 = 0.01	mg/liter	N_2 = 0.02	mg/liter
B_0 = 0.01	mg/liter	N_t = 0.1	mg/liter
P_t = 0.05	mg/liter	D = 0.05	/day

13.6 Growth Inhibition by Sub-Lethal Concentrations of Toxicants

As a first approximation, it should be possible to model the effect of sub-lethal concentrations of certain toxic agents on the growth of microorganisms by using the analogous equation for the inhibition of enzyme reaction rates.

Enzymes exhibit several types of inhibition depending on how V_{max} and K_m are affected. It is likely that toxic agents would exhibit similar effects on the growth kinetics of microbes. Such considerations suggest that an equation of the following type may have some potential for modeling toxicity:

$$\mu = \left(\frac{\bar{\mu}}{1 + i/K_i}\right)\left(\frac{S}{K_s + S}\right) \qquad (13.61)$$

This equation is analogous to the equation for non-competitive inhibition of enzymes. In this equation, $\bar{\mu}$ is changed by a factor of $1/(1 + i/K_i)$. Here, i is the toxicant (inhibitor) concentration, and K_i is a constant equal to the inhibitor concentration which causes a halving of the apparent value for $\bar{\mu}$. S is the concentration of the limiting nutrient, and K_s is the half-saturation constant for the nutrient.

A second possibility for a model is based on the equation for competitive inhibition of enzymes. This equation should model a system in which the toxic agent competes with an essential nutrient for its uptake and utilization.

$$\mu = \bar{\mu} \cdot \frac{S}{S + K_s(1 + i/K_i)} \qquad (13.62)$$

Here the effect of the inhibitor is to increase the concentration of nutrient necessary for half-saturation by a factor of $(1 + i/K_i)$. K_i is equal to the inhibitor concentration which will cause a doubling of the apparent K_s.

A third type of equation which one might be able to employ under special circumstances involves inhibition by high concentrations of nutrient. Two 'nutrients' which might demonstrate this effect are oxygen and light. An equation analogous to the substrate inhibition equation is

$$\mu = \bar{\mu}/\left(1 + \frac{K_s}{S} + \frac{S}{K_i}\right) \qquad (13.63)$$

Here K_i is the constant for nutrient inhibition. An equation of this form was suggested in Chapter 10 as a model for saturation and inhibition of photosynthesis by high intensities of light (refer to Equation 10.56).

<u>Exercise 13.7</u>
Use one of the equations in Section 13.6 and other necessary equations to model the effect of some hypothetical toxic agent on microbial growth and nutrient utilization. Set up the simulation model so that it is possible to vary nutrient concentration. Graph data in an appropriate form to show the effects of toxicity on growth.

Conclusions

A variety of simple models of the Monod type have been described in this chapter. One of their major attributes appears to be simplicity since on close scrutiny, they are often rather unrealistic. Numerous authors have shown that microbial populations often deviate markedly from the growth patterns predicted by the Monod model. Some of the simplifying assumptions involved clearly do not hold. For example, there is a difference between the composition of cells under different growth regimes. The concentration of nutrient in cells is not constant. Some nutrients affect growth of biomass, and others affect cell division. The 'luxury consumption' of nutrients such as phosphorus is not accounted for by this model.

Caperon and Meyer (1972) developed a model equation which places emphasis on the internal nutrient reservoir, that portion of the nutrient per cell which is greater than the amount which makes up biomass and is thus available for growth. This adaptation of the Monod equation involves transport of materials across cell membranes, the concept of pacemaker reactions, and other interesting aspects of the growth process which will not be taken up until the final chapter in Part II.

None of these criticisms in any way detracts from the value of modeling or the use of the Monod model when applicable. Quite the contrary, the Monod model has served to focus attention on details of the growth process which probably would have gone unnoticed for some time without the aid of simulation. Generally, as we become more sophisticated about a system, it is necessary to modify our models and in some cases replace them with new models. In many cases, the new level of sophistication is the direct result of trying to use overlysimplistic models and noting the discrepancies between simulation data and experimental data. The basic Monod concept still appears to be sound, as is the use of the hyperbolic or saturation type equation. Both of these topics will receive further consideration in later chapters.

CHAPTER 14
COMPARTMENT MODELS IN PHYSIOLOGY

In physiology, modeling and simulation techniques have focused on two chief areas, material transport processes and physiological control mechanisms. Because of the close relationship of these two fields to analogous areas of engineering, physiological simulation has made extensive use of existing techniques in control theory and systems analysis. These and related topics are an area of primary interest to the emerging field of bioengineering. The area of material transport makes use of compartment models not unlike those which were employed in Chapter 12 for the transport of energy and material in an ecosystem.

For purposes of modeling, the compartment was defined by Milhorn (1966) as follows:

> "If a substance is present in a biological system in several distinguishable forms or locations, and if it passes from one form or location to another form or location at a measurable rate, then each form or location constitutes a separate compartment for the substance."

Material may be transported from one physical location or compartment to another by several mechanisms. It may diffuse across a membrane, it may be actively transported across a membrane, or it may be carried by fluid flow. In addition, material may be converted from one form (compartment) to another form by chemical reactions of the type discussed in Chapter 6.

14.1 Transport by Simple Diffusion**

In compartment models involving simple diffusion, the following assumptions are usually made (Milhorn, 1966):

1. The volume of the compartment remains constant.

James D. Spain, BASIC Microcomputer Models in Biology

ISBN 0-201-10678-7

2. Any substance entering the compartment is instantaneously distributed throughout the compartment.

3. The rate of diffusion from the compartment is proportional to the concentration of the substance in the compartment.

Equations of the following type are employed to describe the loss of Q from compartment i by forward diffusion, f:

$$\frac{dQ_i}{dt_f} = \frac{-k \cdot Q_i}{V_i} \qquad (14.11)$$

Q_i is the quantity of some diffusing substance in compartment i, k is the rate constant, and V_i is the volume of compartment i.

If compartment i is connected to some other compartment, j, then there will be diffusion in the reverse direction when a significant concentration of Q exists in that compartment.

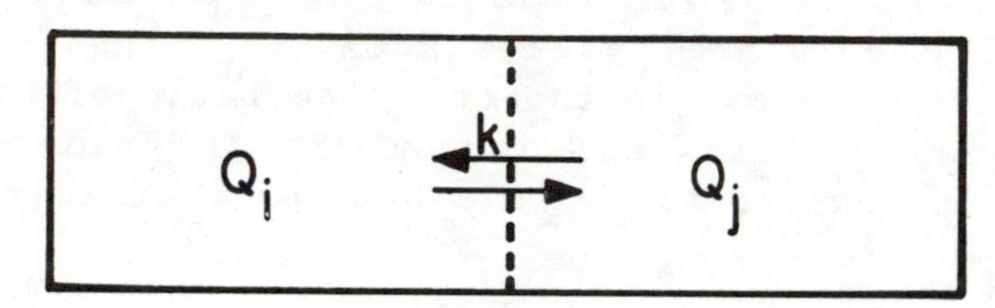

The rate of diffusion in the back direction (b) will be described by the equation

$$\frac{dQ_i}{dt_b} = \frac{k \cdot Q_j}{V_j} \qquad (14.12)$$

Net diffusion is therefore

$$\frac{dQ_i}{dt_{net}} = \frac{dQ_i}{dt_f} + \frac{dQ_i}{dt_b} = k\left(\frac{Q_j}{V_j} - \frac{Q_i}{V_i}\right) \qquad (14.13)$$

The change in each compartment may be described by difference equations.

$$\Delta Q_i = k\left(\frac{Q_j}{V_j} - \frac{Q_i}{V_i}\right)\Delta t \qquad (14.14)$$

$$\Delta Q_j = k\left(\frac{Q_i}{V_i} - \frac{Q_j}{V_j}\right)\Delta t \qquad (14.15)$$

The new states are obtained by update expressions.

$$Q_i \longleftarrow Q_i + \Delta Q_i \qquad (14.16)$$

$$Q_j \longleftarrow Q_j + \Delta Q_j \qquad (14.17)$$

Exercise 14.1**
Set up a simulation of passive diffusion between two compartments, one of which receives an initial quantity of some substance. Employ the following input parameters:

$$(Q_j)_0 = 0,\ (Q_i)_0 = 100,\ V_i = 10,\ V_j = 50,\ k = 0.8.$$

Graph the concentration of the substance in each of the two compartments as a function of time.

14.2 Osmotic Pressure Model

Osmosis is a special case of diffusion in which the solvent, usually water, is the primary diffusing substance. The following osmotic pressure model is based on the classical osmosis experiment, illustrated in Figure 14.1.

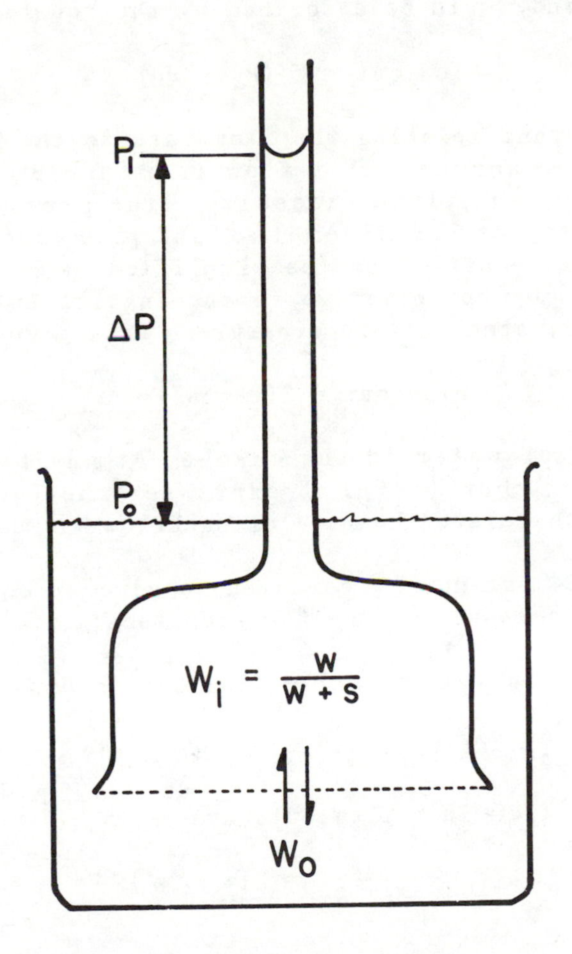

Figure 14.1. Diagram of the classical osmotic pressure experiment.

The movement of water during osmosis is simply diffusion from high concentration of water towards lower concentration of water (high solute concentration). The solute is restricted to the osmometer compartment because of the semipermeable nature of the membrane. The rate of water diffusion may be described by the equation

$$\text{Flow in} = k(W_0 - W_i) \qquad (14.21)$$

where k is a diffusion rate constant which is a function of the properties of water in relation to the membrane, the thickness of the membrane, and the area of the membrane. W_0 is the water concentration outside the membrane expressed as a mole fraction, usually 1, which would be pure water. W_i is the mole fraction of water inside the membrane which equals $w/(w + s)$, where w is the number of moles of water and s is the number of moles of solute per unit volume inside the compartment. Equation 14.21 then becomes

$$\text{Flow in} = k(W_0 - \frac{w}{w + s}) \qquad (14.22)$$

The backflow of water across the membrane is the result of a difference in hydrostatic pressure, and would be described by the equation:

$$\text{Flow out} = K'(P_1 - P_0) \qquad (14.23)$$

where K' is a flow constant relating the flow rate to the hydrostatic pressure difference across the membrane. This flow is a function of membrane area, thickness and porosity, and fluid viscosity. The pressure difference is a function of the geometry of the manometer, and perhaps other factors. For practical purposes, this equation may be simplified by basing the flow on the amount of water in the chamber relative to the initial amount, which will be assumed to have been adjusted to zero pressure. This gives the following:

$$\text{Flow out} = K(w - w') \qquad (14.24)$$

where w is the amount of water in the chamber at any time t, and w' is the initial amount of water when $P = P_0$. K expresses the factors listed above, plus other factors which relate the rate of backflow to the pressure produced.

Equations 14.22 and 14.24 can now be combined to give an equation for net flow across the membrane as a result of the two processes:

Net flow = flow in (diffusion) - flow out (hydrostatic flow)

$$\frac{\Delta w}{\Delta t} = k(W_0 - \frac{w}{w + s}) - K(w - w') \qquad (14.25)$$

At osmotic equilibrium, flow in = flow out, and

$$k(W_0 - \frac{w}{w + s}) = K(w - w') \qquad (14.26)$$

<u>Exercise 14.2</u>
Use Equation 14.25 to program the osmotic pressure model. Use the following parameters to implement your simulation:

$$k = 20,\ K = 0.3,\ W_0 = 1,\ w = 55,\ s = 1.$$

The backflow may be converted to pressure by multiplying by a factor 62.199

which will give pressure in atmospheres, and this may be converted to inches of water by multiplying by a factor of 414.

The simulation should be allowed to run until net flow across the membrane is near zero, at which point the pressure is the osmotic pressure.

14.3 Model of Countercurrent Diffusion

Countercurrent diffusion has evolved in a variety of organisms in order to transport materials or heat from one fluid vessel to another in an efficient manner. One example is found in the special anatomical relationship between the loops of Henle in the kidney nephron and the medullary interstitial fluid (Guyton, 1971). Another example is found in the transfer of heat between arterial and venous circulation in the extremities of whales, seals, and birds of cold regions (Hardin, 1966). This system, called the rete mirabile, is designed for the efficient conservation of heat. The common feature of these examples is the close proximity of two counter flowing vessels between which materials (or heat) are exchanged through simple diffusion as a result of a concentration (or temperature) gradient. The efficiency of the countercurrent system results from maximizing the intensity gradient at all points where exchange is taking place.

An effective model of countercurrent diffusion may be based upon two series of n parallel compartments separated by a membrane which permits transport. The following diagram, Figure 14.2, illustrates the model system:

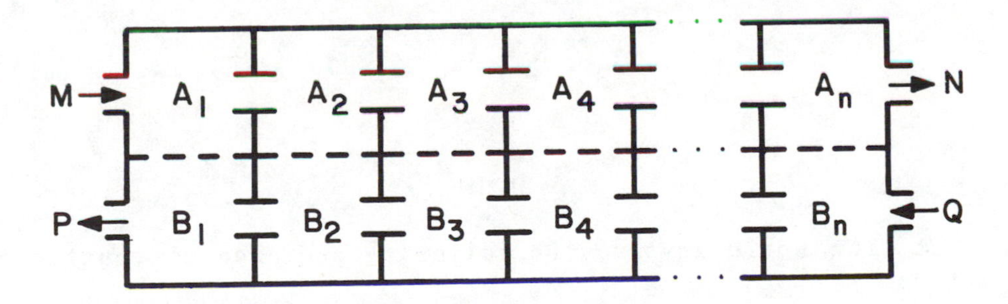

Figure 14.2. Conceptual diagram of a countercurrent system

The diffusion between any pair of compartments A_j and B_j would be described by Fick's law, which states that the rate of diffusion across a thin membrane is proportional to the area of membrane, the difference in concentrations, and a permeability constant.

$$\frac{\Delta X}{\Delta t} = PA(A_j - B_j) \tag{14.31}$$

where ΔX is the amount of X diffusing in the time Δt, P is the permeability constant, A is membrane area, A_j is the concentration of X in compartment A_j, and B_j is the concentration of X in compartment B_j.

In order to simplify this system, the following assumptions are made:

1) The membrane area is the same for all compartments.
2) The permeability constant is the same for all compartments.
3) The volume of each compartment is constant (unit volume), so that amounts diffusing and concentrations are equivalent.
4) The movement of material in the vessel occurs by simple plug flow.

Plug flow is modeled by having the material in each of the A compartments move one compartment to the right, and under countercurrent conditions the material in each B compartment move one compartment to the left. In concurrent flow, the latter movement would be to the right. Two sets of equations are involved. The diffusion equations first involve calculating the differences.

$$\Delta A_j = k(B_j - A_j)\Delta t \quad (14.32)$$

$$\Delta B_j = k(A_j - B_j)\Delta t \quad (14.33)$$

Then, each compartment is updated.

$$A_j \longleftarrow A_j + \Delta A_j \quad (14.34)$$

$$B_j \longleftarrow B_j + \Delta B_j \quad (14.35)$$

The flow equations are set up in the following sequence for the A compartments:

$$N = A_n \quad (14.36)$$

$$A_j = A_{j-1}$$

$$A_1 = M$$

Countercurrent flow would involve the following sequence of equations for the B compartments:

$$P = B_1 \quad (14.37)$$

$$B_{j-1} = B_j$$

$$B_n = Q$$

Concurrent flow would involve the reverse sequence from 14.37, starting with

$$Q = B_n \quad (14.38)$$

Exercise 14.3

Program the preceding simulation model to compare the efficiency of countercurrent flow with that of concurrent flow. The best way to program this multicompartment system, involving perhaps 20 compartments, is to utilize the subscripted variable technique. This would permit you to program a loop which repeats Equations 14.32 - 14.35 for each of the 20 compartments. In a similar manner other loops may be programmed to perform the sequences initiated by

Equations 14.36, and 14.37 or 14.38. Make sequences 14.37 and 14.38 interchangeable, so that it is possible to run the simulation in either the concurrent or countercurrent mode. Use the following input parameters:

Make initial values of A_1 to $A_n = 0$ and B_1 to $B_n = 0$, M = 100, Q (or P) = 0, k = 0.1.

Allow the simulation to generate data for at least 25 time intervals, and produce some output under both types of flow.

The diffusion rate constant, k, may be varied in order to observe its effect on the efficiency of the extraction process. Note that the k values lie between 0 and 0.5, with 0.5 being equivalent to 100 percent equilibrium at each stage (compartment) and 0 being equivalent to no transport.

How could you modify the model and the program to simulate the effect of a layer of interstitial tissue separating the two counter-flowing vessels?

14.4 A Model of Active Transport

A simple model for the active transport of some metabolite, B, into a cell may be developed from the following assumptions based on the diagram in Figure 14.3:

1) Assume two compartments separated by a membrane with differential permeability for a substance B, and a closely related compound C, in which C is the more permeable of the two.

2) Assume an enzyme on one side of the membrane which catalyzes the conversion of B to C. This might occur by a reaction such as the following:

$$B + ATP \rightleftharpoons C + ADP$$

   in which the equilibrium lies far to the right, K>>1 and $\Delta G^o << 0$.

3) Assume there is another enzyme which catalyzes the conversion of C to B. For example,

$$C + H_2O \rightleftharpoons B + Pi$$

   Again the equilibrium is assumed to lie far to the right and $\Delta G^o << 0$.

4) Assume that the concentrations of enzymes on each side of the membrane remain constant throughout.

If each of the two compartments involved had a unit volume, and the membrane had a unit area, then the rate of transport due to diffusion would be described by the equations:

$$\Delta B_d = k_b(B_i - B_0)\Delta t \qquad (14.40)$$

$$\Delta C_d = k_c(C_i - C_0)\Delta t \qquad (14.41)$$

where k_b and k_c are diffusion rate constants.

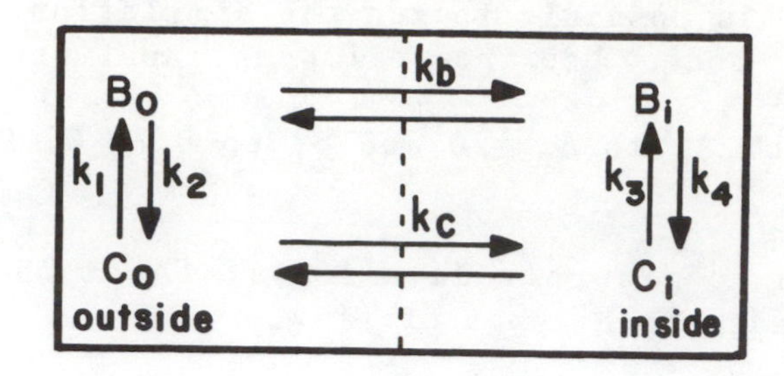

Figure 14.3. Compartment model of active transport process

The concentration changes resulting from enzyme-catalyzed reactions will be assumed to involve first order rate equations, which would be the case if the other reactants involved were maintained at constant concentrations, and enzymes were well below saturation by substrate. The following difference equations would result:

$$\Delta B_i = (k_3 C_i - k_4 B_i)\Delta t \quad (14.42)$$

$$\Delta C_i = (k_4 B_i - k_3 C_i)\Delta t \quad (14.43)$$

$$\Delta B_0 = (k_1 C_0 - k_2 B_0)\Delta t \quad (14.44)$$

$$\Delta C_0 = (k_2 B_0 - k_1 C_0)\Delta t \quad (14.45)$$

where k_1, k_2, k_3, and k_4 are rate constants for the reactions.

The update expressions for numerical integration of this process become:

$$B_i \leftarrow B_i + \Delta B_i - \Delta B_d \quad (14.46)$$

$$B_0 \leftarrow B_0 + \Delta B_0 + \Delta B_d \quad (14.47)$$

$$C_i \leftarrow C_i + \Delta C_i - \Delta C_d \quad (14.48)$$

$$C_0 \leftarrow C_0 + \Delta C_0 + \Delta C_d \quad (14.49)$$

With selection of appropriate rate constants, this simulation should proceed to some steady-state concentration gradient with respect to compound B.

<u>Exercise 14.4</u>

Use the equations in Section 14.4 to develop a simulation model of active transport. Use the following input parameters to implement this simulation:

B_0 = 50 mM $\quad B_i$ = 50 mM $\quad C_0$ = 1 mM $\quad C_i$ = 1 mM

k_2 = 0.005/min $\quad k_3$ = 0.5/min $\quad k_b$ = 0.001/min

k_2 = 0.5/min $\quad k_4$ = 0.005/min $\quad k_c$ = 0.1/min

Increment time in minutes. Allow the system to approach steady-state. Graph the concentration of species B, the molecule which is being transported against the diffusion gradient, as a function of time.

14.5 Simple Approach to Active Transport**

In the previous section, we discussed a possible mechanism for active transport, which resulted in a steady-state concentration gradient across the membrane. This same end result would be obtained by simply employing different diffusion rate constants for each direction across the membrane.

The rate of diffusion then becomes

$$\frac{dQ_i}{dt} = k_1 \cdot \frac{Q_j}{V_j} - k_2 \cdot \frac{Q_i}{V_i} \tag{14.51}$$

where Q_j and Q_i are quantities, V_j and V_i are volumes, and k_1 is greater than k_2 if active transport is toward compartment i. At equilibrium, the rate in each direction is equal. Hence

$$k_1 \cdot \frac{Q_j}{V_j} = k_2 \cdot \frac{Q_i}{V_i} \tag{14.52}$$

so that

$$\frac{k_1}{k_2} = \frac{Q_i \cdot V_j}{Q_j \cdot V_i} = \frac{C_i}{C_j} \tag{14.53}$$

Thus, the ratio of the two constants is equal to the equilibrium (or steady-state) ratio of the two concentrations, C_i and C_j.

Neither of the above active transport models account for mediated or facilitated transport processes. These are membrane transport processes, either active or passive, which show saturation type kinetics because only a limited number of transport sites exist. In addition, they may show specificity for a particular chemical species being transported. Because these two properties are similar to those of enzymes, mediated transport systems are sometimes called translocases (the -ase suffix denoting that they behave like enzymes).

The distinction between simple diffusion and mediated-transport processes is best illustrated in Figure 14.4. This diagram shows that the carrier or transport sites of the mediated-transport become saturated at high concentrations of the diffusing substance, and that the rate does not exceed T_{max}.

Mathematically we may represent this by an equation of the Michaelis-Menten type.

$$T_c = \frac{T_{max} \cdot C_i}{K_c + C_i} \tag{14.54}$$

where T_c is the rate of carrier-mediated diffusion, T_{max} is the maximum rate, C_i is the concentration of diffusing substance, and K_c is the half-saturation constant.

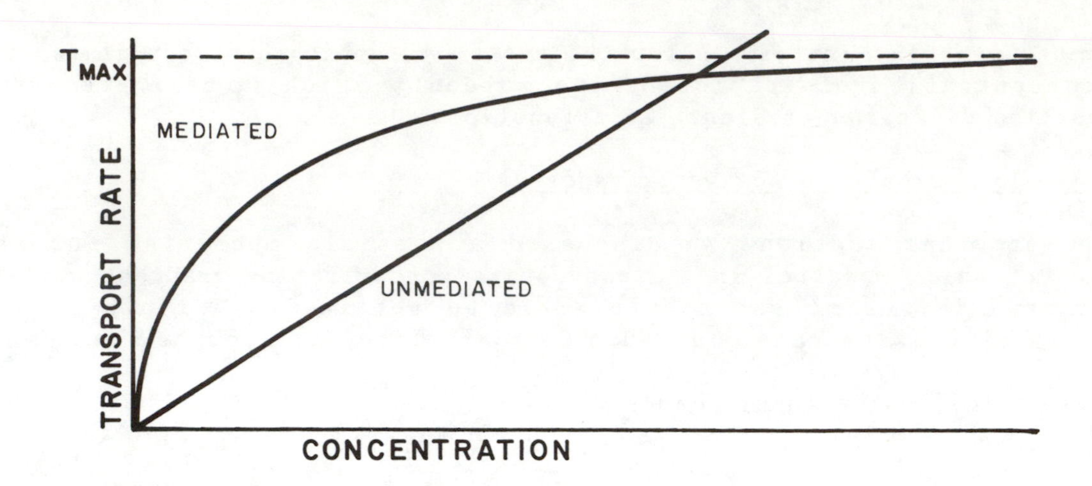

Figure 14.4. Comparison of mediated transport with simple diffusion. Based on a graph in Lehninger (1975).

14.6 Extracellular Space Determination**

One example of a compartment model which involves both passive diffusion and active transport is the simulation of the determination of extracellular fluid space with radioactive sodium, ^{22}Na. The compartments involved would be the vascular compartment, the interstitial space compartment, and the intracellular space compartment. The ^{22}Na is added to the vascular space in a single slug, and will be lost in the urine at an assumed constant rate which is dependent on ^{22}Na concentration in the plasma. The block diagram for this model is as follows:

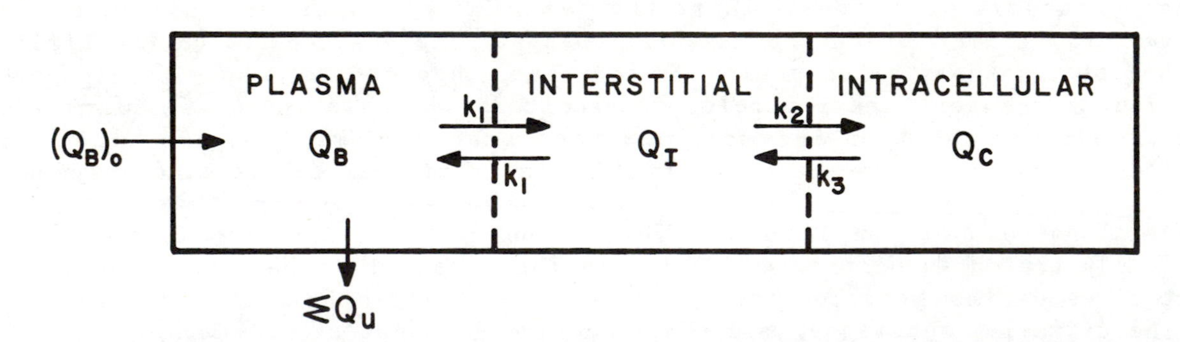

Figure 14.5. Conceptual diagram for the determination of extracellular space.

C_B, C_I, and C_C representing concentrations of ^{22}Na in blood, interstitial fluid and intracellular fluid respectively.

V_B, V_I, and V_C are the volumes of the three compartments.

Q_B, Q_I, and Q_C are the amounts of ^{22}Na in each compartment, $(Q_B)_0$ the amount injected, and Q_u the amount lost in the urine during each time increment where total lost in urine is ΣQ_u.

k_1, k_2, and k_3 are diffusion rate constants, and k_4 is the excretion rate constant, all expressed in units of inverse minutes.

The differential equations which describe the system are as follows:

$$\frac{dQ_B}{dt} = -k_1C_B + k_1C_I - k_4C_B \tag{14.61}$$

$$\frac{dQ_I}{dt} = k_1C_B - k_1C_I - k_2C_I + k_3C_C \tag{14.62}$$

$$\frac{dQ_C}{dt} = k_2C_I - k_3C_C \tag{14.63}$$

$$C_B = \frac{Q_B}{V_B}, \quad C_I = \frac{Q_I}{V_I}, \quad \text{and } C_C = \frac{Q_C}{V_C} \tag{14.64}$$

These equations may be employed to develop a reasonably good simulation of this three-compartment system.

Exercise 14.5**
Use the information given in Section 14.6 to develop a simulation of the ^{22}Na transport that occurs during the measurement of extracellular fluid space. First draw up a flowchart of the simulation, and then program it so that the concentrations of ^{22}Na in each of the fluid spaces are recorded for each time interval. Implement your simulation model with the following parameters:

$k_1 = 0.08/\text{min}$, $k_2 = 0.01/\text{min}$, $k_3 = 0.14/\text{min}$, $k_4 = 0.005/\text{min}$,

$V_B = 3$ liters, $V_I = 12$ liters, $V_C = 25$ liters, $(Q_B)_0 = 100$ μ-curies ^{22}Na.

Increment time in minutes. Graph data as a function of time for each 10 minute interval until the system is well stabilized.

Exercise 14.6
Modify the above simulation model by including an additional compartment for bone and connective tissue which would be exchanging sodium with the interstitial space compartment. Assume the exchange involves a simple diffusion with an apparent rate constant, $k_5 = 0.0005$ per minute, and an apparent volume of 9 liters. Draw the modified block diagram. Then, program the simulation model and see how the resulting simulation data differs from that of the simpler model.

14.7 Modeling of Fluid-Flow Processes**

The basic fluid-flow model is diagrammed in Figure 14.6.

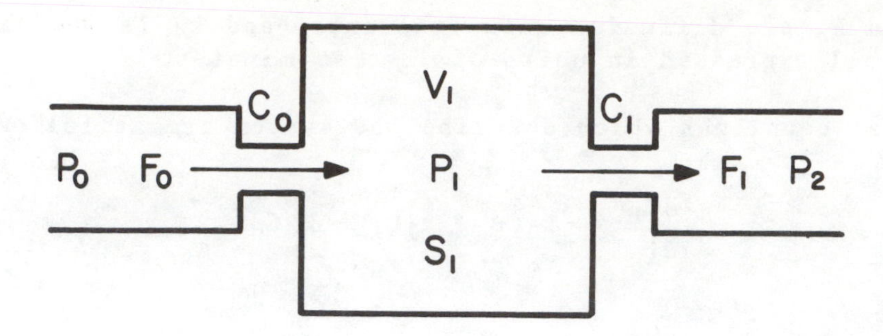

Figure 14.6. Compartment model of a generalized fluid-flow process

This model represents a chamber with a variable volume, V_1, which is supplied by a single input having a conductance, C_0, which under pressure, P_0, results in an input flow, F_0. It has a single output having a conductance, C_1, which results in an output flow of F_1. The pressure in the chamber, P_1, is the result of the volume of fluid, V_1, acting against the elastic walls having stiffness, S_1.

Chamber pressure is given by the equation

$$P_1 = V_1 S_1 \tag{14.71}$$

This equation assumes that the stretching of the chamber walls is directly proportional to the force applied. If a non-linear relationship is involved, this equation could then assume the form of a polynomial of the type

$$P_1 = A + BV_1 + CV_1{}^2 \tag{14.72}$$

where A, B, and C are empirically-derived stiffness coefficients.

Flow into the chamber is defined by

$$F_0 = C_0(P_0 - P_1) \tag{14.73}$$

Here C_0 refers to conductance, i.e. the flow that results from a unit pressure differential, and P_0 and P_1 are existing pressures. C_0 could also include the effect of a simple passive type valve which had flaps which shut under a given back pressure. Then when $P_0 - P_1$ is less than 0, the valve would be shut and C_0 equals 0.

In the same manner flow out of the chamber is controlled by conductance, C_1, and the pressure differential, $P_1 - P_2$.

$$F_1 = C_1(P_1 - P_2) \tag{14.74}$$

The fundamental mass balance equation results from numerical integration of the equation for net flow to the chamber

$$\frac{dV}{dt} = F_0 - F_1 \tag{14.75}$$

The difference expression for simple Euler integration is the following:

$$V_1 \leftarrow V_1 + (F_0 - F_1)\Delta t \tag{14.76}$$

Exercise 14.7
Model the simple fluid-flow process described above. Use the following input parameters to implement your simulation.

P_0 = 15 mm Hg, P_2 = 6 mm Hg, C_0 = 0.8 ml/min·mm Hg, C_1 = 1 ml/min·mm Hg, $(V_1)_0$ = 50 ml, S_1 = 0.15 mm Hg/ml

Graph the input and output flows, F_0 and F_1, as well as pressure, P_1, as a function of time to observe how this simple flow model approaches a steady-state condition.

14.8 Mechanical Operation of the Aortic Segment of the Heart

An interesting simulation of ventricular and aortic blood pressure has been developed by Sias and Coleman (1971). In this simulation, the aorta is assumed to be a single chamber of variable volume. Blood flows into this chamber through the aortic valve when ventricular pressure is greater than aortic pressure. Blood flows out of the chamber (aorta) at a rate which is a function of aortic pressure. The driving function of the system is the blood pressure in the left ventricle of the heart. This pressure is periodic in nature with a frequency of 72 beats per minute, and may be simulated by a truncated sine curve with a maximum of 120 mm Hg pressure at peak systole. Aortic pressure is a function of blood volume in the aortic compartment. The aortic volume in milliliters is obtained by numerically integrating the equation for net flow into and out of the aorta.

Ventricular pressure is defined by the following equation:

$$P_v = 120\left[\sin\left(\frac{6.28t}{0.0139}\right)\right] \tag{14.80}$$

where time t is expressed in minutes. The constant 0.0139 is the length of one heart beat (1/72). Only positive pressures are assumed to occur. Hence, if P_v is less than zero, then P_v equals zero .

Ventricular output (flow) is described by the equation

$$F_v = (P_v - P_a)C_a \tag{14.81}$$

where P_a is the aortic pressure and C_a is the aortic valve conductance which has a value of 1000 ml/min·mm Hg when the valve is open, and a value of zero when it is closed, i.e. when $P_v - P_a$ is less than zero.

Peripheral blood flow is a function of aortic pressure and is defined by an empirical equation of the following form:

$$F_p = A + BP_a + CP_a^2 + DP_a^3 \tag{14.82}$$

Aortic volume is obtained by numerical integration of the equation for volume change based on net inflow.

$$\frac{dV}{dt} = F_v - F_p \tag{14.83}$$

By simple Euler integration this becomes

$$V_a \longleftarrow V_a + (F_v - F_p)\Delta t \tag{14.84}$$

Aortic pressure is a function of the aortic volume and the stiffness of the walls of the aorta. Because of the non-linear nature of the stiffness, it is modeled by an empirical equation of the type

$$P_a = R + SV_a + TV_a^2 + UV_a^3 \tag{14.85}$$

Typical graphic output of this siumlation may be seen in Figure 14.7. The model embodies many basic ideas of physiological simulation, including the response of systems to various inputs. For example, a given aortic volume results in a particular aortic pressure, and this in turn results in a particular peripheral blood flow. The simulation of the system as a whole is the result of the interaction or coupling of a group of subsystems. As such, it provides some understanding of how these systems operate in the coupled phase.

Exercise 14.8
Develop the simulation model for flow in the aortic segment using the information in Section 14.8. Implement the simulation using the suggested input paramters. The following data from Sias and Coleman (1971) should be used to determine the necessary coefficients for the empirical equations describing peripheral blood flow and aortic pressure:

Aortic Pressure (mm Hg)	20	50	100	150	200
Peripheral Blood Flow (ml/min)	0	1000	5000	10000	15000

The following data relates aortic volume to aortic pressure:

Aortic Volume (ml)	0	50	100	150	200
Aortic Pressure (mm Hg)	0	20	70	175	220

Graph the simulation data in the form shown in Figure 14.7. As a special project, you might see if this simulation is improved by using a more sophisticated numerical integration technique such as the second order Runge-Kutta.

14.9 The Iodine Compartment Model

In Chapter 6 we discussed the basic technique for dealing with chemical reaction kinetics. In the present example, taken from Riggs (1970), a slightly different approach is employed. Each chemical form of iodine is assumed to represent a hypothetical compartment, and a chemical conversion from one form to another is dealt with as if it were a physical transport process.

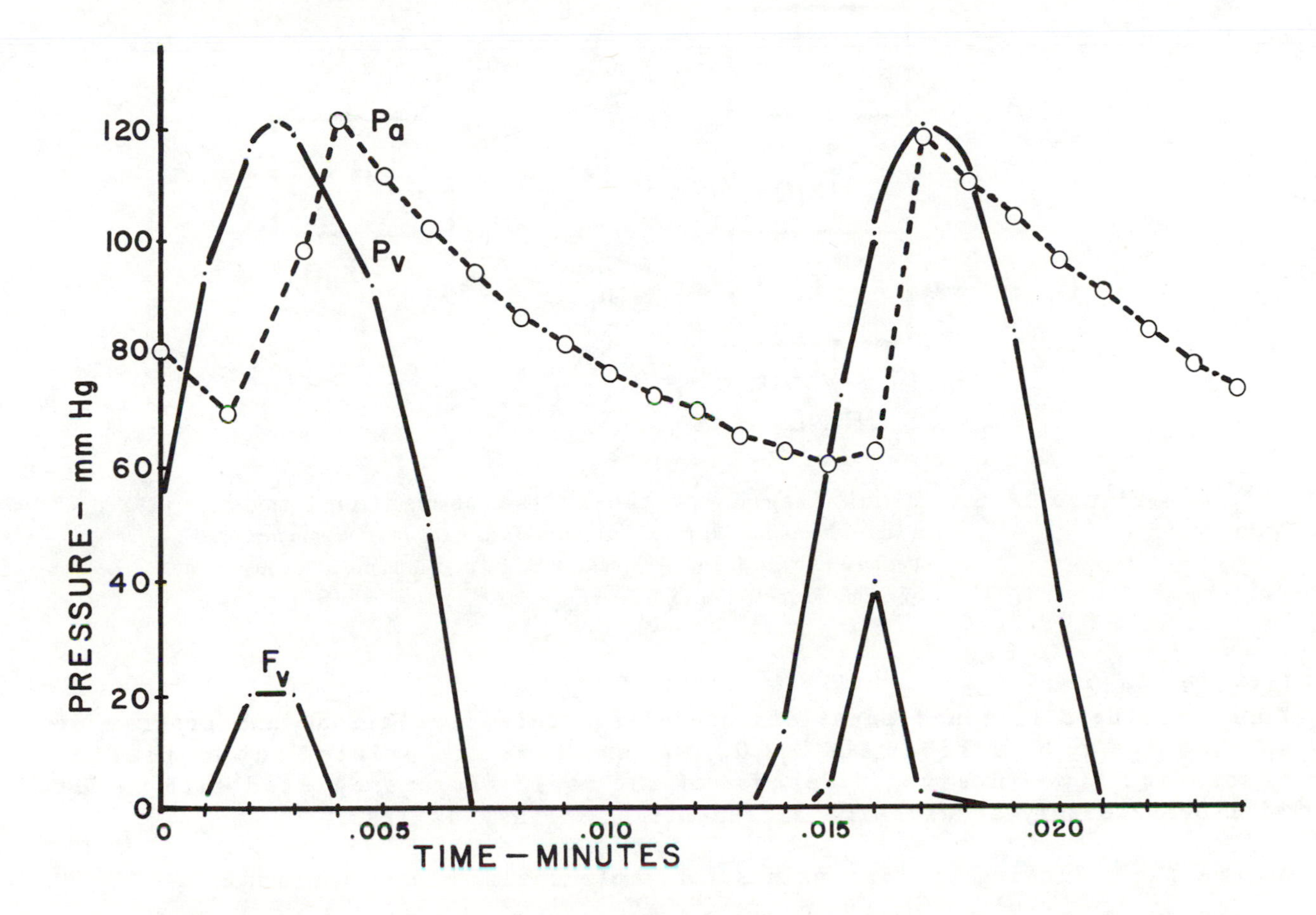

Figure 14.7. Simulation of aortic pressure and aortic blood flow.

Conversion rates are assumed to be dependent upon the amounts of material in each compartment. Despite the questionable validity of this latter assumption, this technique has been employed successfully in simulating processes involving this type of chemical conversions.

The distribution of iodine in the human body involves a three compartment system. Assume I to represent the inorganic iodide compartment, with an amount of iodine represented by Q_i. G represents the stored organic iodine compartment (thyroid gland) with an amount of iodine, Q_g. H represents hormonal iodine in extra thyroid tissue, with an amount of iodine, Q_h. D_i is the rate of input of exogenous iodine. The block diagram for the system is shown in Figure 14.9.

The differential equations describing this system are as follows:

$$\dot{Q}_i = D_i + k_3Q_h - k_1Q_1 - k_4Q_i \qquad (14.91)$$

$$\dot{Q}_g = k_1Q_i - k_2Q_g \qquad (14.92)$$

$$\dot{Q}_h = k_2Q_g - k_3Q_h - k_5Q_h \qquad (14.93)$$

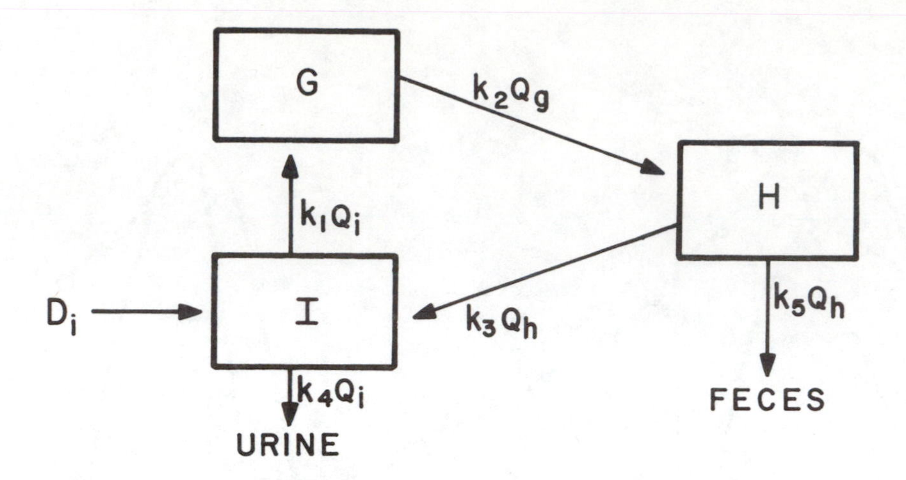

Figure 14.8. Block diagram of the iodine compartment model. Adapted from Figure *a*., p. 232, in *Control Theory and Physiological Feedback Mechanisms* by D. S. Riggs, Williams & Wilkins Co., Baltimore, 1970. Reprinted by permission of Williams & Wilkins Co.

Exercise 14.9

Work out the difference equations needed for this simulation, and program it so that D_i may be varied, and Q_i, Q_g, Q_h, and time are printed out or plotted after each time interval. Because of the rapid flux associated with k_4, it will be necessary to use time increments of 0.1 day or less.

Assume the following initial values for implementing your simulation:

Q_g = 6821 μg k_1 = 0.84/day

Q_h = 682 μg k_2 = 0.01/day

Q_i = 81 μg k_3 = 0.08/day

D_i = 150 μg/day k_4 = 1.68/day

k_5 = 0.02/day

After the system achieves steady-state conditions with respect to the quantity of iodine in each compartment, reduce D_i from 150 μg/day to 15 μg/day, thus simulating the effect of moving a person to a region of intense iodine deficiency.

Conclusion

This section has introduced you to some representative examples of physiological compartment models. This is a general approach with a wide range of applications in the area of physiology. For additional examples of simple compartment models see Simon (1972). For many examples of more sophisticated physiological models, see Riggs (1970), Milsum (1966), and Milhorn (1966).

CHAPTER 15
PHYSIOLOGICAL CONTROL SYSTEMS

Much of the theory concerning feedback control has been developed as a result of the basic need of engineers to control technological systems. Since many of the same principles are applicable to biological systems, there is a large body of terminology and mathematical methodology which should be mastered by anyone wishing to specialize in physiological control. Milsum (1973) presents a wide range of examples of biological control mechanisms, all with excellent diagrams and lucid descriptions. What follows is a very brief introduction to the modeling of physiological control systems.

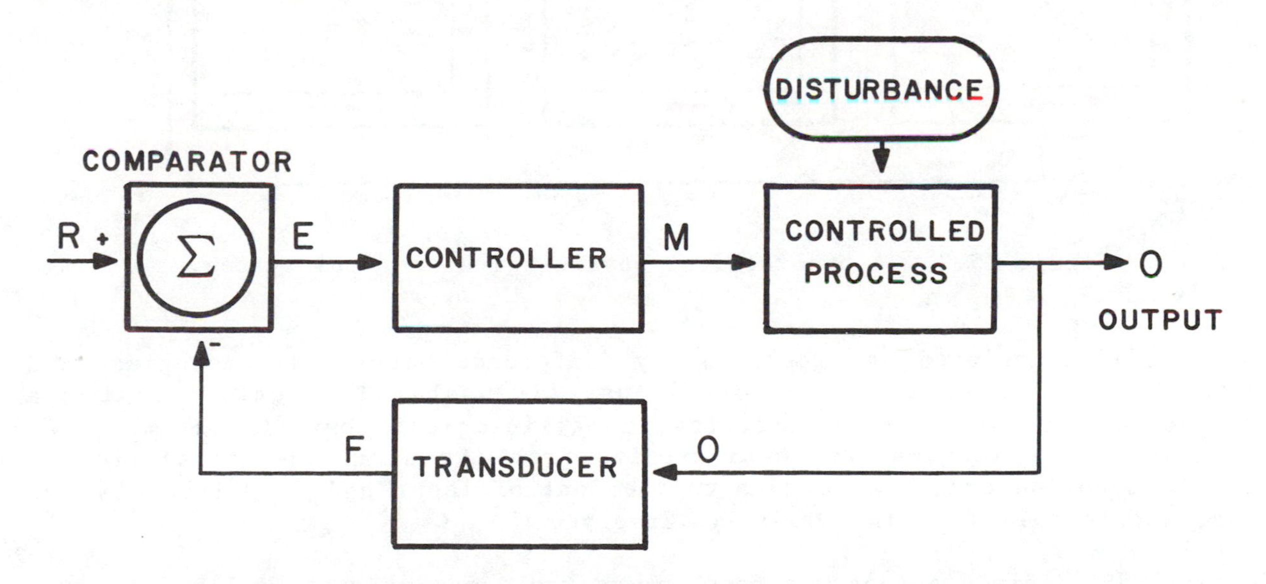

Figure 15.1. General diagram of a technological control system.

James D. Spain, BASIC Microcomputer Models in Biology ISBN 0-201-10678-7

15.1 The Generalized Feedback Control System**

The block diagram describing the general technological feedback control system is shown in Figure 15.1. In this diagram, there is a controlled process, such as heat production, or fluid flow, with an output, O. The intensity of this output is sensed by a transducer which transmits a feedback signal, F, to the comparator. This signal is a function of output intensity. The comparator sums the feedback signal with the reference or set point value, R, and transmits the sum to the controller as an error signal, E. The controller then converts the error to a manipulator or modulator function, M, which causes modulation of the output. In modeling such a system, the reference value, R, is usually a constant, and O, F, E, and M are variables.

In physiological control systems the general relationship between system components is expressed by function blocks which graphically describe the input-output behavior of one or more system components. See Figure 15.2.

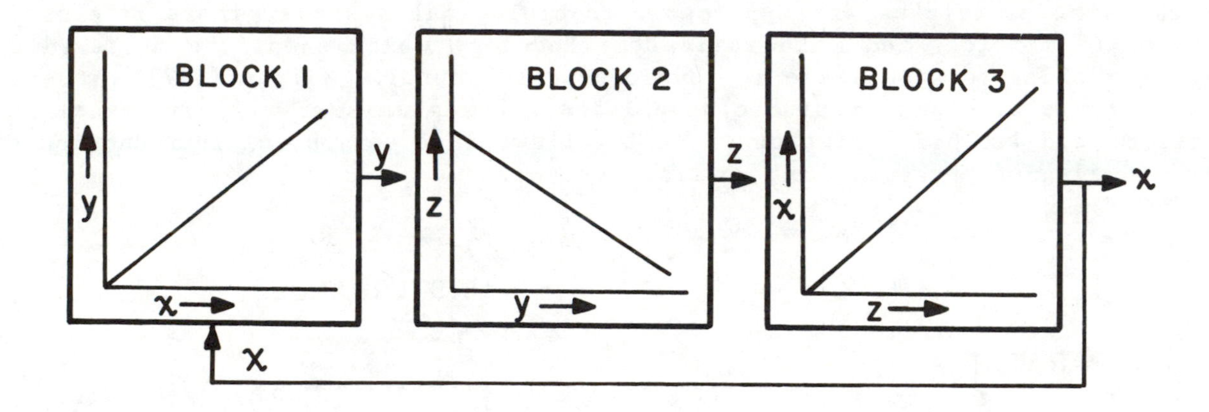

Figure 15.2. A hypothetical physiological control system.

Milhorn (1966) pointed out that the key difference between technological and physiological control is that one is usually unable to identify either a reference point or an error detector in physiological control systems. The equilibrium or steady-state point results from the simultaneous solution of all the function equations making up the control loop, and is defined by the steady-state values of the variables involved.

Jones (1973) defines a negative feedback control system in the following way:

1) It is composed of a set of processes each with its own input and output variables.

2) The processes are coupled in a sequential manner such that the output of one provides the input for the next to form a complete loop.

3) There must be an odd number of processes (usually one) which exhibit sign reversal, i.e. a function equation having a negative slope.

Notice that the example given in Figure 15.2 satisfies all of these requirements. Notice also that there is no reference point specified. Instead, it is defined by the steady-state relationship of all the function equations acting in concert. This may be shown graphically by combining the curves of the above example in the following manner.

First, we lump together the entire system which is external to the function block having the negative slope, in this case block 2. Lumping of the equations in blocks 1 and 3 give a curve relating y to z. This curve has the form shown in Figure 15.3(a).

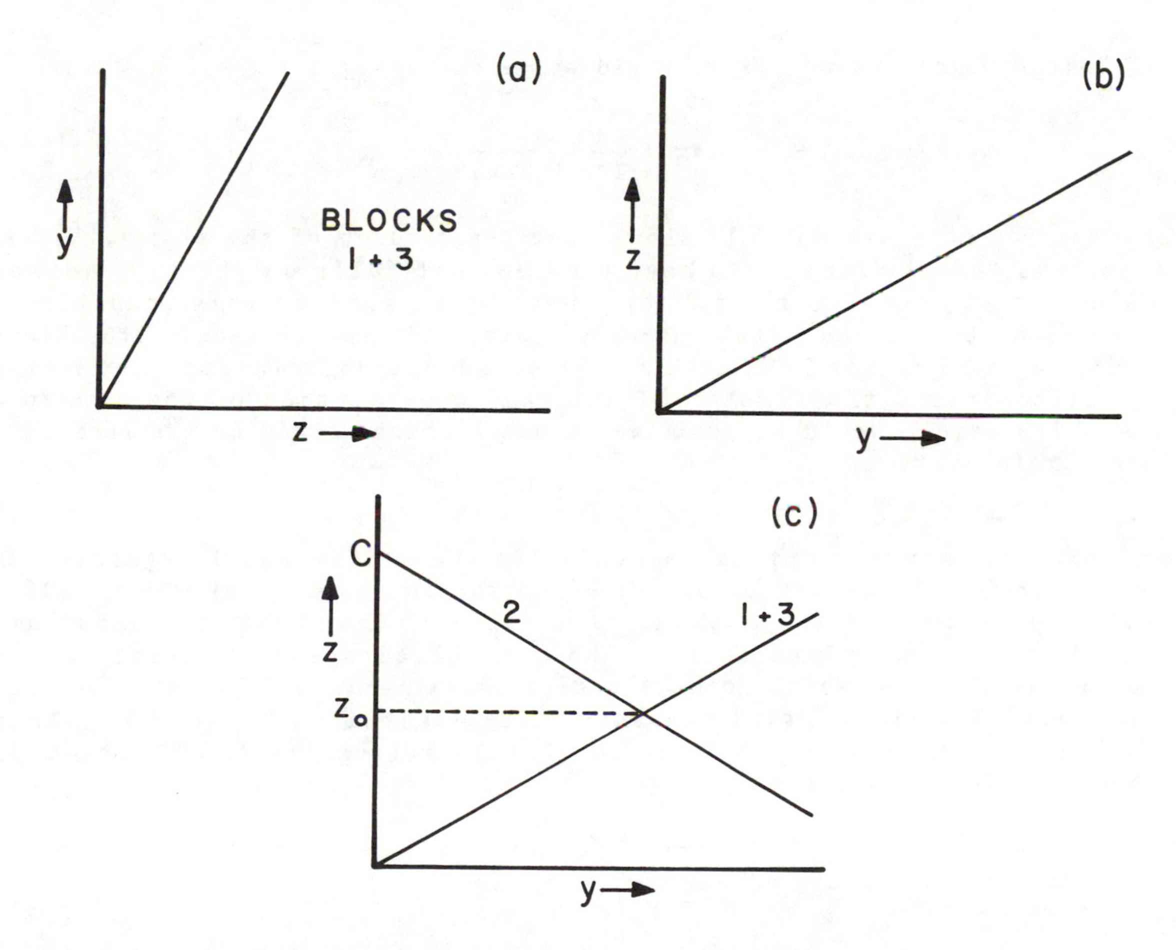

Figure 15.3. Graphical evaluation of the equilibrium point for the hypothetical system in Figure 15.2. See text for description.

This curve is then inverted and plotted on the same coordinate axes as the curve for function block 2, the curve with the negative slope. This results in curve (b) shown in Figure 15.3. The steady-state point for the system is then defined by the intersection of the two curves, as shown in graph (c) of Figure 15.3.

Mathematically, the steady-state point is defined by the simultaneous solution of all the function equations. As an example, assume that the following three equations define the functions for the system diagrammed in Figure 15.2.

$$\text{Equation 1:} \quad y = k_1 x \tag{15.10}$$

$$\text{Equation 2:} \quad z = k_2 y + C \quad (k_2 < 0) \tag{15.11}$$

$$\text{Equation 3:} \quad x = k_3 z \tag{15.12}$$

The steady-state value for z would be given by:

$$z_0 = \frac{C}{1 - k_1 k_2 k_3} \tag{15.13}$$

And the steady-state value for x, would be given by:

$$x_0 = \frac{k_3 C}{1 - k_1 k_2 k_3} \tag{15.14}$$

An examination of Equation 15.13 shows that the product of the slopes, $k_1k_2k_3$, must be less than 1 if z_0 is to have a value that falls on the line between $z = C$ and $z = 0$. See Figure 15.3 (c). One way to achieve this condition is to have one of the slopes with a negative value. If the product of the slopes in Equation 15.13 is not less than 1, the steady-state solution of the equations falls outside the first quadrant, and thus outside of the domain of control. The result would be positive feedback which characterizes runaway or vicious circle systems.

Exercise 15.1**

An example of a simple control mechanism is the system which regulates the concentration of CO_2 in the blood. The controller consists of the medullary respiratory center and the mechanical portion of the chest and lungs which produce the alveolar ventilation. The controlled system consists of the portion of the system which forms the relationship between alveolar ventilation and arterial CO_2 partial pressure. The diagram in Figure 15.4, taken from Milhorn (1966), demonstrates the feedback loop and the input-output relationships for this system.

The two input-output functions may be described mathematically by the following equations:

$$V_a = P_a(0.25) - 5 \tag{15.15}$$

$$P_a = \frac{200}{V_a} \tag{15.16}$$

Here, V_a is alveolar ventilation rate, and P_a is partial pressure of CO_2 in arterial blood.

A very simple simulation of this cybernetic system may be developed by repeatedly taking the output of one of these equations and using it as the input of the other. The system is quite unstable unless one averages the old P_a value with the new P_a value to simulate the damping effect of the body. The results demonstrate that after some oscillation and overshoot the values for

P_a and V_a converge on the steady-state point. This point also can be obtained by the simultaneous solution of the two equations, and results in $P_a = 40$ and $V_a = 5$. The overshoot observed is characteristic of control systems with some lag in the response.

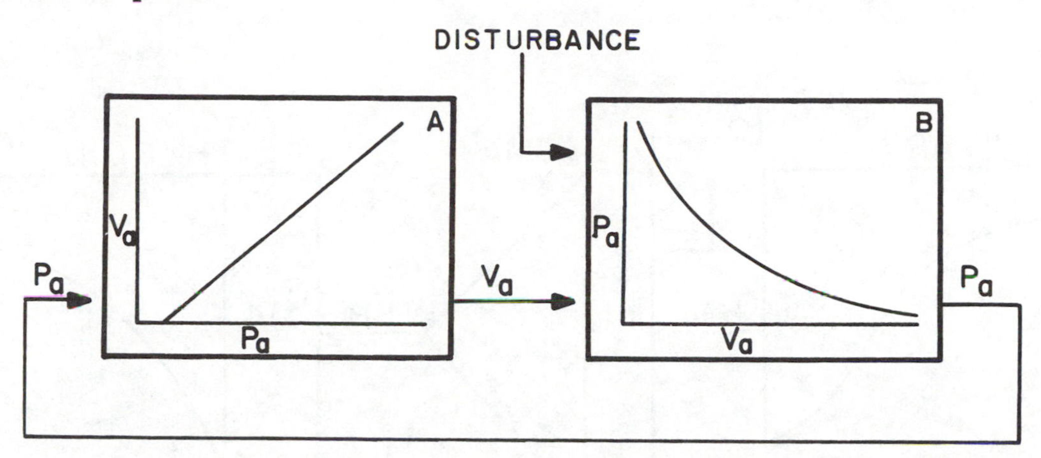

Figure 15.4. Regulation of ventilation rate (V_a) by the partial pressure of carbon dioxide (P_a). Based on Milhorn (1966).
Adapted from Figure 8-3, p. 115 in *The Application of Control Theory to Physiological Systems* by H. T. Milhorn, W. B. Saunders Co., 1966. Reprinted by permission of W. B. Saunders Co.

This very simple model of respiratory control lumps several components of the real system into function block A. It also ignores the contribution of blood oxygen partial pressure and pH to respiratory control. For a complete discussion of the more complex respiratory control model, refer to Milhorn (1966).

15.2 Regulation of Thyroxine Secretion by the Pituitary

The following model provides a good example of a simple feedback control system in which the primary function blocks describe changes in variables rather than the static control described in the previous example.

The rate of thyroxine, TH, secretion from the thyroid gland is regulated by the amount of thyroid stimulating hormone, TSH, in the plasma. At the same time, the release of TSH by the pituitary is regulated by the amount of TH in the plasma. Both TH and TSH are slowly degraded to inactive products so that the concentration of each in the plasma at any time is the integral of the net change (gain minus loss).

In this simple model, we are neglecting the effects of the related hormone, triodothyronine, and assuming only a single form of thyroxine in the plasma. We are also assuming that the sole effect of TSH is on thyroxine secretion, neglecting the role of the hypotholamus. This very simplified model is diagrammed in Figure 15.5.

In this model, the input to the pituitary block is TH concentration and the output is the amount of TSH released. The TSH released is summed with the loss due to degradation and integrated to give the plasma concentration of TSH. The TSH concentration then becomes the input to the thyroid block, where the amount of TH secreted is determined. The net change in TH is integrated to obtain the plasma concentration, thus completing the loop.

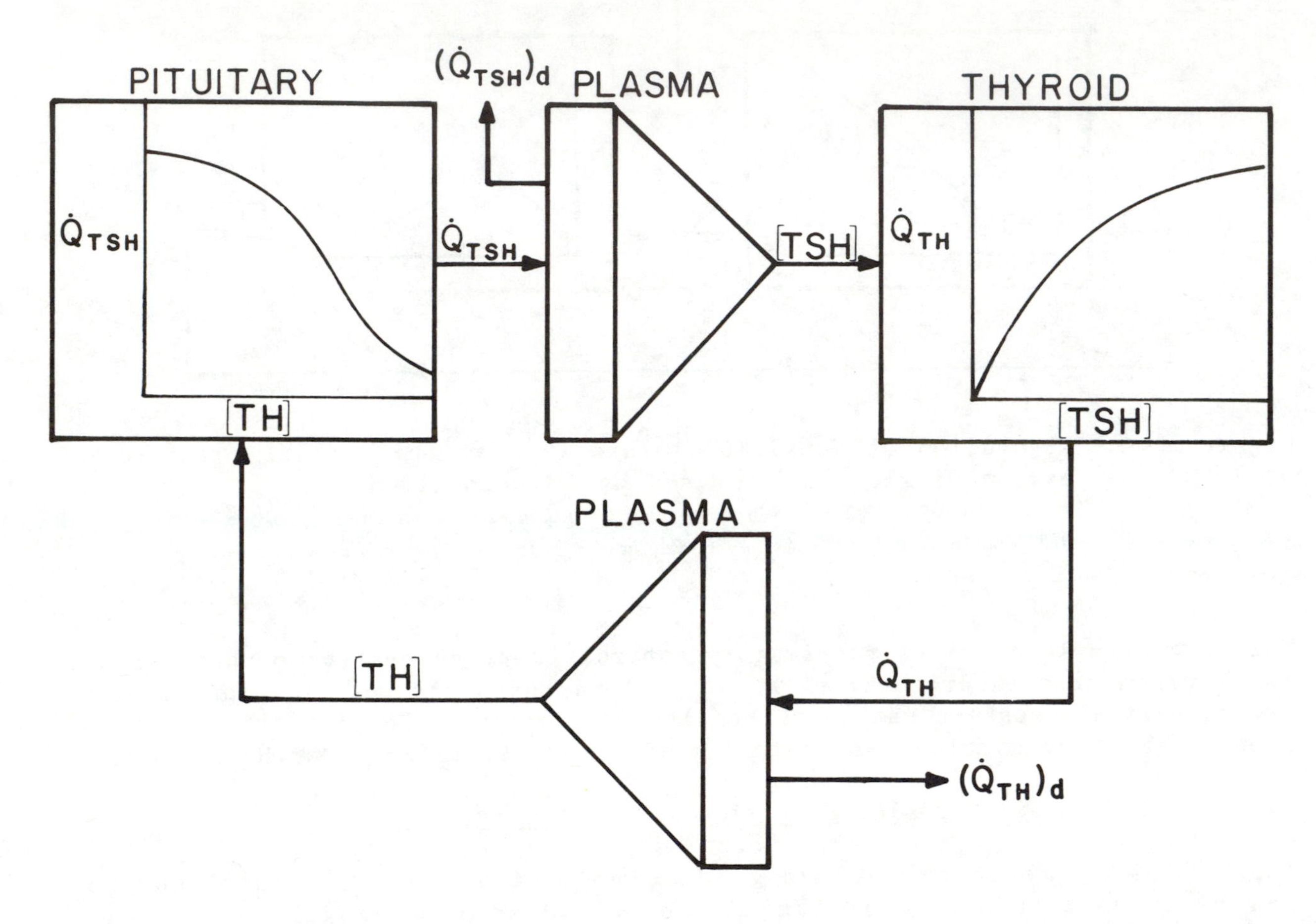

Figure 15.5. Regulation of thyroid hormone concentration, [TH], by thyroid stimulating hormone, [TSH].

The following equations describe the process. The net change in plasma TSH is defined as

$$\Delta Q_{TSH} = (A/(1 + B[TH]^n) - C_{TSH}\cdot[TSH])\cdot\Delta t \tag{15.21}$$

where A, B, and n are parameters in the reverse sigmoid function equation, [TH] is the concentration of thyroid hormone, C_{TSH} is the first order decay coefficient (clearance) for TSH, and [TSH] is the concentration of TSH.

After one time interval, the new concentration of TSH in plasma becomes

$$[TSH]_{t+\Delta t} = [TSH]_t + \Delta Q_{TSH}/V \tag{15.22}$$

where V is the plasma volume.

The net change in the amount of circulating thyroid hormone is given by

$$\Delta Q_{TH} = [R[TSH]/(K + [TSH]) - C_{TH}[TH]]\cdot\Delta t \qquad (15.23)$$

where R and K are parameters in the hyperbolic function equation, and C_{TH} is a first order decay coefficient (clearance) for TH.

The concentration of TH in the plasma after one time interval is given by

$$[TH]_{t+\Delta t} = [TH]_t + \Delta Q_{TH}/V \qquad (15.24)$$

The model, as described, goes to a steady-state condition with very little oscillation.

Exercise 15.2

Program the model describing the regulation of thyroxine by pituitary TSH, and implement it with the following parameters:

[TSH]	= 2 μg/liter	R =	100 μg/day
[TH]	= 30 μg/liter	K =	1.5 μg/1
C_{TSH}	= 2.5 liters/day	A =	8 μg/day
C_{TH}	= 1 liter/day	B =	0.8×10^{-5} $(1/\mu g)^3$
V	= 10 liters	N =	3

Increment time in days and generate about 50 days of simulated data to evaluate the behavior of the model.

15.3 Temperature Control Through Sweating

Under heat stress, the human body reaches a point where it is unable to maintain control simply by convection and radiation. Above this point, control of body temperature is primarily accomplished by sweating. The general feedback control pathway for the process is diagrammed in Figure 15.6.

Metabolic heat production, ΔQ_i, may be assumed to be a function of body temperature. It can be modeled by the following equation:

$$\Delta Q_i = 75\left(Q_{10}^{(T_B - 37)/10}\right)\Delta t \qquad (15.31)$$

where 75 is the metabolic heat production in kcal/hr at 37°C, T_B is the body temperature, and Q_{10} is the factor by which heat production increases for a 10° rise in temperature. See Section 11.3. A typical value for Q_{10} is 2.

Convective and radiative heat loss, ΔQ_r, is a function of body temperature and environmental temperature.

$$\Delta Q_r = k(T_B - T_A)\Delta t \qquad (15.32)$$

where k is the heat loss coefficient, T_B is body temperature, and T_A is environmental temperature. A typical value for k would be 10 kcal/hr C°.

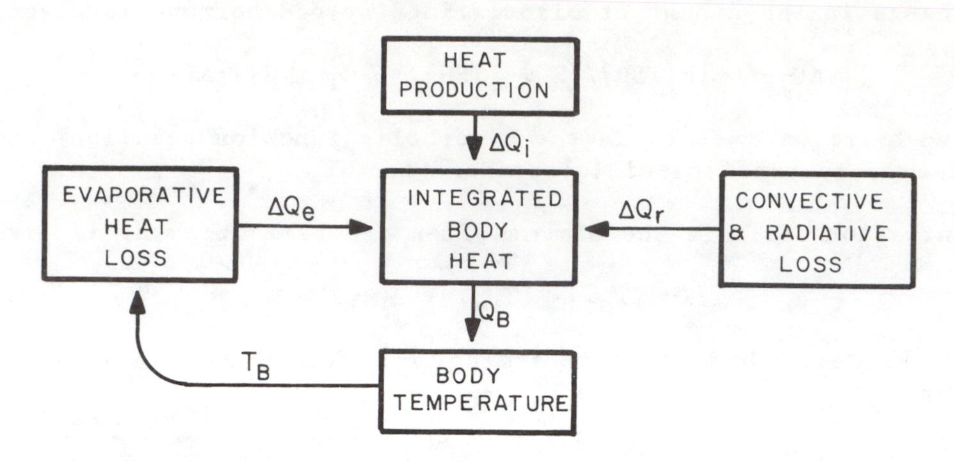

Figure 15.6. Function block diagram for body temperature control above thermal neutrality.

Assume evaporative heat loss is described by the following equation:

$$\Delta Q_e = \frac{E_{max} \cdot T_d \cdot (1 - R_H) \cdot \Delta t}{K + T_d} \tag{15.33}$$

where E_{max} is the maximum evaporative heat loss of 1400 kcal/hr, T_d is the number of degrees of body temperature above 37°C (positive values only), R_H is the relative humidity expressed as a decimal fraction, and K is the half maximum parameter in the hyperbolic function employed. K is approximately 0.2 C°. Because of insensible evaporation, ΔQ_e has a minimum value of 5 kcal/hr. Thus if ΔQ_e is less than 5 from Equation 15.33 then ΔQ_e is assumed to equal 5.

The total change in heat content is given by

$$\Delta Q_B = \Delta Q_i - \Delta Q_r - \Delta Q_e \tag{15.34}$$

The heat balance expression is therefore

$$Q_B \leftarrow Q_B + \Delta Q_i - \Delta Q_r - \Delta Q_e \tag{15.35}$$

The heat capacity of the body is assumed to be equal to that of water, 1 kcal/kg, and the mass of the body is assumed to be 70 kg. Hence, the temperature of the body is given by the following equation:

$$T_B = Q_B/70 \tag{15.36}$$

The initial value for Q_B is 2590 kcal which corresponds to a body temperature of 37°C.

Exercise 15.3

Use the above information to develop a simulation model of the temperature control mechanism above thermal neutrality. Assume the constants given above and observe the approach to steady-state when the environmental temperature is each of the following values: 30°, 35°, 40°, 45°, 50°, 55°, 60°C.

Assume that the initial body temperature in each case is 37°C, and determine the steady-state body temperature for each environmental temperature. Assume an initial relative humidity of 50 percent, $R_H = 0.5$. Repeat the simulation with relative humidities of 80 and 90 percent. Assume that body temperatures over 41.4° normally result in death.

15.4 A Model of Temperature Control Below Thermal Neutrality

When a small mammal, such as a bat, is allowed to cool below its thermal neutral point, two factors combine to resist body temperature change. These factors are embodied in the following assumptions.

Heat loss will be described by the equation for Newton's law of cooling.

$$\text{Heat loss} = k(T_B - T_A) \qquad (15.41)$$

where T_B is body temperature, T_A is ambient temperature, and k is thermal conductance. The latter is dependent on body surface area and insulation effect, and varies in the bat according to the following experimentally determined values relative to ambient temperature (Holyoak, 1969).

T_A(C°)	35	30	25	20	15	10	5
k(cc O_2/g·hr)	1.00	.70	.60	.50	.46	.44	.42

Note that k is expressed in cc O_2, energy equivalent, per gram of body weight per hour. This dependency is best modeled by an empirical equation of the polynomial type.

Chemical thermogenesis appears to vary as a straight line function of body temperature.

$$\text{Chemical thermogenesis, } Q_r = 2 + 2(T_N - T_B) \qquad (15.42)$$

where T_B is the body temperature, and T_N is the body temperature at thermal neutrality, 35°C for bats.

Peak metabolic rate, or maximum rate of metabolism, was found to be about 10 cc O_2/g·hr. The model should be designed to increase chemical thermogenesis up to 10 cc O_2/g·hr., but when Q_r exceeds 10 then set Q_r equal to 10.

Effect of temperature on reaction rates, and hence heat, production is assumed to decrease by a factor of two for each 10 degrees. This effect would be modeled by an equation of the type

$$Q_r' = Q_r e^{-A(35-T_B)} \qquad (15.43)$$

The computer may be programmed to continue making corrections on body temperature until the difference between heat loss and heat production is zero and steady-state exists. Then the computer should be instructed to print out body temperature and heat loss, which equals heat production. This simulation is diagrammed in Figure 15.7.

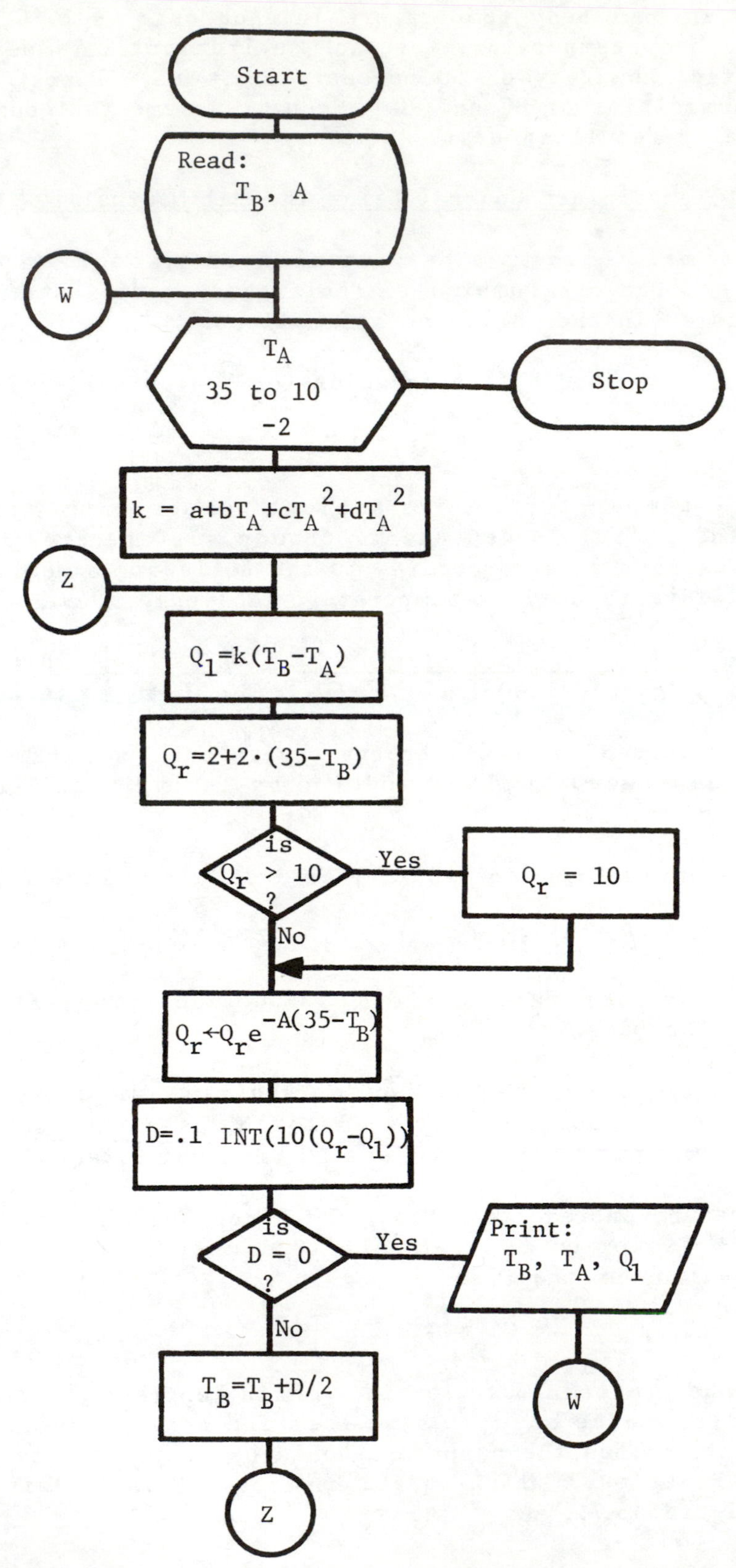

Figure 15.7. Flowchart for model of temperature below neutrality.

With this program it is possible to demonstrate the following principles concerning temperature control below thermal neutrality:

1) A plateau region of thermoregulation where body temperature is relatively constant with declining ambient temperature above the critical temperature.
2) Ambient temperatures below the critical temperature result in positive feedback because of the Q_{10} effect on thermogenesis. The organism therefore comes to equilibrium at a body temperature considerably below that of the thermoregulatory zone.
3) In the zone of hypothermic cooling, the body temperature declines rapidly with ambient temperature maintaining a slight differential all the way down to T_A equaling 0°C.
4) Heat production approaches the limit of 10 cc O_2/g·hr as one approaches the critical temperature from the high side.
5) The warming curve follows a different pathway from the cooling curve due to a hysteresis effect.

A typical simulated cooling curve appears in Figure 15.8.

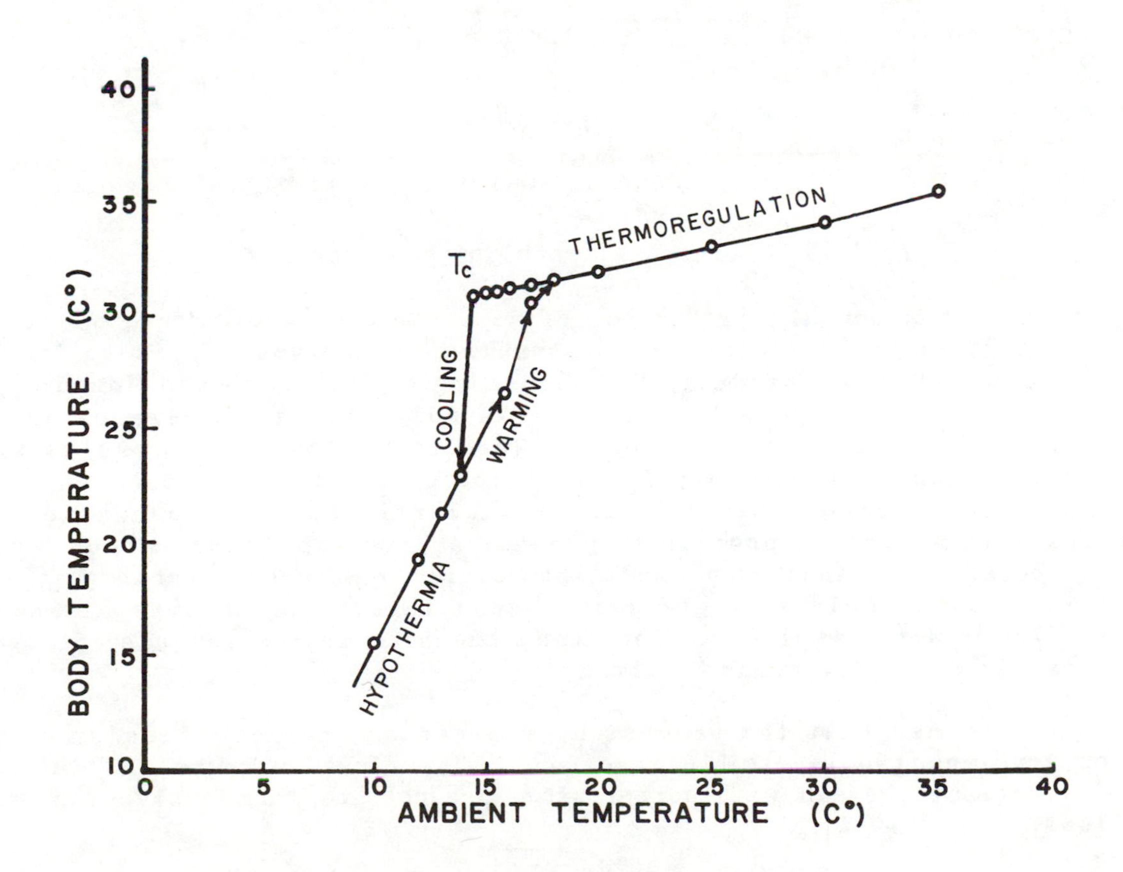

Figure 15.8. Simulated cooling curve for the bat.

Exercise 15.4
Use the above information and the flowchart given in Figure 15.7 to develop a model of temperature control below thermal neutrality. The program should vary ambient temperature from 35°C to 0°C in 2° increments. For each ambient temperature, allow the system to iterate to a steady-state condition, and then print out steady-state values for body temperature and heat loss. Graph your results as shown in Figure 15.8. Assume A is equal to 0.0693.

15.5 The Control of Protein Synthesis by Gene Repression

A simple model for the control of protein synthesis by gene repression was proposed by Maynard Smith (1968). His model is based on the conceptual model of Jacob and Monod (1961), which is diagrammed in Figure 15.9.

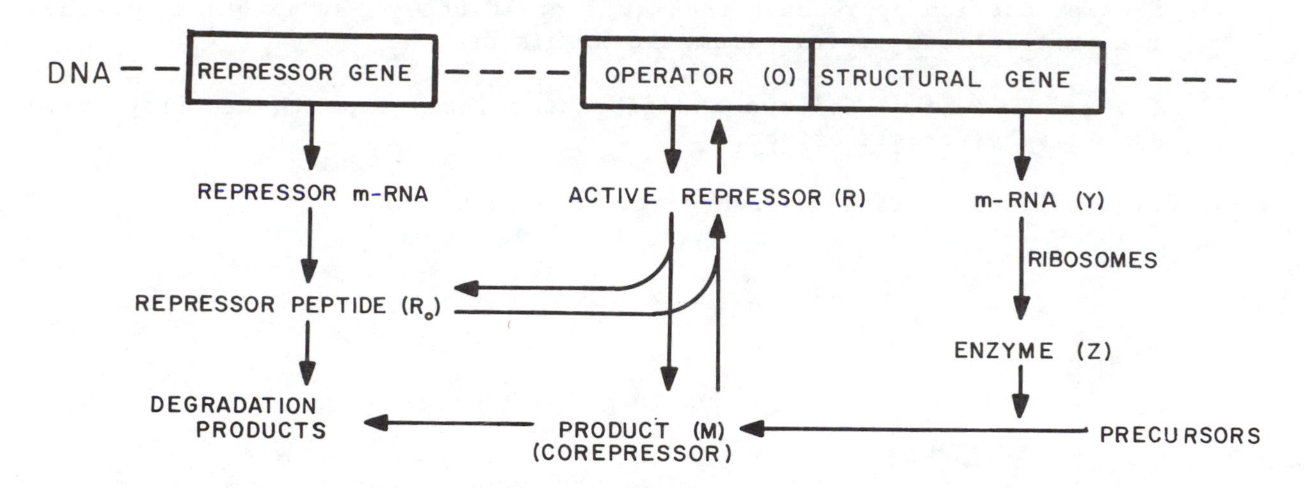

Figure 15.9. Conceptual model of gene repression.

The model explains how the production of ribose nucleic acid (m-RNA), designated in equations by Y, needed for the synthesis of enzyme, Z, is controlled by the concentration of product, M. According to the Jacob and Monod hypothesis, the structural gene produces m-RNA only when the operator is not interacting with active repressor, R. The active repressor results from interaction of repressor protein, R_0, with the corepressor molecules. Hence, when the product (corepressor) is in sufficiently high concentration, it interacts with repressor protein to produce active repressor, which 'turns off' the operator and thus stops m-RNA production by the structural gene. In bacteria, there is only a single gene responsible for each type of enzyme protein. For additional information about the gene repression concept, consult a text in cell biology or biochemistry.

It may be assumed that the process of repression and derepression of the operator involves two reversible reactions. The first concerns the interaction of repressor protein with corepressor (product) to form active repressor as follows:

$$M + R_0 \rightleftharpoons R$$

M is the metabolite or corepressor, R_0 is the inactive repressor protein, and R is active repressor.

This reaction may be represented by the following equilibrium expression:

$$K = \frac{[R]}{[R_0][M]} \quad \text{and} \quad [R] = K[R_0][M] \tag{15.51}$$

Assuming repressor protein is in steady-state between production and degradation, its concentration would be constant. The value for [R] then becomes a direct function of [M].

$$[R] = b[M] \tag{15.52}$$

where b represents the combined constants from Equation 15.51.

A second reaction describes the interaction of active repressor with the operator to form the repressor-operator complex, [OR].

$$R + O \rightleftharpoons [OR]$$

The reversible reaction involves two processes, repression and derepression. By using steady-state assumptions similar to those employed in Chapter 2, especially in Section 2.3, it may be shown that the fraction of time that the structural gene exists in the unrepressed condition, $\emptyset$, would be described by an equation having the following form:

$$\emptyset = \frac{a}{a + b[M]} \tag{15.53}$$

Here a and b are constants, and [M] is the concentration of repressor. The rate of m-RNA synthesis is assumed to be directly proportional to the fraction of time that the gene is unrepressed and thus to be described by the differential equation

$$\frac{d[Y]}{dt_s} = \frac{c}{a + b[M]} \tag{15.54}$$

where c is the product of a and the proportionality constant for RNA production.

There is a continuous breakdown or decay of m-RNA which is assumed to be described by a second differential equation.

$$\frac{d[Y]}{dt_d} = -k[Y] \tag{15.55}$$

where k is a first order decay constant. The net change in m-RNA is therefore given by the overall equation:

$$\frac{d[Y]}{dt} = \frac{c}{a + b[M]} - k[Y] \tag{15.56}$$

The rate of enzyme formation and breakdown is similarly described by a differential equation of the form

$$\frac{d[Z]}{dt} = e[Y] - f[Z] \qquad (15.57)$$

where e is an m-RNA dependent enzyme synthesis rate constant, and f is a first order enzyme decay constant.

If the enzyme is assumed to be saturated with the precursor, P, required for the formation of corepressor, M, then the rate of formation is a direct function of enzyme. The breakdown or conversion of M to something else is assumed to be directly dependent on its concentration. The differential equation for M is therefore given as

$$\frac{d[M]}{dt} = g[Z] - h[M] \qquad (15.58)$$

where g is the formation rate constant for M, and h is a first order decay constant for M.

Exercise 15.5

Use the above information to write the difference equations required for the Euler integration of Equations 15.56 to 15.58. Then program the model and implement it with the following input parameters:

$$a = 5,\ b = 0.00025,\ c = 1,\ e = 2,\ f = 0.1,$$
$$g = 100,\ h = 0.1,\ k = 0.1,\ Y = Z = M = 0.$$

These constants are appropriate for steady-state concentrations of 1μM for m-RNA, 20 μM for enzyme, and 20000 μM for the repressor.

Graph the three variables of this model as a function of time. The result should be an oscillating system.

Conclusion

This chapter has attempted to present some of the basic concepts involved in feedback control of physiological systems. Further discussion of this topic as it applies to feedback inhibition of biochemical reaction pathways will be reserved for Chapter 17. The stochastic nature of the Jacob and Monod model for repression of enzyme synthesis will be discussed in Chapter 22.

CHAPTER 16

THE APPLICATION OF MATRIX METHODS TO MODELING

Up to this point, all of our computer models have employed what might be called the direct or brute-force approach. It has been my belief that this approach is most easily understood by individuals who are just learning how to develop computer models. However, the direct approach can become rather cumbersome when dealing with large numbers of variables and the equations which define them. This problem was exemplified by the age-class models in Chapter 8, and the compartment models encountered in Chapters 12 and 13. As you will see, some of the exercises presented in these chapters are much more conveniently programmed using matrix methods.

Matrix algebra provides an efficient way of manipulating large amounts of data. Although the matrix technique predates the development of the high speed computer (MacDuffee, 1933), it is especially valuable when employed with the computer. A few simple instructions can result in the simultaneous solution of an almost unlimited number of related linear equations. The effect is to greatly facilitate computer programming. What is presented below is a brief description of a few matrix techniques which have direct application to the modeling exercises provided. For additional information, refer to the text on matrix algebra for biologists by Searle (1966) or to the simplified discussion of matrix methods written by Poole (1974).

16.1 The Matrix Approach to Linear Transport Phenomena**

A matrix is an array of numbers or symbols. It is usually named by a single bold-faced letter. Assume we have a 3 compartment system in which all possible transfers are defined by transfer coefficients a_{21}, a_{12}, ... , a_{ij}, as diagrammed in Figure 16.1.

James D. Spain, BASIC Microcomputer Models in Biology

ISBN 0-201-10678-7

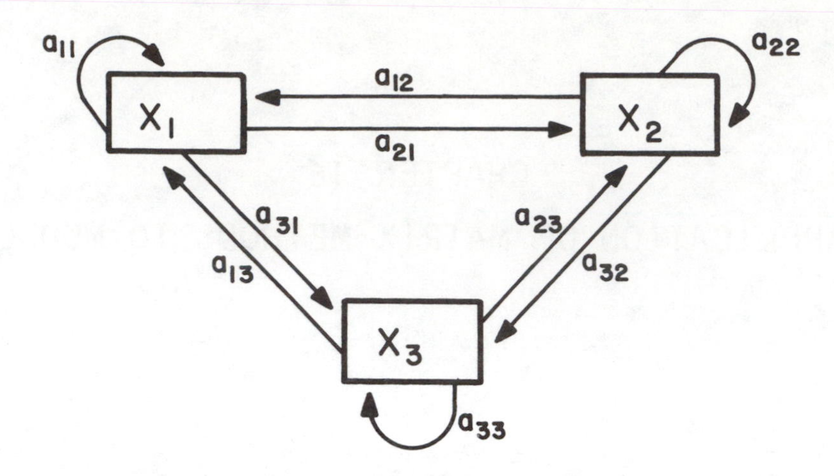

Figure 16.1. Interactions within a three compartment system.

The transfer coefficients may be assigned as elements of a 3 x 3 matrix **A**

$$\mathbf{A} = \begin{bmatrix} a_{11} & a_{12} & a_{13} \\ a_{21} & a_{22} & a_{23} \\ a_{31} & a_{32} & a_{33} \end{bmatrix}$$

In order to conform with matrix terminology, we are now designating the receiving compartment first in the subscripts, i.e. a_{21} refers to the flow of energy or material to compartment 2 from compartment 1. This matrix contains three columns (running up and down) and three rows (running across). The diagonal from top left to bottom right often plays an important role in the matrix. In this example, the diagonal contains the coefficients which express the effect of each compartment on itself.

The variables describing the state of all compartments in energy or mass units are typically contained in a one-dimensional matrix or column vector having the following form:

$$\mathbf{X} = \begin{bmatrix} x_1 \\ x_2 \\ x_3 \end{bmatrix}$$

If we assume that we are dealing with an unforced linear (first order) system, then the change in each of the compartments is defined by the equations

$$\dot{x}_1 = -x_1a_{21} - x_1a_{31} + x_1a_{12} + x_3a_{13} \tag{16.10}$$

$$\dot{x}_2 = x_1a_{21} - x_2a_{12} - x_2a_{32} + x_3a_{23} \tag{16.11}$$

$$\dot{x}_3 = x_1a_{31} + x_2a_{32} - x_3a_{13} - x_3a_{23} \tag{16.12}$$

By combining terms which have the same state variable we obtain

$$\dot{x}_1 = x_1(-a_{21} - a_{31}) + x_2 a_{12} + x_3 a_{13} \qquad (16.13)$$

$$\dot{x}_2 = x_1 a_{21} + x_2(-a_{12} - a_{32}) + x_3 a_{23} \qquad (16.14)$$

$$\dot{x}_3 = x_1 a_{31} + x_2 a_{32} + x_3(-a_{13} - a_{23}) \qquad (16.15)$$

These equations may be expressed as the product of a redefined matrix **A** and the vector, **X**, as follows:

$$\begin{bmatrix} \dot{x}_1 \\ \dot{x}_2 \\ \dot{x}_3 \end{bmatrix} = \begin{bmatrix} -(a_{21} + a_{31}) & a_{12} & a_{13} \\ a_{21} & -(a_{12} + a_{32}) & a_{23} \\ a_{31} & a_{32} & -(a_{13} + a_{23}) \end{bmatrix} \cdot \begin{bmatrix} x_1 \\ x_2 \\ x_3 \end{bmatrix} \qquad (16.16)$$

Notice that the diagonal now contains the negative sum of the coefficients which relate to the loss of material from each compartment. The matrix designation for this operation is

$$\dot{\mathbf{X}} = \mathbf{A} \cdot \mathbf{X} \qquad (16.17)$$

This matrix operation is described as post-multiplication of a square matrix by a column vector in which the product is a column vector. The three differential equations, 16.10, 16.11, and 16.12, are now expressed by a single matrix equation, 16.17, and the previous definition, of **A** and **X**.

For those who are unfamiliar with the multiplication of a matrix by a vector, this operation consists of the following two steps:

1. Each column of the matrix is multiplied by the analogous vector element. In this example, each element of column 1 is multiplied by x_1, each element of column 2 by x_2, and so on.

2. The multiplication products in each row are summed across to generate the analogous elements of the product vector. In this example, the sum of the products in row 1 = $\dot{x}_1$, the sum of the products in row 2 = $\dot{x}_2$, and so on.

Examine Equation 16.16 to verify that this matrix multiplication accomplishes the same thing as Equations 16.13, 16.14, and 16.15.

To complete the process of Euler integration, the new values for each of the state variables would be obtained by multiplication of the change vector, $\dot{\mathbf{X}}$, by a scalar, Δt. This product is then added to the state variable vector, **X**, as follows:

$$\begin{bmatrix} x_1 \\ x_2 \\ x_3 \end{bmatrix} \longleftarrow \begin{bmatrix} x_1 \\ x_2 \\ x_3 \end{bmatrix} + (\Delta t) \begin{bmatrix} \dot{x}_1 \\ \dot{x}_2 \\ \dot{x}_3 \end{bmatrix} \qquad (16.18)$$

In matrix terminology, the result is

$$\mathbf{X}_{t+1} = \mathbf{X}_t + (\Delta t)\dot{\mathbf{X}}_t \tag{16.19}$$

Environmental inputs, or forcing function, normally make up an additional vector, **F**, which should be added to Equation 16.19 as follows:

$$\mathbf{X}_{t+1} = \mathbf{X}_t + (\mathbf{F} + \dot{\mathbf{X}}_t)\Delta t \tag{16.1A}$$

Vector **F** is given by

$$\mathbf{F} = \begin{bmatrix} F_1 \\ F_2 \\ F_3 \end{bmatrix}$$

where F_1, F_2, F_3 are forcings to compartments x_1, x_2, x_3.

Some BASIC dialects include functions which allow one to accomplish the calculations described by Equations 16.17 and 16.19 (or 16.1A) by two simple MAT statements. Although most microcomputers do not have this capability, it is relatively easy to program matrix operations using the simple subroutines listed in Appendix I. The following coding sequence would accomplish the matrix operations described in Equations 16.16 through 16.1A. This subroutine assumes that preliminary matrix operations such as dimensioning the array and reading data have already been accomplished.

```
190  REM STAGE I...CALCULATE CHANGES
200  FOR I = 1 TO N
210  DX(I) = 0
220  FOR J = 1 TO N
230  DX(I) = DX(I) + A(I,J)*X(J)
240  NEXT J
250  NEXT I
260  REM STAGE II...UPDATE THE VARIABLES
270  FOR I = 1 TO N
280  X(I) = X(I) + (DX(I) + F(I))*DT
290  NEXT I
```

In this sequence, N is one dimension of the square matrix, in this case, three. You can appreciate the power of this method when you realize that the same 10 statements would be just as effective for a 30-compartment system as for a 3-compartment system.

<u>Exercise 16.1</u>**
Carry out Exercise 12.2, the linear Silver Springs model, using the matrix approach. Compare these results with those obtained from the direct approach.

<u>Exercise 16.2</u>
Modify the multicompartment linear diffusion model (Exercise 6.8) so that it may be programmed using matrix algebra. Since transfers are occurring only between adjacent compartments, the coefficient matrix, **A**, takes the form

described below.

$$
A = \begin{bmatrix}
-(a_{21}) & a_{12} & 0 & 0 & \cdots & 0 \\
a_{21} & -(a_{12} + a_{32}) & a_{23} & 0 & \cdots & 0 \\
0 & a_{32} & -(a_{23} + a_{43}) & a_{34} & \cdots & 0 \\
0 & 0 & a_{43} & -(a_{34} + a_{54}) & \cdots & 0 \\
\vdots & \vdots & \vdots & \vdots & & \vdots \\
0 & 0 & 0 & a_{ij} & & -a_{ij}
\end{bmatrix}
$$

Use a k value (a_{ij}) of 0.1 and simulate the diffusion process for 60 time units. Compare the results of the model based on the matrix approach with those of the direct approach.

16.2 Solution of Non-Linear Equations

Matrices are designed to solve systems of linear equations. It is therefore necessary to depart somewhat from simple matrix operations in order to develop models which contain non-linear equations.

Suppose that all fluxes between one compartment and another in Section 16.1 were a function of both the donor and the acceptor compartments. The result would be the following equations, which define change in state variables.

$$\dot{x}_1 = x_1x_2(a_{12}) + x_1x_3(a_{13}) - x_1x_2(a_{21}) - x_1x_3(a_{31}) \qquad (16.20)$$

$$\dot{x}_1 = x_1x_2(a_{21}) + x_2x_3(a_{23}) - x_2x_1(a_{12}) + x_2x_3(a_{32}) \qquad (16.21)$$

$$\dot{x}_3 = x_1x_3(a_{13}) + x_2x_3(a_{32}) - x_3x_1(a_{31}) - x_2x_3(a_{23}) \qquad (16.22)$$

Factoring out the common state variable from each equation gives

$$\dot{x}_1 = x_1[\quad 0 \quad + x_2(a_{12} - a_{21}) + x_3(a_{13} - a_{31})] \qquad (16.23)$$

$$\dot{x}_2 = x_2[x_1(a_{21} - a_{12}) + \quad 0 \quad + x_3(a_{23} - a_{32})] \qquad (16.24)$$

$$\dot{x}_3 = x_3[x_1(a_{31} - a_{13} + x_2(a_{32} - a_{23}) + \quad 0 \quad] \qquad (16.25)$$

The matrix for the non-linear transfer coefficients therefore becomes

$$
A_n = \begin{bmatrix}
0 & (a_{12} - a_{21}) & (a_{13} - a_{31}) \\
(a_{21} - a_{12}) & 0 & (a_{23} - a_{32}) \\
(a_{31} - a_{13}) & (a_{32} - a_{23}) & 0
\end{bmatrix} \qquad (16.26)
$$

Solving matrices for non-linear systems on the microcomputer involves only a minor modification of the technique used for linear systems. Line 280 of the previous subroutine is simply changed to incorporate the multiplication of the coefficient by the second state variable so that it becomes a second order term.

```
275  REM NON-LINEAR UPDATE EQUATION
280  X(I) = X(I) + (DX(I)*X(I) + F(I))*DT
```

Check Equations 16.23 - 16.25 if you do not fully understand all that statement 280 accomplishes.

Exercise 16.3

Set up the non-linear form of the Silver Springs energy flow model (Exercise 12.3) using the matrix technique. Notice that only the feeding coefficients are non-linear. Respiration and mortality are simple donor-dependent first-order relationships. It is therefore necessary to set up two coefficient matrices. One, $\mathbf{A}_n$, should consist of the non-linear feeding coefficients with zeros making up the other elements, and the other, $\mathbf{A}_1$, should consist of the linear coefficients. Because of the high turnover rate of the decomposer compartment, use the value of 0.1 wk. for Δt, but print out the state of the system only once each week. This will be rather tricky model to accomplish using matrices.

16.3 Multicomponent Interspecific Action

The Lotka-Volterra equations in Section 7.3 are used to describe the interaction of two species. The basic approach may be expanded to account for interaction of an unlimited number of species (or taxa) within an ecosystem by simply adding additional terms to the equations involved.

The basic equations for two-species interaction are as follows:

$$\dot{N}_1 = r_1N_1 - \frac{r_1N_1^{\,2}}{K_1} - \frac{r_1\alpha N_1N_2}{K_1} \quad (16.30)$$

$$\dot{N}_2 = r_2N_2 - \frac{r_2N_2^{\,2}}{K_2} - \frac{r_2\beta N_1N_2}{K_2} \quad (16.31)$$

Factoring out common terms gives

$$\dot{N}_1 = r_1N_1(1 + a_{11}N_1 + a_{12}N_2) \quad (16.32)$$

and

$$\dot{N}_2 = r_2N_2(1 + a_{22}N_2 + a_{21}N_1) \quad (16.33)$$

where

$$a_{11} = \frac{-1}{K_1} \qquad a_{22} = \frac{-1}{K_2}$$

and

$$a_{12} = \frac{-\alpha}{K_1} \qquad a_{21} = \frac{-\beta}{K_2}$$

In this model, the coefficients refer to the influence of one organism on another, or to its ability to utilize the environment, rather than material or energy flux between compartments.

The block diagram which shows all possible interactions between four components is presented in Figure 16.2.

The differential equation expressing the change in the population density of N_1 is

$$\dot{N}_1 = r_1N_1 + r_1N_1^2a_{11} + r_1N_1N_2a_{12} + r_1N_1N_3a_{13} + r_1N_1N_4a_{14} \quad (16.34)$$

$$\dot{N}_1 = r_1N_1(1 + N_1a_{11} + N_2a_{12} + N_3a_{13} + N_4a_{14}) \quad (16.35)$$

By appropriate selection of coefficients, a_{ij}, it is possible to express almost any kind of interaction between four components.

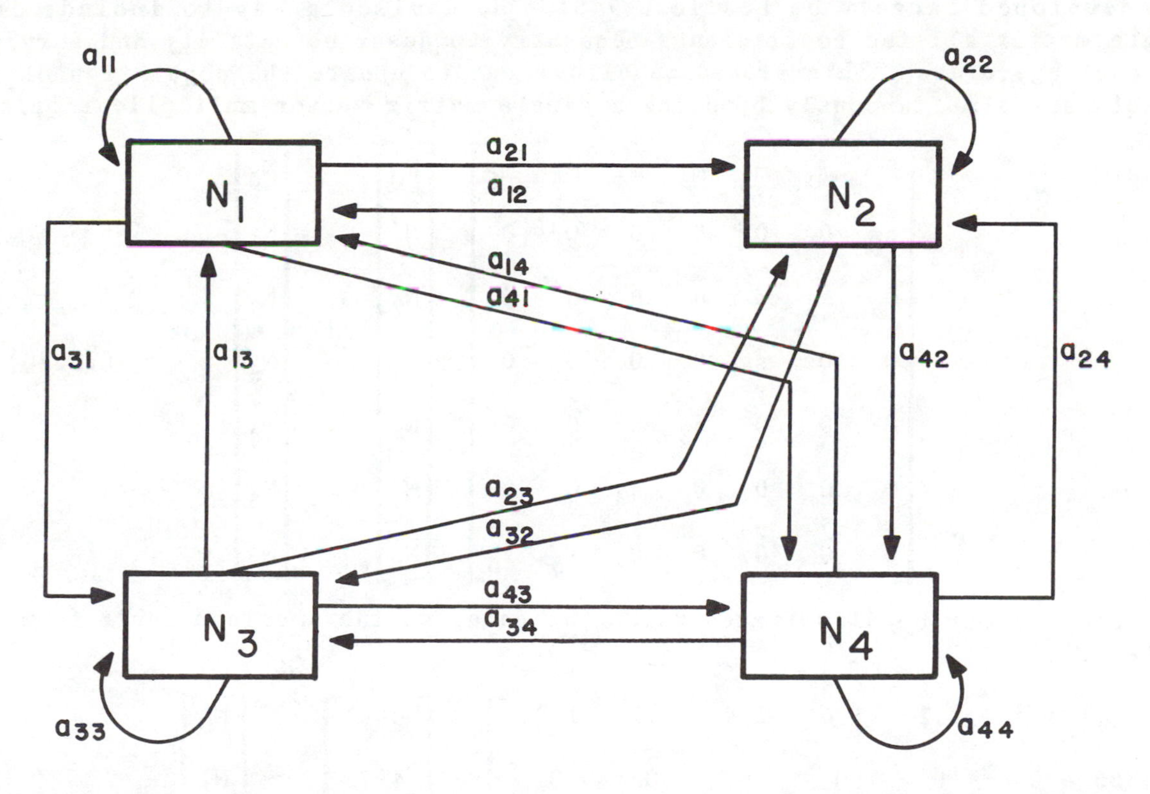

Figure 16.2. Block diagram for the interaction between four components.

<u>Exercise 16.4</u>

The form of Equation 16.35 suggests that this model would lend itself to the matrix approach. Write the four equations required to define the population densities of the four species. Then write the coefficient matrix and the matrix calculations necessary to implement this model. Note that special treatment will be required because of the non-linear nature of these equations.

Use the following values to implement the model:

$N_1 = 10$	$a_{11} = -0.0005$	$a_{21} = +0.000003$	$a_{31} = +0.00007$
$N_2 = 10$	$a_{12} = -0.001$	$a_{22} = -0.1$	$a_{32} = 0$
$N_3 = 10$	$a_{13} = -0.001$	$a_{23} = 0$	$a_{33} = -0.11$
$N_4 = 10$	$a_{14} = -0.001$	$a_{24} = 0$	$a_{34} = 0$
$r_1 = 0.5$	$r_3 = 0.3$	$a_{41} = +0.0001$	$a_{43} = 0$
$r_2 = 0.2$	$r_4 = 0.3$	$a_{42} = 0$	$a_{44} = -0.115$

16.4 The Leslie Age-Class Matrix**

The technique of using matrices for age-class models and population demography was developed largely be Leslie (1945). He devised a way to include in a single matrix all the coefficients necessary to describe natality and survival for each age-class. This procedure allows one to update the population of all age classes simultaneously by using a single matrix-vector multiplication.

$$\begin{bmatrix} 0 & m_1 & m_2 & m_3 & m_4 & m_5 & m_6 \\ s_0 & 0 & 0 & 0 & 0 & 0 & 0 \\ 0 & s_1 & 0 & 0 & 0 & 0 & 0 \\ 0 & 0 & s_2 & 0 & 0 & 0 & 0 \\ 0 & 0 & 0 & s_3 & 0 & 0 & 0 \\ 0 & 0 & 0 & 0 & s_4 & 0 & 0 \\ 0 & 0 & 0 & 0 & 0 & s_5 & s_6 \end{bmatrix} \cdot \begin{bmatrix} N_0 \\ N_1 \\ N_2 \\ N_3 \\ N_4 \\ N_5 \\ N_6 \end{bmatrix}_t = \begin{bmatrix} N_0 \\ N_1 \\ N_2 \\ N_3 \\ N_4 \\ N_5 \\ N_6 \end{bmatrix}_{t+\Delta t} \tag{16.40}$$

This process may be illustrated using the data for the American robin found in Table 8.1.

$$\begin{bmatrix} 0 & 1.4 & 2.1 & 1.6 & 1.4 & 1.1 & 1.0 \\ 0.355 & 0 & 0 & 0 & 0 & 0 & 0 \\ 0 & 0.461 & 0 & 0 & 0 & 0 & 0 \\ 0 & 0 & 0.433 & 0 & 0 & 0 & 0 \\ 0 & 0 & 0 & 0.364 & 0 & 0 & 0 \\ 0 & 0 & 0 & 0 & 0.300 & 0 & 0 \\ 0 & 0 & 0 & 0 & 0 & 0.200 & 0.100 \end{bmatrix} \cdot \begin{bmatrix} 1400 \\ 497 \\ 229 \\ 99 \\ 36 \\ 10 \\ 6 \end{bmatrix} = \begin{bmatrix} N_0 \\ N_1 \\ N_2 \\ N_3 \\ N_4 \\ N_5 \\ N_6 \end{bmatrix}_{t+\Delta t} \tag{16.41}$$

If the Leslie matrix is symbolized by L, and the population vector by **N**, then the matrix equation for this calculation would be the following:

$$\mathbf{N}_{t+1} = \mathbf{L} \cdot \mathbf{N}_t \qquad (16.42)$$

This approach greatly simplifies the programming of age-class models.

Exercise 16.5**
Program the human population model (Exercise 8.3) using the Leslie matrix approach. Compare the results obtained with those with the direct approach.

16.5 Modifications of the Leslie Matrix Model

Usher (1972) has presented an excellent discussion of modifications which can be made to the basic Leslie matrix model. One which we will discuss here involves a coefficient matrix which includes natality rates and survival rates for both males and females. The Leslie matrix then takes the following form:

$$\mathbf{L} = \begin{bmatrix}
0 & m_0 & 0 & m_1 & 0 & m_2 & 0 & m_3 & \cdots & 0 & m_x \\
0 & f_0 & 0 & f_1 & 0 & f_2 & 0 & f_3 & \cdots & 0 & f_x \\
s_0 & 0 & 0 & 0 & 0 & 0 & 0 & 0 & \cdots & 0 & 0 \\
0 & t_0 & 0 & 0 & 0 & 0 & 0 & 0 & \cdots & 0 & 0 \\
0 & 0 & s_1 & 0 & 0 & 0 & 0 & 0 & \cdots & 0 & 0 \\
0 & 0 & 0 & t_1 & 0 & 0 & 0 & 0 & \cdots & 0 & 0 \\
0 & 0 & 0 & 0 & s_2 & 0 & 0 & 0 & \cdots & 0 & 0 \\
0 & 0 & 0 & 0 & 0 & t_2 & 0 & 0 & \cdots & 0 & 0 \\
0 & 0 & 0 & 0 & 0 & 0 & s_3 & 0 & \cdots & 0 & 0 \\
0 & 0 & 0 & 0 & 0 & 0 & 0 & t_3 & \cdots & 0 & 0 \\
\cdot & \cdot & \cdot & \cdot & \cdot & \cdot & \cdot & \cdot & \cdots & \cdot & \cdot \\
\cdot & \cdot & \cdot & \cdot & \cdot & \cdot & \cdot & \cdot & \cdots & \cdot & \cdot \\
0 & 0 & 0 & 0 & 0 & 0 & 0 & 0 & \cdots & s_x & 0 \\
0 & 0 & 0 & 0 & 0 & 0 & 0 & 0 & \cdots & 0 & t_x
\end{bmatrix}$$

where m_x and f_x are rate coefficients for male and female fawn production per year for each age-class, x, and s_x and t_x are the respective survival coefficients of males and females for each age-class.

The population vector, N, now takes the form

$$N = \begin{bmatrix} M_0 \\ F_0 \\ M_1 \\ F_1 \\ M_2 \\ F_2 \\ \cdot \\ \cdot \\ \cdot \\ M_x \\ F_x \end{bmatrix}$$

where M_x is the number of males in age-class x, and F_x is the number of females in age-class x.

It can be readily seen that a population update is accomplished by a simple matrix multiplication as in Equation 16.42. This is an elegant technique, to say the least. For other interesting modifications of the basic Leslie matrix, see Usher (1972).

Exercise 16.6

Use the technique described in Section 16.5 to program the deer management problem described in Exercise 8.5. Make whatever simplifying assumptions you need to adapt it to this procedure.

Conclusion

This chapter has presented a few representative examples of applications of matrix algebra to the field of modeling and computer simulation. Other examples will be included in subsequent chapters.

CHAPTER 17

DYNAMICS OF MULTIENZYME SYSTEMS

Biochemists have traditionally used an analytical approach in which living material is broken down and reactions are studied as separate entities. These reductionist techniques have generated a large amount of data concerning the kinetics of individual enzyme-catalyzed reactions. Over the years, this expanding body of knowledge has led to a number of conceptual models describing the functional relationship between observed system components. Glycolysis, Kreb's cycle, the pentose phosphate cycle, and various biosynthetic pathways are all recognized conceptual models of multienzyme systems, supported by many lines of evidence.

Synthetic methods, in which enzymes and other cell components are recombined to form metabolic systems, have not met with much success. Thus, computer simulation provides a potential mechanism for better understanding the dynamic interaction of components in multienzyme systems.

As in any other field of science, simulation provides biochemists with an unambiguous test of ideas concerning system behavior. It helps to interpret the results of previous experiments and plan new ones. It also provides a way of determining the relative importance of different factors in relationship to observed system behavior. Because of their non-linearity, many multienzyme systems tend to behave in a counter-intuitive fashion. In such cases, simulation can provide an important tool for understanding the complex interactions involved. Thus, simulation can be an important adjunct to the study of biochemical reaction pathways.

Multienzyme systems may involve as many as 50 chemical reactants, often transported between several cell compartments, and catalyzed by 30 or more different enzymes. Each enzyme involved shows different kinetic behavior and may respond to feedback control. Clearly, such systems offer a great challenge to the simulation programmer.

James D. Spain, BASIC Microcomputer Models in Biology ISBN 0-201-10678-7

In this chapter, we will describe some simple examples of modeling techniques which have proved useful in this field. The large models alluded to above are simply extensions of the basic techniques described here. The major limitation to models of this type is simply the time it takes to integrate the large number of rate equations involved.

Much of the early work in biochemical simulation was carried on by Britton Chance and his colleagues at the University of Pennsylvania. Chance, et al., (1960) were the first to develop a simulation of the glycolysis pathway, for example. For an excellent review of these early models, refer to Garfinkel, et al., (1970).

There are two major approaches to the simulation of multienzyme systems. The one employed for much of the earlier work was the direct mass action approach in which enzymes are considered to react with substrates to form complexes which in turn break down to form products. This method might be thought of as a multiple or extended Chance-Cleland model. Its advantage is that it is capable of being used on almost any type enzyme reaction involving one or more substrates, and it can be used for reversible reactions as well as non-reversible reactions.

The second approach involves the use of rate law expressions based on steady-state derivations. Thus, the equations involved have a form similar to the Michaelis-Menten equation. Because it involves fewer mathematical operations, the rate law approach is much faster. The problem with this technique is that it is difficult to obtain valid rate law expressions for certain bisubstrate and trisubstrate reactions. Those that are available often do not show full reversibility, a very desirable characteristic from the standpoint of enzyme theory.

Perhaps the most interesting feature of multienzyme systems from the modeling standpoint concerns the dynamics of control. Models involving enzyme control are explored primarily using the rate law approach. The Monod allosteric enzyme model, presented in Section 2.6, provides a rate law expression which is useful in multienzyme systems involving control.

17.1 Mass Action Model of Glycolysis

Most of the early models of multienzyme systems employed the direct approach of solving each equation based on the law of mass action. Enzymes were included as chemical reactants in the system. One such model was developed by Chance et al., (1960). This model, developed and implemented on UNIVAC I, was able to accept up to 49 chemical components, each of which could participate in one or more first, second, or third order, reversible or non-reversible reactions. The model was used to investigate the control of glucose metabolism as it relates to oxygen, ATP concentration, and other factors. Despite the relatively crude technique employed, it made possible a comprehensive study of the kinetics of a multienzyme system in such a way that the results could be compared with the dynamics of intact cells.

The glycolysis model presented below is based on the simulation developed by

Chance and his colleagues. It has been greatly simplified in order to make it run in a reasonable time using microcomputer BASIC. In order to accomplish this, the model deals only with anaerobic aspects of glycolysis. Table 17.1 gives a symbolic representation of the reactions involved. For a full discussion of the reactions of glycolysis, refer to a biochemistry text such as Lehninger (1975). For a list of abbreviations used, see Table 17.2.

Table 17.1

<u>Reactions used in the glycolysis model</u>

1. GLU + HEX $\xrightarrow{F1}$ CHEX

2. CHEX + A3P $\xrightarrow{F2}$ HEX + H6P + A2P

3. H6P + PFK $\xrightarrow{F3}$ CPFK

4. CPFK + A3P $\xrightarrow{F4}$ PFK + FPP + A2P

5. FPP $\xrightarrow{F5}$ DHAP + GAP

6. DHAP + HNAD $\xrightarrow{F6}$ AGP + NAD

7. AGP + NAD $\xrightarrow{F7}$ DHAP + HNAD

7A. DHAP $\xrightarrow{FA}$ GAP

7B. GAP $\xrightarrow{FB}$ DHAP

8. GAP + CGPDH $\xrightarrow{F8}$ CX + HNAD

9. CX + PI $\xrightarrow{F9}$ DPGA + GPDH

10. NAD + GPDH $\xrightarrow{F\emptyset}$ CGPDH

11. DPGA + A2P $\xrightarrow{FC}$ A3P + P3GA

12. P3GA + A2P $\xrightarrow{FD}$ A3P + PYR

13. PYR + HNAD $\xrightarrow{FE}$ LAC + NAD

14. LAC + NAD $\xrightarrow{FF}$ PYR + HNAD

20. A3P $\xrightarrow{FG}$ A2P + PI

Numbers refer to analogous reactions employed by Chance, et al., (1960). Reaction 7A and 7B have been added by the author. Reaction 20 summarizes the effects of ATP utilization. The symbols F1 through FG refer to the fluxes which are associated with rate constants K1 through KG (see flux equations on following page).

The symbols employed in Table 17.1 conform as much as possible to standard usage. In some cases, they have been modified in order to provide variable names which would be recognized as unique and legal by the microcomputer. Table 17.2 provides a complete list of variable names employed in this model.

Table 17.2
List of variable names and abbreviations used in the glycolysis model

Glucose	GLU	Hexokinase	HEX
Hexose 6-phosphate	H6P	Hexokinase-complex	CHEX
Fructose 1,6-diphosphate	FPP	Phosphofructokinase	PFK
Dihydroxyacetone phosphate	DHAP	Phosphofructokinase-complex	CPFK
α-Glycerophosphate	AGP	Glyceraldehyde 3-phosphate dehydrogenase	GPDH
Glyceraldehyde 3-phosphate	GAP	Glyceraldehyde 3-phosphate dehydrogenase-NAD complex	CGPDH
Diphospho glyceric acid	DPGA	Glyceraldehyde 3-phosphate dehydrogenase-acyl complex	CX
3-phospho glyceric acid	P3GA	Nicotinamide adenine dinucleotide	NAD
Pyruvic acid	PYR	Reduced nicotinamide adenine dinucleotide	HNAD
Lactic acid	LAC		
Inorganic phosphate	PI		
Adenosine triphosphate	A3P		
Adenosine diphosphate	A2P		

For the most part, this glycolysis model has employed the same simplifying assumptions as the original. To reduce the number of equations involved, many enzymes are assumed to be operating well below V_{max} so that simple first order reactions may be employed (see reactions 5, 7A, 7B.) Hexose isomerase and enolase are assumed to be sufficiently rapid that they can be combined with other reactions. Such simplifications reduce the number of calculations by about one third.

The model presented here employs the simple Euler technique to integrate the rate equations involved. The following flux equations are required for Stage I of the integration process:

```
F1 = K1*GLU*HEX*DT
F2 = K2*CHEX*A3P*DT
F3 = K3*H6P*PFK*DT
F4 = K4*CPFK*A3P*DT
F5 = K5*FPP*DT
F6 = K6*DHAP*HNAD*DT
F7 = K7*AGP*NAD*DT
FA = KA*DHAP*DT
FB = KB*GAP*DT
F8 = K8*GAP*CGPDH*DT
F9 = K9*CX*PI*DT
FØ = KØ*NAD*GPDH*DT
FC = KC*DPGA*A2P*DT
FD = KD*P3GA*A2P*DT
FE = KE*PYR*HNAD*DT
FF = KF*LAC*NAD*DT
FG = KG*A3P*DT
```

Note that these equations and those which follow are written as BASIC computer statements to assist you in developing this simulation.

Stage II employs the following update equations to define the concentration of each constituent at time t + Δt:

```
GLU   = GLU - F1
HEX   = HEX - F1 + F2
CHEX  = CHEX + F1 - F2
H6P   = H6P + F2 - F3
PFK   = PFK - F3 + F4
CPFK  = CPFK - F4 + F3
FPP   = FPP + F4 - F5
DHAP  = DHAP + F5 - F6 + F7 - FA + FB
AGP   = AGP + F6 - F7
GAP   = GAP + F5 + FA - FB - F8
CGPDH = CGPDH - F8 + FØ
CX    = CX + F8 - F9
DPGA  = DPGA + F9 - FC
GPDH  = GPDH + F9 - FØ
P3GA  = P3GA + FC - FD
PYR   = PYR - FE + FF + FD
LAC   = LAC + FE - FF
NAD   = NAD - FF + FE - FØ + F6 - F7
HNAD  = HNAD - F6 + F7 + F8 - FE + FF
PI    = PI + FG - F9
A3P   = A3P - FG - F2 - F4 + FC + FD
A2P   = A2P - FD - FC + F2 + F4 + FG
```

The objective of the glycolysis model described here is to demonstrate the direct mass action approach in a system familiar to most biologists. The model simulates anaerobic glycolysis in that the conversion of glucose to pyruvate and lactate is accompanied by an increase in ATP (A3P) with negligible effect on NAD and NADH.

Actually, the model is rather large and cumbersome to be implemented on a microcomputer. You will find it takes about one hour for a significant conversion of glucose to pyruvate to occur. Attempting to speed up the model by changing the time increment usually results in instability, with the result that certain variables assume zero or negative values. If this occurs, the model must be stopped, and the size of the time increment decreased.

<u>Exercise 17.1</u>
Program the glycolysis model described above and run it with the following rate constants and concentrations of intermediates, adapted from Chance, et al. (1960).

K1 = 3×10^{-4}	$(\mu mole \cdot \mu sec)^{-1}$	GLU = 3000	μmoles/liter
K2 = 10^{-4}	''	H6P = 30	''
K3 = 4×10^{-4}	''	FPP = 10	''
K4 = 4×10^{-4}	''	DHAP = 334	''
K5 = 10^{-1}	$(\mu sec)^{-1}$	AGP = 3188	''

K6 = $2x10^{-5}$	$(\mu mole \cdot \mu sec)^{-1}$		GAP = 362	μmoles/liter
K7 = $8x10^{-7}$	''		DPGA = 580	''
KA = 10^{-1}	$(\mu sec)^{-1}$		P3GA = 660	''
KB = 10^{-1}	''		PYR = 540	''
K8 = $6x10^{-5}$	$(\mu mole \cdot \mu sec)^{-1}$		LAC = 2300	''
K9 = $4x10^{-6}$	''		NAD = 143	''
KØ = $6x10^{-5}$	''		HNAD = 33	''
KC = 10^{-4}	''		A3P = 1000	''
KD = $5x10^{-5}$	''		A2P = 500	''
KE = $5x10^{-6}$	''		HEX = 10	''
KF = 10^{-7}	''		PKF = 10	''
KG = 10^{-2}	$(\mu sec)^{-1}$		GPDH = 50	''
			PI = 100	''

As the model runs, keep track of glucose, ATP, and lactate. These variables should be printed or plotted about every 100 time intervals (μseconds) to follow the changes which occur in the system.

The concentrations given above are close to the steady-state concentrations for the rate constants employed. Thus, the major long term changes will be in glucose, ATP, and lactate. Initial concentrations of these three variables may be changed to see the behavior of the system under different conditions.

Another experiment might be to incorporate an additional equation to simulate the reoxidation of NADH (HNAD) by the respiratory chain. A simple reaction which would model this effect is

$$\text{HNAD} \xrightarrow{\text{FH}} \text{NAD}$$

where FH = KH*HNAD*DT. This flux should be included in the update equations for NAD and HNAD. Assume KH equals 0.1.

17.2 General Mass Action Approach for Single Substrate Reactions

The reversible single substrate enzyme catalyzed reaction may be represented by the following steady-state system:

$$S + E \underset{k_2}{\overset{k_1}{\rightleftharpoons}} ES \underset{k_4}{\overset{k_3}{\rightleftharpoons}} E + P$$

where S is substrate, E is enzyme, ES is enzyme-substrate compound, and P is product. The rate constants for the four reactions involved are k_1, k_2, k_3, and k_4. The general mass action approach would treat this as an extended Chance-Cleland model. See Section 6.4 for further details on the basic model. The fluxes involved in a single reaction of this type are defined as follows:

$$F_1 = k_1[S][E]\Delta t \tag{17.21}$$

$$F_2 = k_2[ES]\Delta t \tag{17.22}$$

$$F_3 = k_3[ES]\Delta t \tag{17.24}$$

The Euler integration of these fluxes involves the following four update expressions:

$$S \longleftarrow S + F_2 - F_1 \tag{17.25}$$

$$E \longleftarrow E + F_2 - F_1 + F_3 - F_4 \tag{17.26}$$

$$ES \longleftarrow ES - F_2 + F_1 - F_3 + F_4 \tag{17.27}$$

$$P \longleftarrow P + F_3 - F_4 \tag{17.28}$$

This approach was employed in Exercise 6.5 to model a single enzyme catalyzed reaction. The technique is modified below to model a sequence of single substrate reactions of the following type:

$$S_1 \longrightarrow S_2 \longrightarrow S_3 \longrightarrow S_i \longrightarrow \cdots \longrightarrow S_n$$

The obvious procedure is to re-form the above equations using subscripted variables. Thus, the flux equations may be written in the following form, appropriate for use in a computer program:

```
F1(I) = K1(I)*S(I)*E(I)*DT
F2(I) = K2(I)*ES(I)*DT
F3(I) = K3(I)*ES(I)*DT
F4(I) = K4(I)*E(I)*S(I+1)*DT
```

Notice that the product of the reaction involving S(I) now becomes S(I+1).

The second stage of integration should make use of the following update equations:

```
S(I)   = S(I)   + F2(I) - F1(I)
E(I)   = E(I)   + F2(I) - F1(I) + F3(I) - F4(I)
ES(I)  = ES(I)  - F2(I) + F1(I) - F3(I) + F4(I)
S(I+1) = S(I+1) + F3(I) - F4(I)
```

This simulation should be programmed to increment time, go through all the flux equations for N values of I, to complete stage one, and then go through all the update equations for N values of I, to complete stage two, and finally print or plot the resulting concentrations.

Exercise 17.2

Use the general mass action technique described above to model a hypothetical reaction pathway consisting of five intermediates connected by four single substrate enzyme catalyzed reactions, as diagrammed below:

$$S_1 \xrightarrow{E_1} S_2 \xrightarrow{E_2} S_3 \xrightarrow{E_3} S_4 \xrightarrow{E_4} S_5$$

Use the rate constants and initial concentrations provided in Table 17.3.

Table 17.3

Parameters for a hypothetical reaction sequence

i	1	2	3	4	5
k_1	0.05	0.005	0.05	0.05	---
k_2	0.1	0.1	0.01	0.1	---
k_3	0.1	0.1	0.1	0.1	---
k_4	0.005	0.05	0.005	0.05	---
K_{eq}	10.0	0.1	100.0	1.0	---
E_i	10.0	10.0	10.0	10.0	---
S_i	1000	100	100	100	100
K_s	0.2	2.0	0.02	0.2	---
V_s	1.0	1.0	1.0	1.0	---
V_p	1.0	1.0	0.1	1.0	---

Program the simulation so that the concentration of each intermediate may be observed at the end of each time interval. Write the program so that S_1 and S_5 may be held constant in order to observe the system under steady-state conditions. Use the model to explore the difference in steady-state concentrations of intermediates when operating with fully reversible equations as opposed to the situation when no product is allowed to convert back to substrate. The latter, irreversible condition, would be simulated if all values for k_4 were made equal to zero.

17.3 The Rate Law Approach for Single Substrate Reaction Sequences

The mass action approach described above is quite slow, even when the computer programs are written in compiled languages such as FORTRAN. Simulation times one or more orders of magnitude greater than the reaction times being simulated have been reported. In addition, the non-linear nature of the mass action equations and the size of the coefficients can cause the numerical integration technique to yield rather large errors. Thus, there were good reasons to find more suitable techniques for simulating multienzyme systems.

The rate law approach, first described by Rhodes, et al. (1968) seems to offer an acceptable alternative. Initial trials of the rate law technique by Rhodes proved to be about 100 times faster than a comparable simulation based on the mass action approach. Essentially identical simulation data were generated by both methods.

The rate law technique employs rate expressions like the Michaelis-Menten equation to determine the flux between a given substrate and its product. The

Michaelis-Menten equation is the specific rate law for the non-reversible single substrate reaction. The rate law for reversible single substrate reactions is described by Savageau (1976) as follows:

$$V_{net} = \frac{\frac{V_s}{K_s}[S] - \frac{V_p}{K_p}[P]}{1 + \frac{[S]}{K_s} + \frac{[P]}{K_p}} \qquad (17.31)$$

where V_{net} is the net forward reaction velocity, or flux, for a single substrate enzyme catalyzed reaction. [S] and [P] are concentrations of substrate and product, V_s and V_p are maximum velocities for the forward and reverse reactions, and K_s and K_p are Michaelis-Menten constants for the forward and reverse reactions. Note that if [P] equals 0, Equation 12.31 reverts to the standard form of the Michaelis-Menten equation. The above rate law appears to model most single substrate reactions effectively.

The four constants involved in Equation 17.31 are related to the equilibrium constant for the reaction by the Haldane relationship

$$K_{eq} = \frac{V_s \cdot K_p}{V_p \cdot K_s} \qquad (17.32)$$

Thus, to be internally consistent the four constants usually provided as input to this model are K_{eq}, K_s, V_p, and V_s. To calculate K_p, use the equation

$$K_p = K_{eq} \cdot V_p \cdot K_s / V_s \qquad (17.33)$$

To minimize the number of calculations required within the numerical integration loop, the terms V_s/K_s and V_p/K_p are defined as VF and VR, respectively, before the loop begins. Thus, the following program statements would be employed to read and calculate the parameters for a general simulation involving N reactions in a sequence, as shown in Section 17.2.

```
200 FOR I = 1 TO N                                    (17.34)
210 READ KE(I),KS(I),VS(I),VP(I)
220 KP(I) = KE(I)*VP(I)*KS(I)/VS(I)
230 VF(I) = VS(I)/KS(I)
240 VR(I) = VP(I)/KP(I)
250 NEXT I
```

The sequence of statements required for stage I of the integration would utilize the rate law expression to calculate the net flux for the reaction, F(I).

```
300 FOR T = 1 TO TM                                   (17.35)
310 FOR I = 1 TO N
320 F(I) = (VF(I)*S(I)-VR(I)*S(I+1))/(1+S(I)/KS(I)+S(I+1)/KP(I))*DT
330 NEXT I
340 NEXT T
```

The second stage of numerical integration updates the concentration for each intermediate, S(I), using the following program statements:

```
400 FOR I = 1 TO N                                        (17.36)
410 S(I) = S(I) - F(I)
420 S(I+1) = S(I+1) + F(I)
430 NEXT I
```

The advantage of the rate law approach is obvious when you compare the number of statements required for numerical integration in this simulation with the one described in Section 17.2. The rate law approach required only one flux equation and two update equations to accomplish the same procedure which employed four flux equations and four update equations using the mass action approach. Because the system is much more stable, time increments need not be as small, and thus the time required for a comparable simulation is at least an order of magnitude less.

Exercise 17.2
Write a program to simulate the same hypothetical reaction sequence that was modeled in Exercise 17.2, but this time, use the rate law approach. Note that Table 17.1 provides the necessary values for the steady-state constants, K_s, K_{eq}, V_s, and V_p.

After the simulation is programmed, determine the time required for the system to reach a given condition relative to the steady-state. Compare this time with that required to achieve the same condition using the mass action simulation.

17.4 The Rate Law Approach For Bisubstrate Reactions

Unfortunately, from the standpoint of simulating multienzyme systems, most reaction pathways involve very few single substrate reactions. Although we often focus attention on one substrate or another, the reaction rate, and the rate law expression, is usually dependent upon a second or third substrate, as well as multiple products. In addition, reversible bisubstrate reactions involve a variety of mechanisms, each of which requires a specific rate law. Thus, using the rate law approach to describe reversible reactions in real reaction pathways requires complicated rate law expressions typically involving about a dozen coefficients. This procedure is well beyond the scope of this book and it is likely to have limited success on the microcomputer. For further information on bisubstrate reaction rate laws for reversible reactions, refer to Savageau (1976).

A general rate law expression for non-reversible bisubstrate reactions is the equation described by Cleland (1970).

$$v = V \cdot \frac{[A]}{([A] + K_a)} \cdot \frac{[B]}{([B] + K_b)} \tag{17.41}$$

where v is the reaction velocity, [A] and [B] are concentrations of two substrates required for the same reaction, K_a and K_b are half-saturation constants, and V is the maximum velocity. An equation of this type will be used in Section 17.6.

17.5 The Yates-Pardee Model of Feedback Control

Many metabolic pathways are controlled by an end product metabolite which feeds back to inhibit one of the initial reactions in a sequence. This simple form of control, found to exist in a variety of biosynthetic pathways, was first described by Yates and Pardee (1956).

The Yates-Pardee control process may be represented by the following reaction sequence:

$$S_0 \xrightarrow{E_0} S_1 \xrightarrow{E_1} S_2 \xrightarrow{E_2} S_3 \xrightarrow{E_3} S_4 \xrightarrow{E_4} S_5 \quad (S_4 \cdots\!\!\to E_0)$$

S_0 through S_5 are hypothetical substrates and products which form an unbranched metabolic pathway. E_0 through E_4 are enzymes which catalyze the reactions involved. S_4 is a product which feeds back and allosterically inhibits enzyme E_0. This model is similar to one investigated in detail by Walter (1974).

Assuming that only single substrate reactions are involved, we may simulate this process by employing the rate law approach described in Section 17.3. The rate law equations for E_1 through E_4 would be those described in Equations 17.31 to 17.36.

The rate law expression for E_0 employs an equation based on the concerted model for allosteric control developed by Monod, et al. (1965). This model is described in Section 2.6 of this text. Using a modification of this equation which assumes no allosteric activation, we may describe the reaction rate as follows:

$$F_0 = V \cdot \frac{\alpha(1 - \alpha)^3}{L(1 + \beta)^4 + (1 + \alpha)^4} \cdot \Delta t$$

V is the maximum velocity, α is the substrate concentration in relation to the dissociation constant, $[S_0]/K_0$, L is the equilibrium constant for conversion of enzyme from the active form to the inactive form, and β is the inhibitor concentration in relation to its dissociation constant, $[S_4]/K_i$.

The process is simulated in much the same manner as the sequence in Section 17.3, except that Equation 17.51 is used to determine the flux from S_0 to S_1. The result is an oscillating system due to the lag in information feedback to the control enzyme. Such systems are called phase shift oscillators.

Exercise 17.4

Program the simulation of the Yates-Pardee feedback control model. Use the following parameters for reactions 1 through 5:

$$K_{eq} = 10, \ V_S = 10, \ V_P = 1, \ K_S = 20.$$

The parameters for the controlled reaction are

$$K_0 = 10, \; K_i = 5, \; L = 200, \; S_0 = 100$$

Plot the resulting concentrations of S_4 and S_1 as a function of time. The result should be a stable oscillating system resulting from the lag in information feedback. Compare the oscillations obtained using the reversible rate law expression with those obtained using the irreversible rate law expression.

17.6 Allosteric Control of Phosphofructokinase

The key enzyme for controlling glycolysis is considered to be phosphofructokinase, the enzyme that catalyzes the conversion of fructose 6-phosphate to fructose diphosphate. The key position of the enzyme at the head of the energy producing section of the pathway permits the whole glycolysis pathway to be controlled from this point. Phosphofructokinase activity is inhibited by adenosine triphosphate (ATP) and activated by adenosine monophosphate (AMP). The result is a slowing down of the rate of glycolysis when an adequate supply of ATP energy is available from other sources, such as Kreb's cycle, and a speeding up of glycolysis when ATP concentration is low and AMP concentration is high.

Oscillations of various metabolites associated with glycolysis have been observed experimentally in a variety of tissues and microbes. The present model was developed by Richter and Betz (1976) to determine if the lags in the feedback pathways to phosphofructokinase could account for the magnitude and frequency of natural oscillations.

The system modeled by Richter and Betz is described by the following reaction scheme:

```
              ATP              ADP
   V0          +       Vp       +       Vr               Vd
------>       F6P    ------>   FDP    ------>   ATP   ------>  consumption
               |
               | Vs
               v
            Storage
```

In this scheme, fructose 6-phosphate (F6P) enters the pathway at a constant rate, V_0. F6P reacts with ATP in a reaction catalyzed by phosphofructokinase to form fructose diphosphate (FDP) and adenosine diphosphate (ADP). The rate of this reaction, V_p, is determined by a rate law expression based on the Monod model. The product, FDP, is converted by a sequence of reactions with velocity, V_r, which ultimately yields four ATP and consumes four ADP. ATP is consumed by energy requiring reactions of the cell at a rate described by V_d. Excess F6P is assumed to be stored at a rate determined by V_s.

The velocity for the phosphofructokinase reaction is defined by the following equation:

$$V_p = \frac{V_m \alpha (\alpha+1)^3}{L \cdot \frac{(1+\beta)^4}{(1+\gamma)^4} + (1+\alpha)^4} \cdot \frac{[ATP]}{(K_m + [ATP])} \tag{17.61}$$

The first half of the equation is based on the Monod model for allosteric control discussed in Section 2.6. The last half of the equation is a saturation term for the second substrate of the reaction, ATP. V_m is the maximum reaction velocity for phosphofructokinase, α is [F6P]/K_f, where K_f is the dissociation constant for F6P, β is [ATP]/K_i, where K_i is the dissociation constant for the allosteric inhibitor, and γ is [AMP]/K_a, where K_a is the dissociation constant for the allosteric activator. [F6P], [ATP], and [AMP] refer to concentrations of these intermediates in millimoles per liter.

The velocity for the reactions involved in utilizing FDP to make ATP is described as follows:

$$V_r = V_n \cdot \frac{[FDP]}{(K_n + [FDP])} \cdot \frac{[ADP]}{(K_r + [ADP])} \tag{17.62}$$

where V_n is the maximum velocity for the reaction, K_n and K_r are half-saturation constants, and [FDP] and [ADP] refer to substrate concentrations.

The rate of consumption of ATP is determined by the decay equation

$$V_d = K_d [ATP] \tag{17.63}$$

The storage rate of F6P is described by the equation

$$V_s = K_s [ATP][F6P] \tag{17.64}$$

ATP can react with AMP to yield two ADP according to the following reaction:

$$ATP + AMP \underset{V_2}{\overset{V_1}{\rightleftharpoons}} 2\ ADP$$

The velocities of the forward and reverse reactions, V_1 and V_2, are determined by the equations

$$V_1 = k_1 [ATP][AMP] \tag{17.65}$$

$$V_2 = k_2 [ADP][ADP] \tag{17.66}$$

The total concentration of the three adenosine nucleotides, N_0, is assumed to be constant, so that AMP concentration at any time is defined by

$$[AMP] = N_0 - [ATP] - [ADP] \tag{17.67}$$

By examining these equations, you can see that this simulation involves aspects of both the mass action approach and the rate law approach. The main equations, 17.61 and 17.62, use the rate law approach.

Equations 17.61 to 17.66 all define the fluxes for the system, and thus constitute stage I of the integration process. Stage II would use the following update expressions:

$$[F6P] \longleftarrow [F6P] + V_0 - V_p - V_s \qquad (17.68)$$

$$[FDP] \longleftarrow [FDP] + V_p - V_r$$

$$[ATP] \longleftarrow [ATP] + 4V_r - V_p - V_s - V_d - V_1 + V_2$$

$$[ADP] \longleftarrow [ADP] - 4V_r + V_p + V_s + V_d + 2V_1 - 2V_2$$

$$[AMP] = N_0 - [ATP] - [ADP]$$

<u>Exercise 17.5</u>

Use the information in Section 17.6 to program the simulation of allosteric control of the phosphofructokinase enzyme. Use the following parameters.

k_1 =	100	$(\mu mole\ min)^{-1}$	V_0 =	24	µmole/min·ml
k_2 =	50	''	V_m =	33	''
k_s =	6	''	V_n =	20	''
k_d =	1.0	$(min)^{-1}$	N_0 =	3.3	µmole/ml
K_a =	0.01	mM	L =	250	
K_f =	0.03	''	[ATP] =	2.0	''
K_m =	0.01	''	[ADP] =	1.0	''
K_n =	1.0	''	[AMP] =	0.3	''
K_r =	0.3	''			

Run the simulation for about 5 minutes of simulated time. Plot the concentration of ATP and AMP to observe glycolytic oscillations. It is also interesting to plot the concentration of FDP. Note that all the parameter symbols used in this simulation may be employed directly as program variables, except for ATP. Because AT is a reserved word in BASIC, ATP must be defined by some other variable name such as A3P.

<u>Conclusion</u>

The objective of this chapter has been to introduce you to some representative examples of multienzyme system simulations. Unfortunately, the microcomputer is not ideally suited for implementation of models of this complexity. This is not the result of limited memory, but rather a limitation on the time required to integrate all the differential equations these models typically entail. Presumably, the process of integration could be speeded up by using a more powerful integration technique. However, the programs are already quite complicated, thus the additional complication of an iterated second order Runge-Kutta integration would have made the exercises totally impractical for a text of this level. It is hoped that these examples provided at least an introduction to this potentially important area of simulation.

CHAPTER 18

MULTISTAGE NUTRIENT LIMITATION MODELS

In Chapter 13 we discussed single stage models for describing microbial growth which were based on the concentration of unbound nutrient. The growth of many organisms is adequately described by simple models of the Monod type. However, researchers are finding more and more examples of systems which deviate significantly from Monod-type kinetics, thus indicating that more realistic, higher resolution models are required.

Long before Monod, biologists had known that the conversion of nutrient to cell biomass involved more than a single stage process combining nutrient uptake and growth. Thus, it was not a lack of knowledge about the complexity of the systems involved which led to the use of the Monod single-stage model, but rather a parsimonious desire to keep the models as simple as possible. Clearly, the process of nutrient uptake and growth involves many pathways of the type described in the previous chapter. In fact, it would be hard to find any biochemical pathway in the cell that is not involved in one way or another with cell growth. The question is: what is the simplest model of cell growth which provides the necessary level of realism to simulate the consequences of nutrient limitation?

It is appropriate that this chapter on microbial growth concludes Part II of this book. It is in models of this type that concepts and modeling techniques relating to many previous chapters find a potential application. It is in this area of biology that ecology joins with biochemistry and cell physiology to provide models for the complex systems involved. This relationship is best exemplified by the fact that the major publications of active ongoing research in this subject are found in journals dealing with aquatic ecology on one hand, and biochemical engineering on the other. Very closely related research is carried on in these superficially diverse fields, one a branch of ecology, and the other closely related to biochemistry and microbiology.

James D. Spain, BASIC Microcomputer Models in Biology ISBN 0-201-10678-7

This integration of ecology with biochemistry results from the fact that cell growth and division are, fundamentally, biochemical processes. At the same time, in natural communities, organisms are exposed to other organisms, to abiotic factors such as heat and light, and to varying concentrations of nutrients. Thus, the simulation of microbial growth could conceivably involve sub-models describing various ecological factors as well as biochemical kinetics. It is in multicomponent system models of this type that the holistic nature of biology is most clearly demonstrated.

18.1 The Two Stage Nutrient Uptake Model

A major advance towards higher resolution nutrient limitation models resulted from the recognition by Caperon (1968) and Droop (1968) that the growth of cells was a function of the internal nutrient concentration rather than the external nutrient concentration. Thus, these authors described an internal nutrient pool which represented unassimiliated nutrient available for biosynthesis. This growth dependency was most evident when cell growth was observed during periods of fluctuating concentrations of external nutrient. Thus, it was observed, that growth continued for a period after cells were transferred from optimal nutrient concentrations to sub-optimal nutrient concentrations. Conversely, there was a lag before growth began when cells were transferred from sub-optimal nutrient to optimal nutrient concentrations. These perturbation studies fit with an earlier observation by Ketchum (1939) that nutrients such as phosphate were taken up by certain cells in excess of that required for immediate growth. This type of uptake was described as luxury consumption. Thus, there was good justification for modeling growth as a two stage process. The first stage involves nutrient uptake to form an internal nutrient pool, and the second stage involves the assimilation of nutrient as new cell biomass.

Nutrient uptake is generally carried on by mediated transport processes such as those described in Section 14.5. Therefore, uptake is characterized by an equation of the Michaelis-Menten type.

$$v = V_m \cdot \frac{N}{(K_n + N)} \qquad (18.11)$$

where v is the rate of nutrient uptake per unit of biomass, V_m is the maximum rate of nutrient uptake per unit of biomass, N is the external nutrient concentration, and K_n is the half-saturation constant for nutrient uptake.

Equation 18.11 describes changes in both the concentrations of external nutrient and internal nutrient, q. The change in external nutrient concentration is

$$\frac{dN}{dt} = -vX \qquad (18.12)$$

where X represents the total cell biomass per liter. The change in internal nutrient concentration is described by the equation

$$\frac{dq}{dt} = v - \mu q \qquad 18.13)$$

where dq/dt refers to the change in the amount of internal nutrient relative to the biomass. As such, q may be considered an internal nutrient concentration expressed as grams of nutrient per gram of biomass or µmoles of nutrient per cell. The term q has been called the yield coefficient by Caperon (1972) and the cell quota by Droop (1973).

The Monod growth equation may be redefined using the internal nutrient concentration and the minimum internal nutrient required for growth, q_o. This term also may be called the internal nutrient threshold.

$$\mu = \bar{\mu}\cdot\frac{(q - q_o)}{K_q + (q - q_o)} \qquad (18.14)$$

where μ is the growth rate per unit of biomass resulting from internal nutrient concentration, q, $\bar{\mu}$ is the maximum growth rate, K_q is the half-saturation constant for growth, and $(q-q_o)$ is the internal nutrient concentration available for growth. This latter term constitutes the nutrient pool or unassimilated nutrient. Thus, q_o might be considered to be the assimilated nutrient, not available for stimulating further growth. Figure 18.1 shows a graph of typical data from Caperon and Meyer (1972).

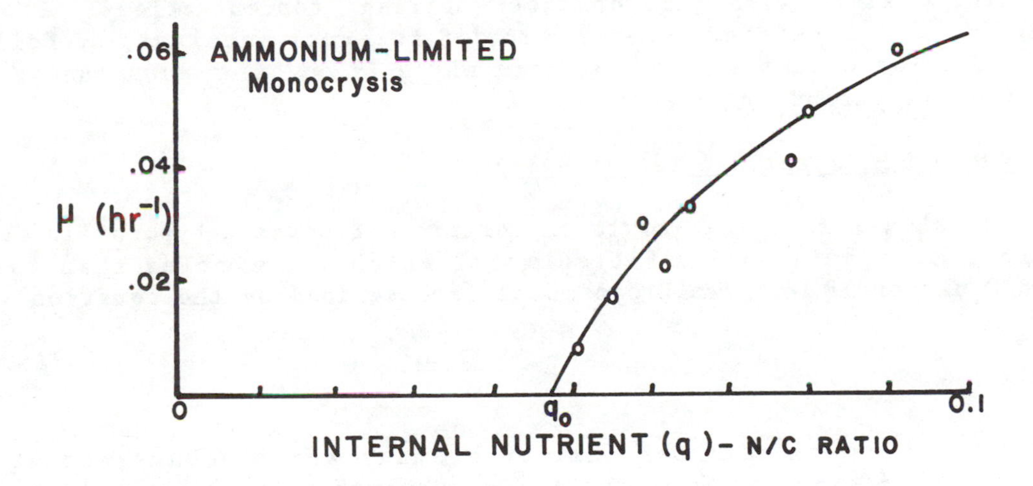

Figure 18.1 Cell growth as a function of internal nutrient concentration. From Caperon and Meyer (1972).

The simulation of batch culture growth by Euler integration involves the following update expressions:

$$N \longleftarrow N + (RqX - vX)\Delta t \qquad (18.15)$$

$$q \longleftarrow q + (v - \mu q)\Delta t \qquad (18.16)$$

$$X \longleftarrow X + (\mu X - RX)\Delta t \qquad (18.17)$$

where R is the respiration/mortality coefficient which describes the rate of nutrient return to the media.

Exercise 18.1
Write a program to simulate microbial growth in batch culture based on the two stage nutrient limitation model. Assume the following growth parameters:

$N = 0.5$ mg/liter	$K_n = 0.05$ mg/liter
$X = 0.01$ mg/liter	$V_m = 0.03$ mgN/mgX·hr
$q_o = 0.02$ mgN/mgX	$\mu = 0.1$/hr
$K_q = 0.03$ mgN/mgX	$R = 0.01$/hr

Plot nutrient concentration and biomass for 240 simulated hours of growth. Compare the resulting growth curve with that which was obtained in Exercise 13.1. Notice particularly the relationship of growth to free nutrient and internal nutrient concentration.

Exercise 18.2
Use Exercise 13.2 as a guide to develop a chemostat simulation based on the two stage nutrient limitation model. Hold the dilution rate at 0.05/hr and vary input nutrient from 0.05 to 0.5 mg/liter. Observe the time required to achieve steady state with each of these nutrient concentrations. Plot biomass, X, and steady-state concentration of internal nutrient, q, following each perturbation of the system. Explain why q is constant even though external nutrient is varied by a factor of 10.

18.2 The Droop Nutrient Limitation Model

Droop (1968) showed that the growth of certain cells was a hyperbolic function of absorbed nutrient, with a fitted model which was simpler than the form shown in Equation 18.14. The Droop model is described by the equation

$$\mu = \bar{\mu}\cdot\frac{(q - q_o)}{q} \tag{18.21}$$

where μ is the growth rate per unit of biomass which occurs for internal nutrient concentration, q. $\bar{\mu}$ is the maximum growth rate which could occur for infinite concentrations of internal nutrient, and q_o is the minimum nutrient required to initiate growth. This model is simpler than Equation 18.14 because it involves only two parameters, $\bar{\mu}$ and q_o, whereas the Monod-Caperon model requires three. DiToro (1980) points out that this simplification results only if $q_o = k_q$ in Equation 18.14. Thus Equation 18.14 simplifies to the Droop model as follows:

$$\mu = \bar{\mu}\cdot\frac{(q - q_o)}{q_o + (q - q_o)} \tag{18.22}$$

$$= \bar{\mu}\cdot\frac{(q - q_o)}{q} \tag{18.23}$$

or

$$= \bar{\mu}\cdot\left(1 - \frac{q_o}{q}\right) \tag{18.24}$$

Although q_o often appears to equal k_q as the Droop equation requires, there seems to be no theoretical justification for this. Because the model involves only two parameters, curve fitting is simplified. Equation 18.21 can be linearized as follows:

$$\mu q = \bar{\mu} q - \bar{\mu} q_o \qquad (18.25)$$

Thus, a plot of μq as a function of q yields $\bar{\mu}$ as a slope and $-\bar{\mu} q_o$ as the intercept. This procedure, called a Droop plot, provides a way to obtain the growth parameters from real data.

Exercise 18.3

The following data was obtained by Droop (1974) for Monochrysis algae when grown in a chemostat under limiting conditions for phosphate.

Internal nutrient (nM[P]/10^6 cells)	Growth rate, μ (day^{-1})
0.398	0.116
0.496	0.123
0.664	0.458
0.695	0.452
0.584	0.459
0.637	0.475
0.803	0.591
0.815	0.587
0.818	0.586
0.917	0.704
1.06	0.709
1.04	0.705
1.02	0.644

Use this data to fit the Droop model, Equation 18.21, and also the Monod-Caperon model, Equation 18.14. To fit the Droop model it will be necessary to modify the CURFIT program slightly . The linear form of the Droop model (Equation 18.25) requires that y be transformed to xy. This transformation may be accomplished by substituting the following statements into the CURFIT program.

```
 785 IF Q = 10 THEN Y = X*Y(D) : REM TRANSFORM Y
1080 IF Q = 10 THEN A = S : B = -I/A : REM A AND B FROM SLOPE AND INT.
1235 IF Q = 10 THEN PRINT " Y = A*(X-B)/X"
1555 IF Q = 10 THEN Y = A*(X-B)/X : REM CALCULATE MODEL Y
2070 IF Q = 10 THEN Y = A*(X-B)/X : REM CALCULATE Y FOR PLOT
```

As you may recognize from these changes, you would be temporarily replacing the statements required for fitting the maxima function with those required for fitting the Droop equation.

The Monod-Caperon model will require an estimate of the x intercept or x offset. This will be your best estimate of q_o. Repeat the curve fitting process with different values for q_o until the best fitting hyperbolic equation is found.

Compare the F statistics for the fitted equations to determine which model provides the best fit. Does K_q turn out to equal q_o? For additional data on microbial growth, see Droop (1974) or Caperon and Meyer (1972).

18.3 Control of Nutrient Uptake by the Internal Nutrient Pool

Rhee (1973) showed that the uptake rate of phosphate by *Scenedesmus* is controlled by the cellular concentration of labile polyphosphates. Thus, the size of an internal nutrient pool appears to cause inhibition of phosphate uptake. He proposed an equation of the form

$$v = V_m \cdot \frac{N}{(K_n + N)} \cdot \frac{K_i}{(i + K_i)} \tag{18.31}$$

where i is the concentration of inhibitor, in this case polyphosphate, and K_i is the inhibitor concentration which causes the apparent maximum uptake to be reduced to one-half.

DiToro (1980) generalized the nutrient uptake equation as follows:

$$v = V_m \cdot \frac{N}{(K_n + N)} \cdot \frac{K_i}{(K_i + q - q_o)} \tag{18.32}$$

This kind of control may account for the variation in apparent values for V_m and K_n noted by Rhee (1978) and others.

Exercise 18.4

Use the modified nutrient uptake equation, Equation 18.32, to model growth of the hypothetical organisms discussed in Exercise 18.1. Use the same growth parameters as provided there, except include control of nutrient uptake with K_i equal to 0.08. How does this affect the growth curves for batch culture and for continuous culture? How would K_i be determined experimentally?

18.4 Multiple Nutrient Limitation

In Chapter 13 we discussed the Chen model for multiple nutrient limitation. This model is also called a multiplicative model because of the relationship of the terms involving nutrient dependency. Using a multiplicative model growth limitation by nitrate and phosphate would be described as follows:

$$\mu = \bar{\mu} \cdot \frac{(q_n - q_{on})}{(K_{qn} + q_n - q_{on})} \cdot \frac{(q_p - q_{op})}{(K_{qp} + q_p - q_{op})} \tag{18.41}$$

where μ is growth rate, $\bar{\mu}$ is maximum growth rate, q_n and q_p are internal concentrations of nitrate and phosphate, q_{on} and q_{op} are minimum internal concentrations of nitrate and phosphate to permit growth, and K_n and K_p are half-saturation constants for growth.

Rhee (1978) carried out a study of *Scenedesmus* growth in a chemostat to explore the limitation of growth by two nutrients. The input phosphate concentration, P, was maintained constant and the input nitrate concentration, N, was varied so that the N:P ratio ranged from 5 to 70. Nitrate was found to be limiting below a N:P ratio of 30 and phosphate was limiting above this level. The primary objective of the study was to determine whether these two nutrients limited Scenedesmus growth in a multiplicative fashion as described by Equation 18.41, or by a single nutrient as suggested by Liebig's law of the minimum. This law states that growth is limited by the nutrient which is in its lowest concentration relative to its need.

The separate equations for single nutrient limitation are as follows:

$$\mu = \bar{\mu}\cdot\frac{(q_p - q_{op})}{(K_{qp} + q_p - q_{op})} \tag{18.42}$$

$$\mu = \bar{\mu}\cdot\frac{(q_n - q_{on})}{(K_{qn} + q_n - q_{on})} \tag{18.43}$$

To simulate nutrient limitation according to Liebig's law, the growth rate is calculated using both Equation 18.42 and 18.43, then the minimum value of μ is used, assuming it to be the growth rate determined by the limiting nutrient.

The uptake equations for nitrate and phosphate are assumed to be

$$v_n = \frac{V_n\cdot N}{(K_n + N)} \tag{18.44}$$

$$v_p = \frac{V_p\cdot P}{(K_p + P)} \tag{18.45}$$

where v_n and v_p are uptake of nitrate and phosphate, V_n and V_p are maximum uptake rates, N and P are nutrient concentrations, and K_n and K_p are half-saturation constants for nutrient uptake.

Rhee (1978), in a very convincing way, showed that Scenedesmus growth conformed very closely to the models described by the dual Equations 18.42 and 18.43, thus confirming Liebig's law.

Exercise 18.5
Use the equations discussed in Section 18.4 to compare the behavior of the multiplicative model with the alternative, dual equation model. Assume that the phosphate concentration of the media coming into the chemostat is 3μM. Allow the nitrate concentration to vary from 20μM to 200μM in steps of 20μM. Assume the chemostat has a constant dilution rate of 0.5. The other necessary growth parameters, based on Rhee (1978) are as follows:

$V_p = 4.36 \times 10^{-9}$ μ moles P/cell·day, $K_p = 0.9$ μM

$V_n = 120 \times 10^{-9}$ μ moles N/cell·day, $K_n = 18$ μM

$\bar{\mu} = 1.16$ /day, $q_{op} = 1.6 \times 10^{-9}$ μ moles P/cell

$q_{on} = 45.4 \times 10^{-9}$ μ moles N/cell

Assume that K_{qp} equals q_{op} and K_{qn} equals q_{on} so that the simpler Droop models may be used in place of Equations 18.41 to 18.43. Set the initial number of cells equal to 1 x 10^8 cells/liter. Allow the model to simulate about 50 time intervals for each nitrate concentration. After you have examined the behavior of the dual equation model, change the program so that growth is determined by the multiplicative model, and repeat the variation of nitrate. What are the major differences in the behavior of these two models?

Conclusion

One of the weaknesses of the simple Monod models discussed in Chapter 13 is that they do not account for physiological changes in cells in response to varying nutrient concentration. Thus, the multistage nutrient limitation models discussed in this chapter are designed to account for such behavior as lag phase growth, luxury consumption, and multiple limiting nutrients.

Several researchers have explored the use of more highly structured models in which growth consists of several parallel processes. Ramkrishna, et al. (1967) divides the biomass into two categories, one the protein-related biomass, and the other the nucleic acid-related biomass. The two pathways of growth are partially dependent, drawing on a common nutrient. The total growth is the sum of the two components. The differential rates of production embodied in these models can account for variations in the nuclear to cytoplasmic ratio often encountered during cell growth.

Shuler, et al. (1979) proposed a highly structured model for growth and division of bacteria. This model involved 14 differential equations for describing changes in concentrations of precursors, proteins, nucleic acids, enzymes, carbohydrates, and waste products. The model is capable of accounting for the effect of cell geometry on growth and predicting temporal events during the cycle of DNA synthesis and cell division.

It is evident that at least some researchers are attempting to combine the multienzyme system approach with nutrient uptake models to describe growth in a very structured way.

DiToro (1980) examined and compared several nutrient limitation models as they relate to phytoplankton growth in aquatic ecosystems. From a practical standpoint, he questioned whether the increased realism that is attained by multistage models is worth the effort required to determine the additional kinetic constants which these models require. This would be true particularly for the mixed populations which occur in nature.

There are several excellent reviews of multistage nutrient limitation models. These include, in addition to DiToro (1980), one by Humphrey (1979) which discusses nutrient limitation models as they relate to bioengineering and industrial fermentation. Dabes, et al. (1973) discuss the Monod model in relation to several other models which provide more realistic growth kinetics, especially as they relate to fermentation processes.

This section of the book began with a discussion of biochemical reaction kinetics, and a promise that the subject of biochemistry was related to ecology. In subsequent chapters, we have seen many instances where concepts and modeling techniques borrowed from one field were applied to another. Thus, it is fitting that this section concludes with a chapter in which the ideas of biochemistry and ecology actually merge into a single subject area, that of nutrient limitation models. It is in models of this type that the holistic nature of biology is most clearly demonstrated.

PART III

PROBABILISTIC MODELING

Up to this point, all the models discussed have been the deterministic type. However, random processes also play an important role in most biological systems. There are several reasons for this.

1) Many important biological processes are the result of discrete events which appear to occur more or less randomly through time. Some examples are birth, death, and genotype determination.

2) Many systems are strongly influenced by environmental factors which are themselves subject to considerable random variation through time and space. Although light intensity and temperature are largely deterministic, fluctuations which behave in a random fashion play an important role in determining the net effect.

3) Individual organisms respond differently to the same stimulus because of differences in genotype and previous environmental conditioning.

4) Although these factors would be averaged out by infinitely large populations living in an average environment, the finite size of real populations, living in a real environment, results in apparent random fluctuations from the normal or deterministic behavior of the system.

The type of model which includes random fluctuations, in addition to deterministic behavior, is called a stochastic or probabilistic model. Stochastic models are generally more complex than deterministic models because of the mathematics involved in describing random processes. However, in the next several chapters we will emphasize the direct approach, that of using the Monte Carlo method for simulating stochastic processes. As with the previous

James D. Spain, BASIC Microcomputer Models in Biology

ISBN 0-201-10678-7

chapters, this must be considered as only a brief introduction to an important area of modeling. Students wishing to pursue the topic further should consult Goel and Richter-Dyn (1974) and Poole (1974).

From a purely philosophical point of view, one may question whether truly random processes really exist. The boundary between what we consider random behavior and what we know to be deterministic has been steadily pushed back throughout the history of science. On close examination, each of the examples given above can be shown to be the result of deterministic causes. They appear to occur randomly because we are unable to predict their occurrence from the information available. Thus, there is a close relationship between unpredictability and what we call randomness. Randomness is a useful conceptual model for characterizing the behavior of unpredictable systems or system components. The stochastic model and the Monte Carlo technique provide an alternate modeling approach which allows us to gain new insight into certain biological systems.

Your objectives in this section should include the following:

1) learning how to use the Monte Carlo technique for modeling random processes,

2) understanding how random processes affect biological systems,

3) learning to apply the Monte Carlo technique to at least one area which interests you directly, and

4) using the stochastic model to gain new insight into certain biological systems.

CHAPTER 19

MONTE CARLO MODELING OF SIMPLE STOCHASTIC PROCESSES

This chapter is concerned with the modeling of simple biological and non-biological processes. The purpose is to introduce you to the Monte Carlo method and give you an opportunity to test it on some familiar systems which readily permit statistical analysis of the simulation data. Subsequent chapters will deal with the use of the Monte Carlo technique to model sampling, random walks, and queueing processes in biological systems.

19.1 Random Number Generators

Monte Carlo simulations depend on the generation of random (actually pseudorandom) numbers by the computer. These numbers are defined as pseudorandom because they depend upon reproducible mathematical operations performed on numbers by the computer. The numbers which result pass most of the tests for randomness. However, the very same set would be produced again if the same starting number (seed) was employed in the same manner. To circumvent this repetition, most programming languages use the time of day or some other variable to initiate the random number generator. This is the purpose of the RANDOM statement in TRS-80 BASIC. True random numbers require some innately random physical process such as radioactive decay, or noise in an electronic circuit for their production. Such a source of random information is a bit more expensive to obtain, and for the purpose of the Monte Carlo simulation employed in this text, is no more effective than the computer-generated pseudorandom numbers.

19.2 Tests of Randomness

In order to appreciate the true nature of random numbers, and also to see just how random they really are, it is quite interesting to perform some standard tests on a set of pseudorandom numbers which have been generated by the

James D. Spain, BASIC Microcomputer Models in Biology

ISBN 0-201-10678-7

computer. The following table resulted from generating and sorting over 1000 pseudorandom numbers.

Table 19.1. Frequencies of ordered pairs of digits from 1012 random numbers.

First	Second digit										
digit	0	1	2	3	4	5	6	7	8	9	Total
0	8	8	1	10	12	11	9	7	12	10	97
1	15	14	12	5	17	10	10	12	8	7	110
2	10	13	9	8	16	11	14	11	11	15	118
3	12	16	8	4	14	10	12	8	14	10	108
4	10	11	7	11	13	12	11	10	13	12	110
5	10	4	12	9	7	11	13	18	8	7	99
6	11	11	5	8	8	12	9	14	7	8	93
7	6	11	7	10	7	8	15	3	9	6	82
8	6	9	6	13	9	10	12	8	10	10	93
9	8	13	8	11	12	10	8	7	10	15	102
Total	96	110	84	89	115	105	113	98	102	100	1012

These numbers were examined by tests similar to those employed by the Rand Corporation (1955) in its classical publication on random numbers. Random numbers are typically in the form of decimal fractions. Since the most important digits from the standpoint of probability simulation are the first and second ones following the decimal, the tests reported here deal exclusively with these two digits. The frequencies of different pairs within the test series of 1012 numbers is presented in the table, where the first digit of the pair is shown in the left column of the table and the second digit is shown at the top. Totals for rows and columns are given along the right side and the bottom.

The following results were obtained from the statistical analysis of the data. The average frequency chi-square, X^2, for all 10 columns was found to be 8.10, which has about 0.53 probability level (P value) for 9 degrees of freedom (d.f.). The X^2 for the column totals was 8.93, and for 9 d.f. this corresponded to 0.45 probability level. The X^2 for row totals was 9.99, which for 9 d.f. corresponded to 0.35 probability level. The frequencies of each of the 10 digits in the 2024 examined gave a X^2 of 7.46, which for 9 d.f. gives a P value of 0.6. Odd digits compared to even digits gave a X^2 of 0.16, which for 1 d.f. gives a P value of 0.65. There were 96 pairs vs. 916 non-pairs which gave a X^2 of 0.30, corresponding to a probability level of 0.6 based on 1 d.f. The unweighted average probability for these 6 frequency tests is 0.53, which suggests that this is about as average a set of random numbers as one is likely to obtain. One final test involved comparison of the observed overall frequency distribution with that expected from the Poisson distribution. The results are given in Table 19.2.

Table 19.2. Frequency distribution.

Frequency	Expected Number* Based on Poisson Series	Observed Number for Each Frequency
1	0.0	0
2	0.2	0
3	0.7	1
4	1.8	2
5	3.6	2
6	6.0	5
7	8.7	8
8	11.0	15
9	12.4	7
10	12 5	17
11	11.5	12
12	9.7	12
13	7.6	6
14	5.5	5
15	3.7	4
16	2.3	2
17	1.4	1
18	0.8	1

*See Appendix 4 for Poisson Distribution Program.

The goodness of fit X^2 for the observed frequency compared to the frequency predicted by the Poisson equation was 7.7 which, for 15 d.f., corresponded to about 0.9 probability level. It would therefore seem that the subroutine for generation of pseudorandom numbers has provided an acceptable source of random numbers for use in various types of Monte Carlo simulations.

Exercise 19.1

Set up a program to generate 1000 random numbers and sort them as in Table 19.1. This can readily be accomplished using a counting technique which increments elements of a 10 by 10 matrix with subscripts derived from the first and second digits of the random number. It is also possible to include a subroutine which sorts the frequencies of numbers in each element of the matrix, so that these frequencies may be compared with the Poisson distribution of frequencies, as in Table 19.2. Run at least three chi-square tests on the data, in addition to the Poisson distribution test, in order to evaluate your set of random numbers. Discuss the results of your analysis.

19.3 The Basic Monte Carlo Simulation

The Monte Carlo computer simulation uses a random number to decide whether or not a particular event relating to an individual atom, molecule, organism, or gene takes place. This is called the Monte Carlo approach since a random number takes the place of coin flipping, dice throwing, or wheel spinning techniques which have been employed in non-computer simulations of this sort. The computer generated number is usually in the form of a decimal fraction so that it may easily be compared with the probability of a particular event, also typically expressed as a decimal fraction.

The simplest example is a coin flipping simulation in which the probability of heads occurring is 0.5. To use a Monte Carlo simulation to determine whether one gets 'heads' or not, one simply compares the random number with the 0.5 probability value. If the random number is greater than 0.5, the computer records this as 'heads', if it is less than 0.5, it is assumed to be 'tails'. The flow diagram for a model which determines the number of heads in a given number of tosses is presented in Figure 19.1.

Such a technique can be employed with equal ease for a number of yes-no or on-off situations such as

1) whether an organism dies or not during a given time period,
2) whether a heterozygous individual throws an 'A' gamete or an 'a' gamete,
3) whether a rat in a 'T' maze turns right or left at the next corner.

Note that the probability value which one compares with the random number may be fixed at some value, or it may change in some mathematically predictable fashion.

In other situations, the random number may be employed to decide between a number of alternative courses, or it may be employed to provide numbers along a continuum between two limits. For example, if you wished to simulate a dice game, the random number, R, may be converted into digits between 1 and 6 by the simple algorithm, D = INT(R*6 + 1). For dice, the results should represent the sum of two random operations, one for each die.

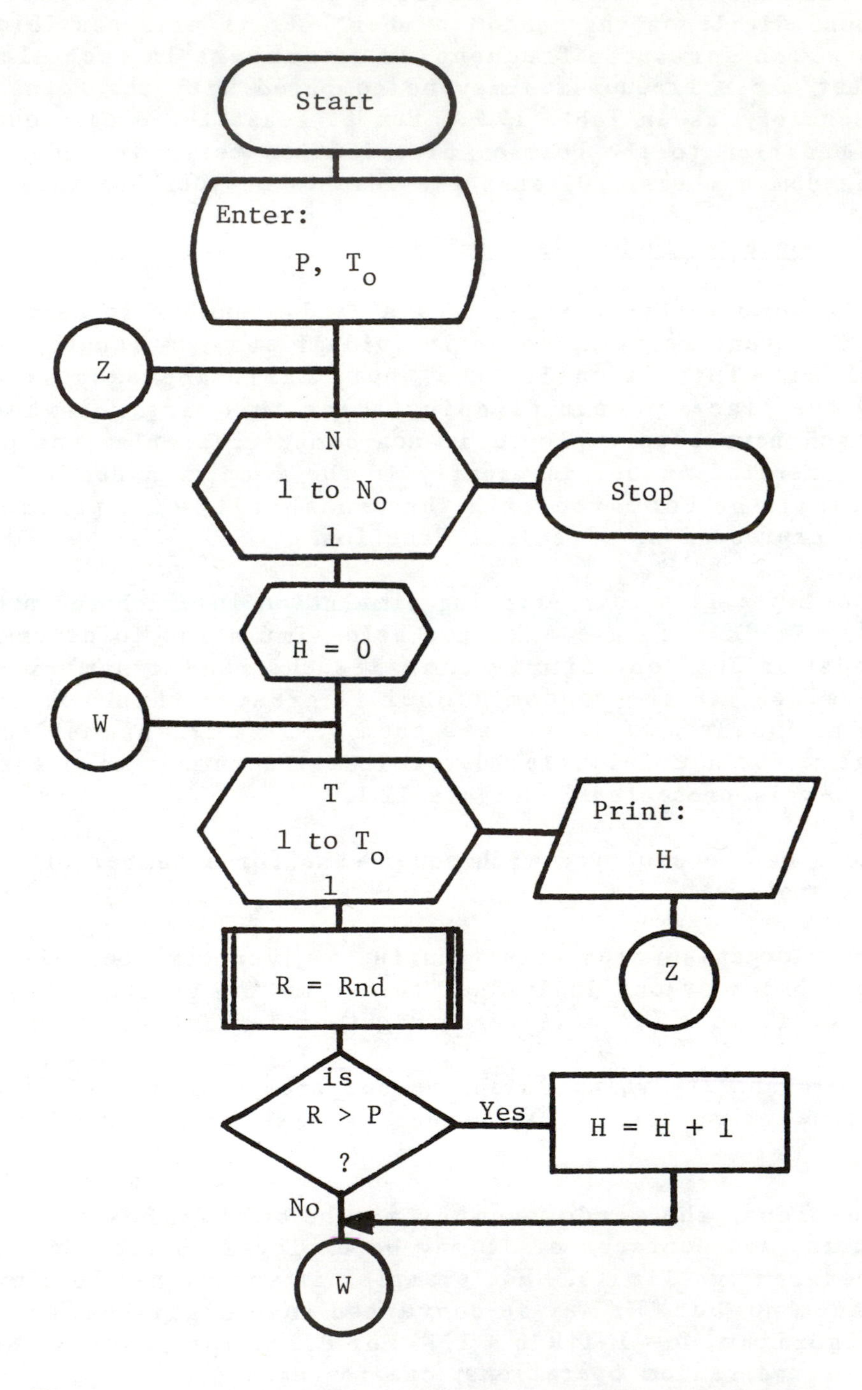

Figure 19.1. Flowchart of coin tossing simulation.
R is the computer generated number.
P is the probability of heads.
T_0 is the number of tosses per simulation.
N_0 is the number of simulations being run.

The outcome of such a model can be identical to that of a real dice game, perhaps better since these 'dice' cannot be manipulated.

Exercise 19.2
Write a program for the coin-flipping simulation described above. Set it up so that it does not print out until it has completed a pre-selected number of tosses.

Test your simulation (and the randomness of the numbers) by carrying out one of the following tests after repeating the simulation about 100 times. Use the same number of tosses each time. One test is the obvious one of running chi-square on the total number of heads and total number of tails. The expected value is of course one half heads, one half tails. The other test is to compare the frequency of occurrence of different numbers of heads with the binomial distribution for the number of tosses you employed. The binomial distribution can be obtained either from Pascal's triangle or from the binomial distribution equation. Use the chi-square test to determine how closely the simulation data fits the expected binomial distribution pattern.

Exercise 19.3
Write a program which will simulate the throwing of dice in pairs. Use this program to generate enough data so that the frequency of 2's, 3's, 4's, etc. may be compared with the expected frequency. Those who are familiar with the game of craps probably know that there is only 1 way to throw either 2 or 12, 2 ways to throw 3 or 11, 3 ways to throw 4 or 10, 4 ways to throw either 5 or 9, 5 ways to throw 6 or 8, and 6 ways to throw 7. Since there are 36 possible ways to throw the dice, the expected frequency for 2's is 1/36, for 3's, 2/36, and so on. Run a chi-square test on the observed frequency and make a statement about goodness of fit.

Exercise 19.4
Use the Monte Carlo method to simulate the movement of a particle in two dimensional space, a Brownian movement or random walk simulation. One of the simplest ways to approach this is to take a random number and subtract 0.5. to get a number between -0.5 and +0.5. Let this number equal the distance a particle moves on the x-axis. Then take a second random number and subtract 0.5 to give the position of the particle on the y-axis. Continue doing this, summing up the x's and the y's after each movement and printing out the position after each movement. Connecting the coordinate points with straight lines will provide a typical two dimensional random walk pattern. Note that this simulation lacks realism because of the unnatural distribution of the distances moved during each time interval since they are uniform between -0.5 and +0.5.

Exercise 19.5
Modify the preceding simulation to examine the positions of a number of points (particles) after several random movements from a point source (0.0). This technique provides a reasonable simulation of diffusion from a point source. The straight line distance from the point source may be obtained by the Pythagorean Theorem.

$$d = \sqrt{x^2 + y^2}$$

19.4 Stochastic Simulation of Radioactive Decay

Radioactive disintegrations proceed at a rate which cannot be modified by any known chemical or physical means. For a given element this rate at any instant is proportional only to the number of the nuclei present, N, and is described by the differential equation

$$-\frac{dN}{dt} = kN$$

The integration of this equation, yielding the classical decay equation, was discussed in Chapter 1.

Since the atoms act independently of one another during radioactive decay, one may employ a model which simulates one atom at a time. Each atom has a certain probability of decay during any one time interval. The Monte Carlo technique is used to decide whether or not an atom is going to decay during a given time interval. If a random number is greater than the probability of decay, the computer can be programmed to add one number to the time counter and repeat the loop, which will generate a new random number to test against the probability of decay. If the random number is less than the probability, the atom is considered to have decayed, and the computer is programmed to record the number of cycles or time intervals that the random number has exceeded the probability. This process simulates the life of one atom. The computer then starts over and determines the life of the next atom by Monte Carlo simulation. After the computer has generated the 'life times' of approximately 200 atoms, the frequency of each life time may be printed out in the form of a histogram to show the nature of the radioactive decay process. By converting the resulting data into an expression of the fraction of atoms remaining undecayed after each time interval, the results may be compared directly with those of the classical model (See exercise 1.2).

Exercise 19.6

Use the above information to program the stochastic simulation for radioactive decay. Implement the model using a k value obtained from one of the following half-lives ($t_{1/2}$) where:

$$k = \frac{0.69315}{t_{1/2}}$$

Element	$t_{1/2}$	k
C^{11}	20.5 min	$0.33812/10^1$ min
C^{14}	6000 yr	$0.11553/10^3$ yr
N^{13}	10.0 min	$0.6863/10^1$ min
Na^{24}	14.8 hr	$0.4683/10^1$ hr
I^{124}	4.0 days	0.1733/day

Collect simulation data for at least 200 atoms. Compare the results obtained with those of the deterministic model by plotting the fraction remaining undecayed as a function of time. How does the half-life compare with the value you started with? Note that the units of time for k have been adjusted so that you have numbers which are convenient to work with. For example, 6000 yrs is expressed as 6×10^3 yrs, and k is expressed as $0.1155/10^3$ yr.

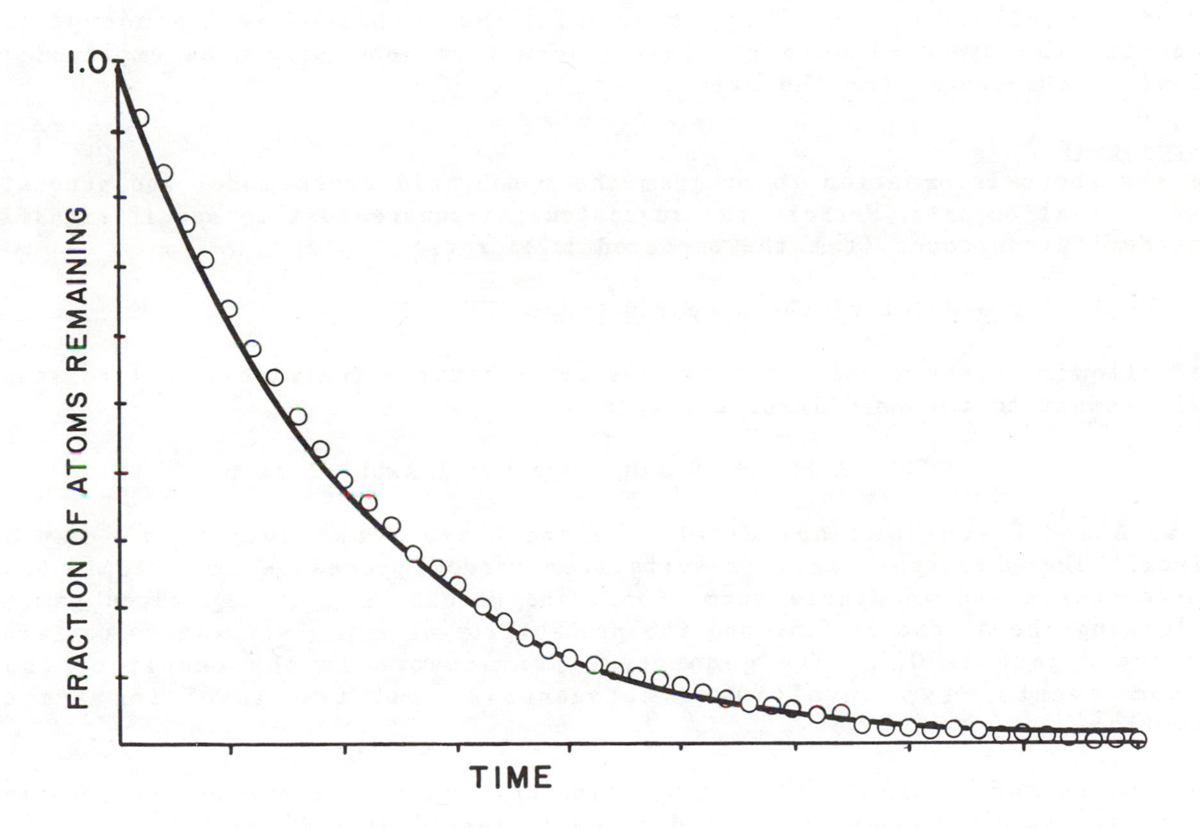

Figure 19.2. Simulation of radioactive decay. Circles represent the fraction of atoms remaining undecayed at each time period in a total of 1000 replications of the Monte Carlo simulation. The line curve is that of the classical equation for the decay process.

19.5 Monte Carlo Model of a Monohybrid Cross

The monohybrid cross may be diagrammed as follows:

$$Aa \times Aa \longrightarrow 1\ aa:\ 2\ Aa:\ 1\ AA$$

This 1:2:1 ratio for the genotypes of the offspring is subject to the random processes involved in gametogenesis and zygote formation. When a single gene locus is involved, there will be only two kinds of gametes formed, each occurring with a probability of 0.5.

This situation may be simulated by generating a uniform random number, R, and testing it against 0.5. If R is greater than 0.5 assume that the gamete from the first parent contains the allele A, if less than 0.5, assume it contains the allele a. Then examine a second random number representing the second parent. If R is greater than 0.5, again we will say that this gamete contains the A allele, and if less than 0.5, the a allele. Continue to examine pairs of random numbers in this manner, keeping track of the number in which both were greater than 0.5 (AA zygotes), neither were greater than 0.5 (aa zygotes), and one was greater and one less than 0.5 (Aa and aA zygotes). At the end of a specified number of zygotes, print the number of each genotype and calculate the expected number. This information should then be employed to calculate chi-square for the data.

Exercise 19.7
Use the above information to program the monohybrid cross model and generate some simulation data. Perform the suggested chi-square test to see if significant deviations occur from the expected 1:2:1 ratio.

19.6 Monte Carlo Model of the Dihybrid Cross

The following diagram characterizes the cross between individuals heterozygous with respect to two non-linked genes.

AaBb x AaBb ⟶ 9 AxBx: 3 aaBx: 1 Axbb: 1 aabb

where A and B are dominant alleles, a and b are recessives, and x may be either. The phenotypic ratio results from random processes involved in both gametogenesis and in zygote formation. The probability of any given gamete containing the A gene is 0.5, and the probability of any given gamete containing the B gene is 0.5. The genotype of any zygote is the result of four random events, two involving gametogenesis, and two involving zygote formation.

The process may be simulated by generating uniform random numbers and testing these against 0.5. From the flow diagram in Figure 19.3 it will be seen that the genotype aabb (phenotype ab) results when four random numbers in succession are less than 0.5. Since it is only necessary to have one A gene to express the A phenotype, it is not necessary for the random number to exceed 0.5 more than once to fall into the Ax category. The same is true for the B gene, and the Bx category.

Exercise 19.8
Use the above information to program the dihybrid cross model. This exercise should be designed to determine the expected phenotypic ratio, and to evaluate the simulation data by calculating chi-square.

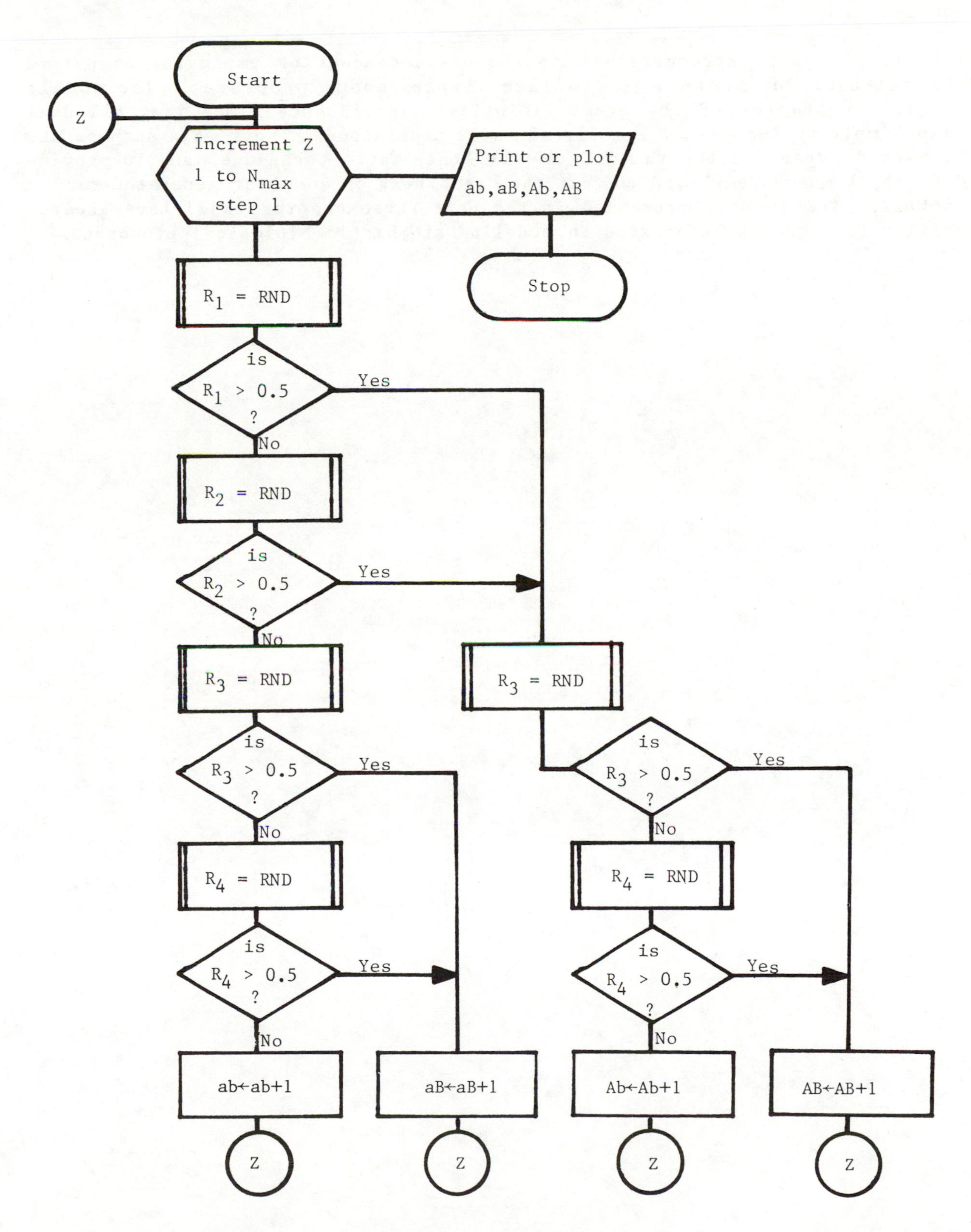

Figure 19.3. Flowchart of dihybrid cross model.

Conclusion

This chapter has introduced you to the basic concept of employing computer-generated random numbers to simulate simple random processes. The models themselves are not of any great biological significance since they all deal with simple systems which are already well understood. Their main purpose has been to demonstrate the validity of the Monte Carlo technique, and to provide data which may be employed to test the randomness of computer-generated random numbers. The models presented in the next three chapters will have greater utility to students interested in modeling stochastic biological processes.

CHAPTER 20

MODELING OF SAMPLING PROCESSES

Of the many activities carried on by a biologist, the one most subject to random fluctuations is the sampling process. Samples are taken in order to draw conclusions about an entire population of potential measurements of some variable of interest. Although it may be theoretically possible to sample the entire population, this is usually not practical, and in some cases, where the population is imaginary it would not even be possible. Therefore, one or more samples are taken so that estimates of population parameters such as the mean, the median, the range or the standard deviation can be made. An estimate of a population parameter is called a statistic. Although statistics will vary between samples taken from a given population, the population parameters are constant. Since one seldom has data for entire populations, it is usually impossible to compare sample statistics with population parameters in order to obtain information about their validity. Such a comparison may be readily accomplished, however, by modeling the sampling process using Monte Carlo methods. Much of what we know about various probability distributions such as the chi-square distribution or the normal distribution has resulted from long tedious hours of investigating the nature of random numbers, largely without the aid of the high speed computer.

Biological variables may exist in two distinct forms. A continuous variable is one which has an infinite number of values within its range. Examples of continuous variables include weights, lengths, volumes, rates, concentrations of chemical constituents and durations of time. On the other hand, some variables may be expressed only as integer values. Such variables are termed discrete variables. Examples include the number of organisms in a population, the number of births and deaths, and the number of leaves on a branch. In the previous chapter, most of the models were concerned with discrete events like tossing of coins, decay of atoms, and interaction of gametes. The random walk model (Exercises 19.4 and 19.5) was the only example which simulated a continuous variable, i.e. the distance moved in one time interval.

James D. Spain, BASIC Microcomputer Models in Biology

ISBN 0-201-10678-7

20.1 Sampling a Community of Organisms

An ecologist is often called upon to make an estimate of the number of organisms of each type in a given community, or to obtain an estimate of the population density of certain organisms in a particular ecosystem. Data may be employed to prepare tables showing percent relative composition, or to calculate one of the indices of diversity. In any case, one of the questions which arises is, how large a sample must be taken in order to have a valid measure of the community composition? One of the problems with real communities is that variability between samples may be the result of a non-uniform environment. Every environment is composed of a mixture of microhabitats which might not be evident to the sampler. Another problem is that communities are not unlimited in size, so that it is possible to sample a real community out of existence. Simulation, in this case, provides a means of evaluating the effect of sample size on estimates of population parameters for a model community which is homogeneous in composition, and unlimited in size. The following describes a simple discrete model of the sampling of a community.

Assume we have a community with a known fraction of organisms of five different types. For this simple example, we will assume the fraction of each of the five types is the same, namely 0.2. The fraction of a given organism in the community is also the probability that any organism drawn at random from the community is a representative of that type. Therefore, we can simulate the drawing of a sample from the community by examining a series of random numbers, equal to the sample size and comparing each random number with the cumulative probabilities of each type. For example, if a uniformly random number, R_u, is less than 0.2, then it will be assumed to represent an organism of type A. If R_u is between 0.2 and 0.4 then we shall assume that it represents an organism of type B. If R_u is between 0.4 and 0.6 then it represents an organism of type C, and so forth. In this way each of the random numbers can be made to represent one of the five types of organisms. The computer can be programmed to keep track of the number of organisms of each type obtained in the simulated sample. When the sampling process is completed, various types of derived data and sample statistics may be calculated and compared with the known population parameters.

Exercise 20.1

Use the techniques described above to model the sampling of a hypothetical community composed of 8 species present in the following proportions:

Species A	0.495	Species E	0.032
B	0.255	F	0.016
C	0.124	G	0.009
D	0.061	H	0.008

Simulate the collection of at least 3 samples each composed of 200 organisms. Calculate the variance of the estimated frequency for each organism. How well does the sample estimate predict the true population frequency? To what extent is the estimate improved by pooling the three samples?

20.2 The Effect of Sample Size on Diversity Index

A parameter often used to compare one community with another is the diversity index. Most diversity indices provide a single parameter to express both the number of species and the relative abundance of each species. The Shannon-Weaver index, one commonly employed to express community diversity, is based on the equation

$$H = -\sum_{i=1}^{s} p_i \cdot \ln(p_i) \qquad (20.20)$$

where H is the estimate of community diversity, p_i is the proportion of the total number of individuals making up the ith species, and s is the total number of species. Although the Shannon-Weaver diversity index is not directly dependent upon sample size, its accuracy is improved by increasing sample size. See Poole (1974) or Pielou (1977) for further discussion about the theory and application of diversity indices.

A question that often arises when calculating diversity is: How large a sample must one take to obtain an accurate measure of diversity? This question can easily be answered by simulating the sampling process on a previously measured community of organisms.

Table 20.1. Dominant Phytoplankton taxa taken from an offshore station in Lake Michigan. From Schelske, et al. (1971).

No.	Taxa	Cells	Frequency
1	*Cyclotella stelligera*	622	.5644
2	*Dinobryon divergens*	125	.1134
3	*Tabellaria fenestrata*	86	.0780
4	*Fragilaria crotonensis*	81	.0735
5	*Cyclotella michiganiana*	75	.0681
6	Unidentified flagellates	66	.0599
7	*Cyclotella ocellata*	13	.0118
8	*Rhizosolenia eriensis*	12	.0109
9	Unidentified coccoid green	11	.0100
10	*Rhisosolenia gracilis*	2	.0018
11	*Anabaena flos-aquae*	9	.0082
	Total of above taxa	1102	1.0000

Exercise 20.2

Use the sampling simulation to determine the effect of sample size on the accuracy of the Shannon-Weaver diversity index. To make this evaluation, you should write a program which will allow you to vary sample size, and simulate the taking of samples from a community with a known composition. Use the results of the sampling simulation to calculate the diversity index according to Equation 20.20. Compare the sample diversity values with the diversity calculated for the population as a whole. The community of aquatic organisms described in Table 20.1 above will serve as the sample reservoir.

After you have completed the simulation draw some conclusions about the effect of sample size on the diversity index. How does the error in diversity index vary as a function of sample size?

20.3 Mark and Recapture Simulation

The mark and recapture method of estimating populations of organisms involves taking a moderately large sample, marking the organisms in some manner, and returning them to the original group. At some subsequent time, additional samples are withdrawn and inspected to determine the fraction of marked individuals. The Peterson equation for estimating the population size is

$$p = \frac{S \cdot M}{R} \qquad (20.31)$$

where p is the population estimate, S is the number sampled, R is the number of originally marked organisms which were recaptured, and M is the total number of marked organisms in the population.

This process may be simulated by using the Monte Carlo technique. The total population size, the number of originally marked organisms, and the sample size are necessary inputs. The computer is programmed to set up a ratio between the marked organisms and the total population. It is then programmed to simulate the sampling process by generating a series of random numbers equal to the sample size. Each random number is checked against the ratio between marked and total population. The computer keeps track of the number of times the random number is greater than that ratio, and records these as unmarked organisms. Those random numbers which are less than the ratio are recorded as marked organisms. The ratio is continually modified to reflect the changes in marked and unmarked organisms in the population as a whole. After each sampling the unmarked organisms are assumed to be returned to the population as marked organisms. At the end of the simulated sampling process, the computer can be programmed to print out the number of marked and unmarked organisms in the sample, the total number of marked organisms in the population, and the population estimate. In addition to the Peterson estimate, two other estimates are useful, particularly if the computer is to simulate a series of samples on the same population.

The two estimates are the Schnabel estimate

$$p = \frac{\Sigma S \cdot M}{\Sigma R} \qquad (20.32)$$

and the Schumacher-Eschmeyer estimate

$$p = \frac{\Sigma M^2 \cdot S}{\Sigma M \cdot R} \qquad (20.33)$$

By simulating repeated samplings it is possible to observe the effect on the estimates which one obtains, compared to the assumed population size. The flowchart for the mark and recapture simulation is shown in Figure 20.1.

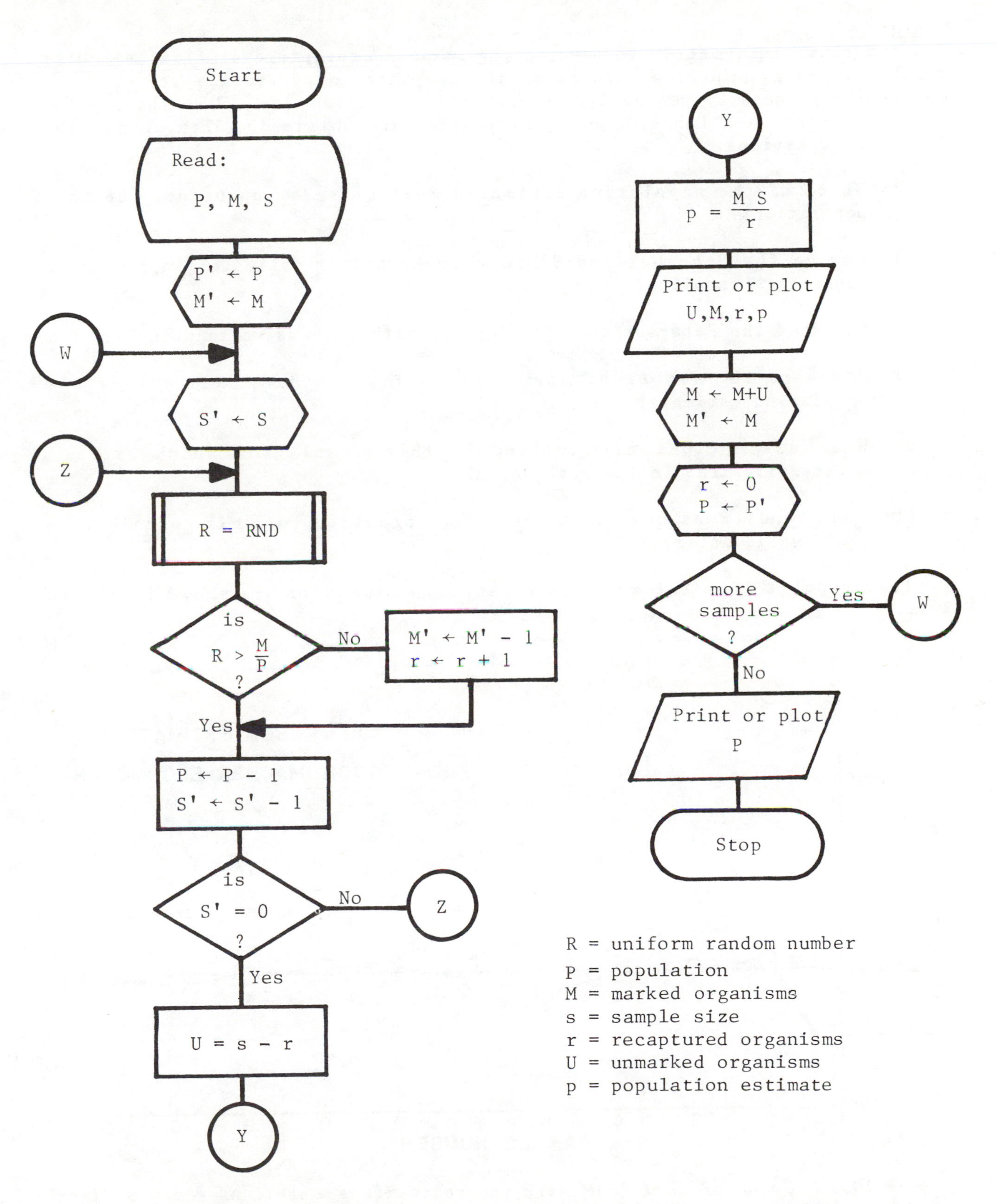

Figure 20.1. Flowchart of the mark and recapture simulation.

<u>Exercise 20.3</u>
Use the above information to develop the mark and recapture simulation. Make the necessary assumptions, and allow the computer to generate a set of data. Repeat the process by requesting several additional samples. Plot the population estimates as a function of total sample size obtained. Then answer the following questions:

a) Which of the population estimates most closely approaches the true populations?

b) Why do the Schnabel and Schumacher-Eschmeyer estimates improve with successive sampling?

c) Why does the Peterson estimate improve with successive sampling?

d) How does the number of marked organisms in the sample affect the population estimate?

e) What assumptions are implied in this simulation which are not necessarily true in the real world?

f) What conditions must exist in actual practice for this technique to yield valid estimates?

Typical results from the mark and recapture simulation are shown in Figure 20.2.

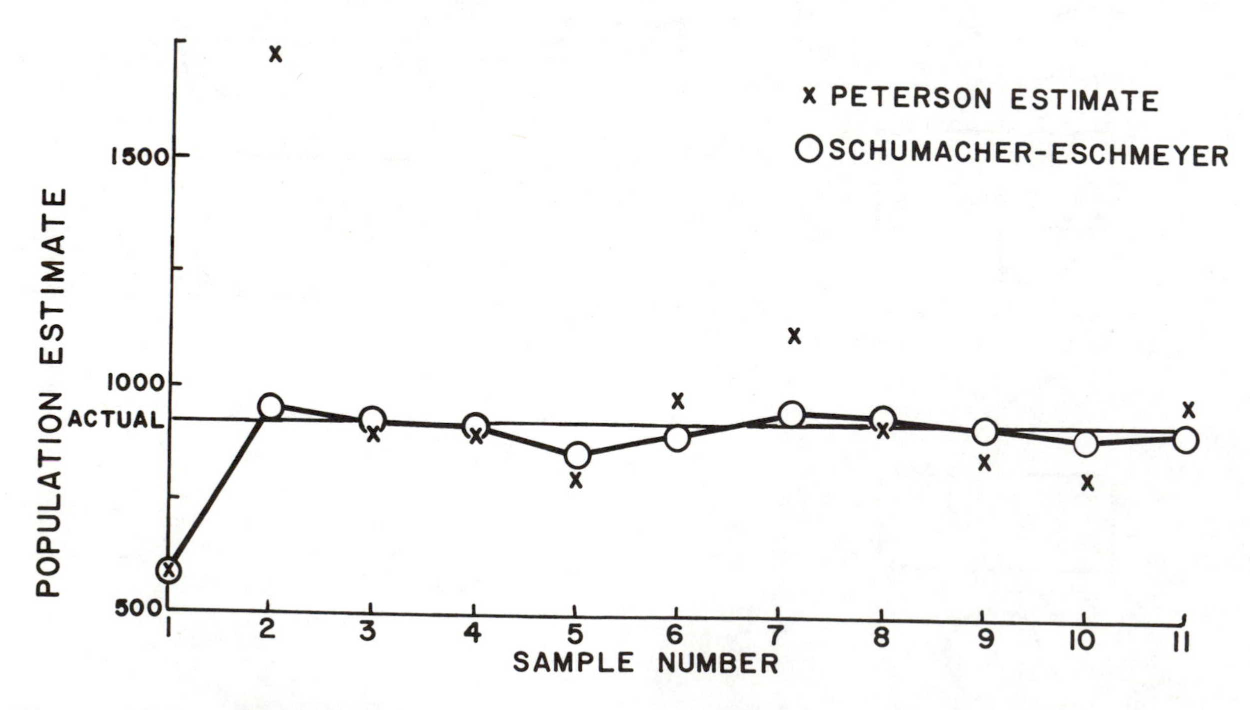

Figure 20.2. Graph of data from mark and recapture simulation. Adapted from Spain (1973).

20.4 Spatial Distribution of Organisms

The distribution of organisms in two dimensional space may be simulated by generating uniform random numbers to define the coordinates of each organism.

Locating organisms in this way assumes that each is totally independent of the other, neither attracted nor repelled by any other. It is useful to count the number of organisms at each location and store this information in a matrix as well as displaying the location of each organism graphically This information will make it possible to evaluate the results statistically.

One way to analyze the results of such a simulation is to find out how many matrix locations end up with no organisms, how many have only one organism, how many have two, and so on. This process results in a frequency distribution which may be compared with the Poisson distribution. The Poisson distribution is obtained when articles are randomly scattered in either space or time, each acting totally independently of the other. The equations that follow may be used to predict the Poisson distribution

$$P_0 = 1/e^M \qquad (20.41)$$

where P_0 is the probability that a unit of space (or a matrix location) will contain no articles when the mean number of articles per unit space is M.

The probability of finding one article in a unit space is

$$P_1 = P_0 \cdot M \qquad (20.42)$$

the probability of finding 2 articles in a unit space is

$$P_2 = \frac{P_1 \cdot M}{2} \qquad (20.43)$$

and the probability of finding x articles in a unit space is

$$P_x = \frac{P_{x-1} \cdot M}{x} \qquad (20.44)$$

In more formal terms, the Poisson equation is the following:

$$P_x = \frac{M^x}{e^M \cdot x!} \qquad (20.45)$$

Because it is necessary to determine the factorial of x, the process is usually calculated in a recursive manner by using Equations 20.41 - 20.44. To express the probabilities in terms of the number of unit spaces showing a given frequency, each of the above equations would be multiplied by the total number of unit spaces (or elements of the matrix), N.

<u>Exercise 20.4</u>
Write a computer program to simulate the random placement of 1000 organisms in space. Use the graphics screen to observe the locations of the simulated

organisms. As the organisms are being plotted on the video screen, have the computer convert the values of x and y to integers which may be summed and stored in a 10 x 10 matrix. After all simulated organisms are located on the screen and stored in the matrix, program the computer to calculate the number of spaces having each possible number of organisms from 0 to 25. These frequencies should be summed in a second array. Subsequently, this frequency distribution should be compared with the expected frequency distribution obtained from the Poisson equation. A chi-square test may be used to compare the observed distribution with the expected one. See Zar (1974) for additional information on this analysis.

The real challenge of this exercise is to build the simulation and the statistical analyses all into the same program. One of the rules of good programming is to do everything in the computer. The Poisson distribution program provided in Appendix 5 may be used to assist you in writing this program.

20.5 Sampling from the Poisson Distribution

The Poisson distribution describes the distribution of individual random occurrences in either space or time. The preceding section showed one way to simulate the Poisson distribution. Another technique which may be used to convert uniform random numbers to numbers having a Poisson frequency distribution was developed by Wiebe (1971). This technique uses the Poisson equations to calculate and sum the probabilities for each frequency until the sum is greater than a uniform random number previously generated for this purpose. The program is best described by the flowchart presented in Figure 20.3.

Still another technique involves what is called a table-lookup method. This is a much faster technique because it minimizes the number of equations involved. The trick is to generate an array of numbers using the Poisson equations. The computer is then programmed to generate a random number which is used to enter the array and access the numbers which fall on the Poisson distribution. Typically, the array is set up with 1000 compartments. The Poisson equations define how many compartments contain zeros, ones, twos, threes, etc. By tricky use of FOR-NEXT loops, and by rounding frequencies to give integer values for numbers of compartments, the table look up method can be programmed with relatively few steps.

Exercise 20.5

Use one of the two techniques described above to model the time required for the germination of seeds. Assume 100 seeds are planted and that all germinated. The mean germination time was 7 days. Graph the resulting germination times on a histogram.

20.6 Generating Normally Distributed Random Numbers Using the Table-lookup Method

The normal or Gaussian distribution is a continuous, symmetrical distribution, which may be fully described by a mean and a standard deviation. This is the sampling distribution which is used to simulate measured parameters such as height, weight, and concentration of chemical components.

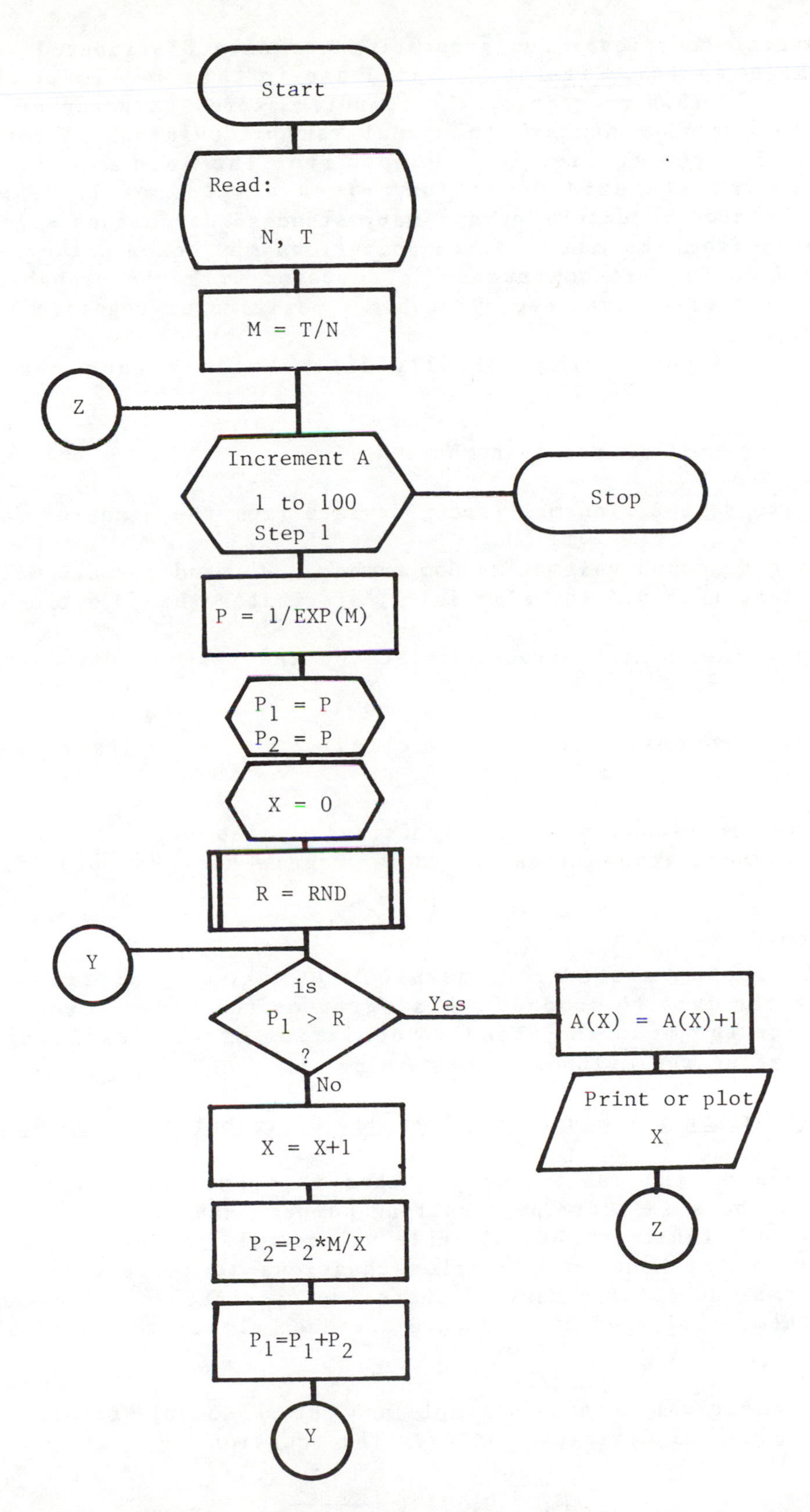

Figure 20.3. Flowchart for sampling from the Poisson distribution.

There are several techniques for generating normally distributed random numbers from the computer. The one we will use in this section is the table-lookup method. In this procedure, the computer stores an array of data which converts uniform random numbers to normal random deviates. A normal random deviate describes the distance of a number from the mean in units which are normalized to the standard deviation, given a value of 1. These random deviates may be used to determine how many standard deviations a given random number deviates from the mean. Since deviations may be on either side of the mean, a second uniform random number is compared with the probability of 0.5 to determine whether a given deviation has a positive or negative sign.

The procedure for generating normally distributed random numbers is as follows:

1) Generate a uniform random number, R_u.

2) Use R_u to select a normal random deviate from the array of data.

3) Generate a second uniform random number, R_u', and compare with 0.5. If R_u' is less than 0.5 the sign is −1, if greater than 0.5 the sign is +1.

4) Multiply the normal random deviate by the assumed standard deviation and the sign.

5) Add this product to the mean to obtain a normally distributed random number.

Because of complications involved in setting up the array to provide better accuracy for higher random deviates, this program has been provided for you. See Figure 20.4.

Exercise 20.6

Use the table-lookup method to generate 1000 normally distributed random numbers. Use the data to produce a histogram on the video screen. In addition, calculate the mean and standard deviation of the resulting data as a means of validating the method.

20.7 Empirical Method for Generating Normally Distributed Random Numbers

One disadvantage of the table-lookup method for generating normally distributed random numbers is that the resulting numbers are not continuous, having been produced by a finite number of table values. To get around this problem, there are several methods which employ equations to convert uniform random numbers to normal deviates. Many of these employ rather complicated and time consuming mathematical procedures, hence are much slower than the table-lookup method.

The following empirical method is simple and yet reasonably accurate over most of the range of normal deviates. It uses the equation

$$N_d = 0.603 \cdot \ln\left(\frac{R_u}{1 - R_u}\right) \qquad (20.71)$$

where N_d is a normal random deviate in the approximate range −3 to +3, and R_u is a uniform random number between 0 and 1. The normal deviate is transformed to a normally distributed random number by the following equation:

$$R_n = N_d \cdot s + M \qquad (20.72)$$

where R_n is the normally distributed random number, s is the standard deviation, and M is the mean.

```
100  REM   NORMAL DISTRIBUTION USING THE TABLE-LOOKUP METHOD
105  DIM R(102),N(102)
106  DIM A(130)
110 SD = 10
120 M = 60
130  FOR I = 1 TO 100
140  READ R(I)
150  NEXT I
160  FOR I = 0 TO 99
170  READ N(I)
180  NEXT I
190  FOR N = 1 TO 100
200 U =  INT ( RND (1) * 500 + 1)
205 S = 1
210  IF  RND (1) > .5 THEN S =  - 1
220  IF U < 400 THEN NR = S * SD * R(U / 5) + M: GOTO 250
230 NR = S * SD * N(U - 400) + M
250 SU = SU + NR
260 S2 = S2 + NR © 2
280  PRINT N,NR
300  NEXT N
310 MEAN = SU / 100
320 ST =  SQR ((S2 - SU © 2 / 100) / 99)
330  PRINT MEAN,ST
350  END
500  DATA  .01,.03,.04,.05,.06,.08,.09,.1,.11,.13
510  DATA  .14,.15,.16,.18,.19,.2,.21,.23,.24,.26
520  DATA  .27,.28,.29,.31,.32,.34,.35,.36,.37,.39
530  DATA  .40,.42,.43,.44,.45,.47,.48,.50,.51,.53
540  DATA  .54,.56,.57,.59,.60,.62,.63,.65,.66,.68
550  DATA  .69,.71,.72,.74,.76,.78,.79,.81,.83,.85
560  DATA  .86,.88,.90,.92,.94,.96,.98,1.0,1.02,1.04
570  DATA 1.06,1.08,1.10,1.13,1.15,1.18,1.20,1.23,1.26,1.29
580  DATA  1.31,1.34,1.37,1.41,1.44,1.48,1.52,1.57,1.60,1.65
590  DATA 1.70,1.75,1.81,1.88,1.96,2.06,2.17,2.33,2.58,3
600  DATA 1.29,1.295,1.30,1.305,1.31,1.315,1.32,1.325,1.33,1.335
610  DATA 1.34,1.345,1.35,1.355,1.36,1.367,1.373,1.38,1.39,1.40,1.41,1.416
     ,1.422,1.428,1.434
620  DATA 1.44,1.448,1.456,1.464,1.472,1.48,1.488,1.496,1.504,1.512
630  DATA  1.52,1.528,1.536,1.544,1.552,1.56,1.568,1.576,1.585,1.592
640  DATA  1.6,1.61,1.62,1.63,1.64,1.65,1.66,1.67,1.68,1.69
650  DATA  1.7,1.71,1.72,1.73,1.74,1.75,1.76,1.77,1.78,1.8
660  DATA  1.81,1.82,1.84,1.85,1.87,1.88,1.9,1.91,1.93,1.95
670  DATA  1.96,1.98,2,2.02,2.04,2.06,2.08,2.1,2.13,2.15
680  DATA   2.17,2.2,2.23,2.26,2.29,2.33,2.38,2.43,2.48,2.53
690  DATA  2.58,2.65,2.75,2.88,3.08
```

Figure 20.4. A program for generating normally distributed random numbers by the table-lookup method.

<u>Exercise 20.7</u>

Use the empirical equation described above to generate about 20 sets of data, each containing 100 normally distributed random numbers, having a mean of 50 and a standard deviation of 10. Program the computer to calculate the mean, the standard deviation, and the standard error of each data set. Evaluate this method by counting how many numbers in a given sample fall within one standard deviation. Find out how many of the means fall within one standard error. Look up one of the standard techniques for assessing deviations from normality in a standard statistics book, such as Zar (1974), and use it to evaluate the data.

20.8 Sampling from the Exponential Distribution

The exponential distribution is a continuous distribution in which the origin is at the start of the frequency curve. This is the frequency distribution which resulted in the exponential decay curve discussed in 19.3. Random distributions of this type characterize processes which have a certain probability of failure. For example, it would be used to describe the lifetimes of organisms having a constant probability of death, i.e. type II survival shown in Figure 8.2.

One way to generate this distribution is to assume a probability of death, failure or decay, and then generate a series of uniform random numbers, checking each number against the probability until one is produced which is less than the probability. The number of random numbers generated is then equal to the lifetime being simulated. This was the technique employed in Section 19.3.

The following equation may be used to convert a single uniform random number into an exponentially distributed random number:

$$R_e = -M \cdot \ln(R_u) \qquad (20.81)$$

where R_e is the exponentially distributed random number, M is the mean, and R_u is a uniform random number.

Exercise 20.8

Assume a certain species of carp exhibited a type II survival pattern and that the mean survival in a particular Japanese temple pond was 14.3 years. Use Equation 20.80 to simulate the lifetimes of 200 carp placed in this pond in 1900. According to this simulation, would any of them still be alive?

Conclusion

This chapter has presented a number of Monte Carlo models which describe different aspects of the sampling process. The techniques for generating specific sampling distributions will be used in subsequent chapters.

For additional information of the techniques discussed in this chapter, refer to Martin (1968), Zar (1974), and Poole (1974).

CHAPTER 21

RANDOM WALKS AND MARKOV CHAINS

There are many processes in biology which involve discrete changes in some state variable occurring as a result of random fluctuations. Processes of this nature may be viewed as random walks in state space. One of the fundamental characteristics of processes of this type is that although the present state is arrived at randomly, it is determined by the immediate past state, and the probabilities involved. This condition results in part from the limited state changes which can occur during any one time interval. Such processes are called Markov chains after the nineteenth century Russian mathematician who first described them. A Markov chain is a stochastic process involving a finite number of states, in which the probability of being in a particular state at time t+1 depends only on the state occupied at time t, and not on the state occupied during any previous time interval. Each of the biological processes discussed in this chapter shows the properties of a Markov chain. For additional information on the theory of Markov processes, refer to Olinick (1978), Feller (1968), or other books on probability theory.

21.1 The Classical Random Walk Model**

The name random walk comes from a classical problem involving an inebriated person who had an equal probability of staggering either one step to the left or one step to the right. Once in his new position, he continued to move one step at a time either to the right or to the left. Each new position was arrived at with a probability of 0.5 from the previous one.

The same sort of result is obtained if we keep track of the total number of heads and tails that occur as we flip a series of coins. The typical random walk pattern is obtained when we plot the difference between the number of heads and tails as a function of time. Sometimes there are more heads than tails, sometimes there are more tails than heads. In a typical run, the lead

James D. Spain, BASIC Microcomputer Models in Biology ISBN 0-201-10678-7

may change several times. Many people are surprised to see that the difference is often found to be quite far removed from the zero or equalization point.

One interesting experiment which can be run using a coin toss simulation involves counting the number of times the difference equals zero during a run of one thousand simulated coin tosses. Another interesting statistic to obtain from a coin toss simulation is the toss number of the last crossover or equalization point before the end of the run. This number can be expressed as the ratio of the toss number for the last crossover, k, to the total number of tosses, n. It can be shown that the last crossover is just as likely to come in the first half of the run as in the last half. Thus, the average value of k/n for many runs would be close to 0.5. When deviations cause this system to move away from the zero point, it is just as likely to deviate further from the mean as to return to it. This fact is obvious when you consider that at any time there is only a 50 percent chance of moving towards the average.

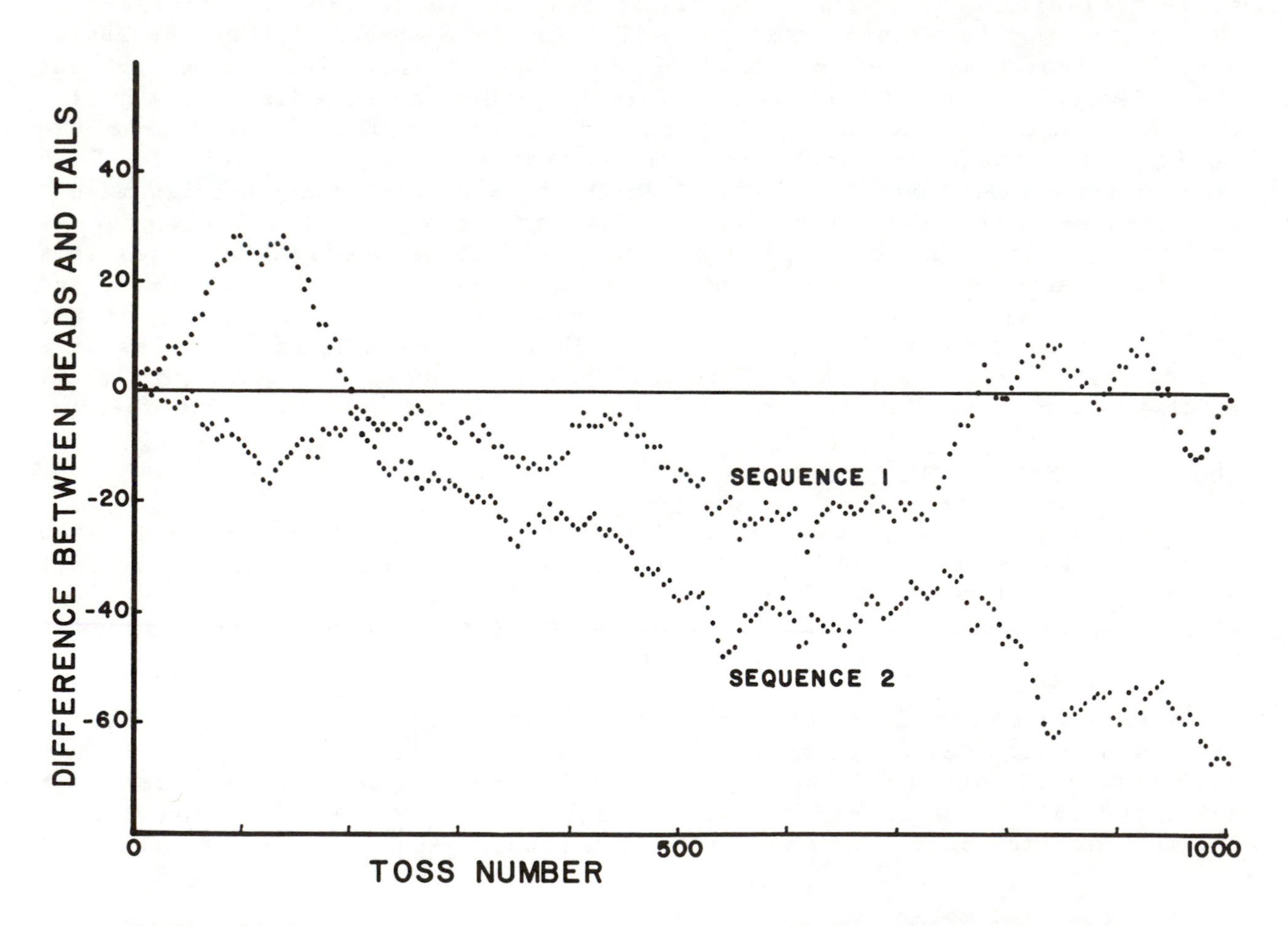

Figure 21.1. An example of the coin toss random walk simulation.

The use of coins in this example may seem rather unrelated to biology, but the principle illustrated here has broad applications to population growth, diffusion, movement of organisms in space, changes in gene frequencies, and random processes in general. For further discussion of the classical random walk simulation, refer to Feller (1968).

Exercise 21.1**
Write a program to implement the classical coin toss model described above. Set it up to simulate 1000 coin tosses per run. Plot representative data for each run relative to a zero line as shown in Figure 21.1. Have the computer keep track of the total number of crossovers, and the toss number of the last crossover. After the simulation is running successfully, carry out 10 successive runs, and calculate the average value of k/n.

21.2 Population Growth, the Direct Approach

In this example, we will use the direct approach to examine the effect of random birth and death processes on population size.

Assume a time interval sufficiently short that both the birth rate and death rate per time interval are less than one. under these conditions, the birth rate may also be considered the probability that a given individual will give birth to a new individual during that time interval. In the same manner, the death rate may be considered the probability that an individual will die during that interval. The two probabilities are assumed to be independent of each other, and independent of age and sex. Then, the probability that any individual will cause the population to increase by one, P_i, is defined by the equation

$$P_i = b(1 - d) \qquad (21.21)$$

where b is the birth rate and d is the death rate. This also may be viewed as the probability that an individual will cause a birth, but not a counterbalancing death.

The probability that a given individual will cause the population to decrease by one, P_d, is defined by the equation

$$P_d = d(1 - b) \qquad (21.22)$$

This may also be considered as the probability of causing a death without a counterbalancing birth.

The probability that an individual will cause no change in the population is

$$P_0 = 1 - P_i - P_d \qquad (21.23)$$

or

$$P_0 = db + (1 - b)(1 - d) \qquad (21.24)$$

Notice from Equation 21.24 that this can occur either by a simultaneous birth and death or no birth and no death.

To simulate changes in the population, a number of uniform random numbers equal to the population size are generated and compared with the probabilities of increase and decrease. The sum of the increases and decreases during one time interval is added to the previous population size to give the new population size. The result is a simple Monte Carlo simulation of population growth.

Exercise 21.2**
Write a program for the simple stochastic simulation of population growth described in Section 21.2. Plot the resulting population data as a function of time. Assume that the birth rate, b, is 0.30 and the death rate, d, is 0.27. Allow the model to run for several time intervals in order to evaluate its behavior. If the population drops to zero, and becomes extinct, start a new population. It is also useful to plot the log of population size as a function of time. Allow the simulation to run for about 150 time intervals. The data points should approximate a line having a slope of b - d.

After you have examined the nature of a single population growth curve, modify the program so that you may repeat the simulation several times and obtain a measure of the variance at different time intervals during the course of the simulation. This analysis will entail calculating the sum of the values of N and the sum of the squares of N at specific times for several runs of the simulation. Suggested times are 5, 10, 20, 40, and 80 time intervals. Program the computer to calculate the variance and standard deviation for each time interval after at least ten simulated growth curves. Determine how the variance changes as a function of time.

Use the following equation from Poole (1974) to calculate the predicted value for the variance at each of the above time intervals:

$$\mathrm{Var}(N_t) = \frac{(b+d)}{(b-d)}(e^{(b-d)t} - 1)(e^{(b-d)t})\cdot N_t \qquad (21.25)$$

How closely do the simulated variances compare with the variances which are predicted by Equation 21.25?

Exercise 21.3
Modify the stochastic model for unlimited growth given in Section 21.2, by including a logistic term of the form (2 - N/K) to modify the probability of birth, or birth rate, b. This will cause b to change as a function of population size in relation to the carrying capacity, K. For a discussion of this kind of logistic term, see Exercise 8.2. Implement the model with steady state probabilities for birth and death, b = 0.08 and d = 0.08, and let K = 100. Run the simulation at least 10 times and determine the variance at the same times as in Exercise 21.2. How does the carrying capacity affect the variance?

21.3 An Unlimited Growth Model Based on Variance Estimates**

Another approach employed in the development of stochastic models of population growth uses calculated values for variance to establish the size of random fluctuations. The technique as described by Poole (1974) makes use of variance calculations for pure birth and death processes which are discussed in detail by Goel and Richter-Dyn (1974).

The variance in the estimate for population size when going from N_t to N_{t+1} is described by the following equation from Poole (1974):

$$Var(N_{t+1}) = \frac{(b+d)}{(b-d)}(e^{b-d} - 1)(e^{b-d}) \cdot N_t \qquad (21.31)$$

where b is the birth rate per time interval, and d is the death rate.

The standard deviation for the estimate of N_{t+1} is

$$S_n = \sqrt{Var(N_{t+1})} \qquad (21.32)$$

The estimate of N_{t+1} is obtained deterministically by the process of Euler integration using the equations described in Chapter 7.

Poole (1974) suggests the following technique for developing a Monte Carlo simulation of population growth based on variance calculations.

1) Calculate the expected value for N_{t+1} using Euler integration and the value of N_t.

2) Calculate the variance of the estimate using Equation 21.31.

3) Calculate the standard deviation using Equation 21.32.

4) Generate a uniform random number and use it in Equation 20.71 to produce a normal random deviate.

5) Use the normal random deviate, the standard deviation, and the estimate of N_{t+1} to obtain a possible value for N_{t+1}.

Note that all the terms which precede N_t in Equation 21.31 are constants, so they may be combined into a single constant which we call V_n. Thus, during the simulation, the standard deviation may be calculated by the equation

$$S_n = \sqrt{V_n \cdot N_t} \qquad (21.33)$$

The normal deviate, N_d, obtained from Equation 20.71 may be multiplied times the standard deviation and employed directly in the population update equation as follows:

$$N_{t+1} = N_t + \Delta N_t + N_d \cdot S_n \qquad (21.34)$$

Thus, the term $N_d \cdot S_n$ represents the error in the estimate of N_{t+1}.

Exercise 21.4**
Use the variance calculation technique to develop a stochastic simulation of population growth. Assume that the birth rate is 0.30 per time interval and the death rate is 0.27. This is equivalent to an intrinsic growth rate of 0.03. Run the simulation for about 100 time intervals and compare the results with the direct approach described in Section 21.2.

21.4 Stochastic Simulation of an Epidemic

A discrete time model of an epidemic was developed by Bartlett (1956). The model described below is based on a simplified version described by Poole (1974). The deterministic model of an epidemic which provides much of the basis for the present discussion is described in Section 8.6 of this text.

The rate of formation of new contagious individuals in a population, ΔC, is described by the following deterministic equation:

$$\Delta C = kNC\Delta t \quad (21.41)$$

where k is the infectivity constant, N is the density of non-immune individuals, and C is the density of contagious individuals in the population. The same expression may be used to describe the decrease in non-immune individuals due to infection, ΔN_t.

$$\Delta N_t = -kNC\Delta t \quad (21.42)$$

The stochastic analog of this equation is the following:

$$\Delta N_r = -kNC\Delta t + E_n \quad (21.43)$$

where ΔN_r is the stochastic estimate of the change in density of non-contagious individuals and E_n is the random fluctuation in the number of individuals becoming infected. The stochastic model is based on a Monte Carlo technique involving a variance calculation similar to that which was used in Section 21.3. E_n is a function of the variance in N which is calculated by the equation

$$Var(N_t) = \Delta N_t(1 - kC) \quad (21.44)$$

The standard deviation of the estimate of N_t is

$$S_n = \sqrt{Var(N_t)} \quad (21.45)$$

The stochastic simulation of an epidemic is structured very much like the deterministic one described in Section 8.6. The difference lies in the two update equations which include the E_n term. The calculation of E_n is very much like the error term used in Section 21.3. The process involves generating a uniform random number, converting it to a normal deviate by Equation 20.71, and using it in conjunction with the standard deviation to calculate E_n, where E_n equals $N_d S_n$.

Exercise 21.5

Use the equations described in Section 8.6 and Equations 21.44 and 21.45 to develop a stochastic simulation of an epidemic. Use the following parameters:

N_0 = 2,000,000 people/100 mi^2, S = 0.95, L = 0.05/wk

C_0 = 1 person/100 mi^2, $E_0 = D_0 = I_0 = 0$

k = 0.000001 100 mi^2/people·wk, R = 5000 people/100 mi^2·wk

First run the simulation in the deterministic mode to establish the frequency and intensity of epidemics. Then, implement the stochastic portion of the model and generate at least two sets of data which can be compared with the deterministic data. How does the stochastic component of the model affect a) the onset of the first epidemic, b) the maximum intensity of the first epidemic, and c) the time between recurring epidemics?

21.5 Modeling Markov Processes with a Transition Matrix**

One of the most useful ways of representing certain types of random processes is with the transition or state diagram. This technique is used when there are a finite number of possible states that a given system may assume. The transition diagram uses a location to represent each state. Transition lines are used to indicate all possible transitions which can occur. The probability that each transition occurs during a fixed time interval is written beside each transition line. The sum of the probabilities for all transitions from a particular state must equal 1. A typical transition diagram is illustrated in Figure 21.2.

This transition diagram describes a very simple model of plant succession involving only three states in which a plant community might exist. State 1 represents grasses and other pioneer species. State 2 represents rapid growing trees such as aspen which require full sunlight. State 3 represents hardwoods such as red maple which are capable of growing in partial shade. A set of hypothetical probabilities has been assumed in order to provide a simple illustration of the principle involved. For a detailed application of the probability matrix to modeling tree succession see Horn (1975).

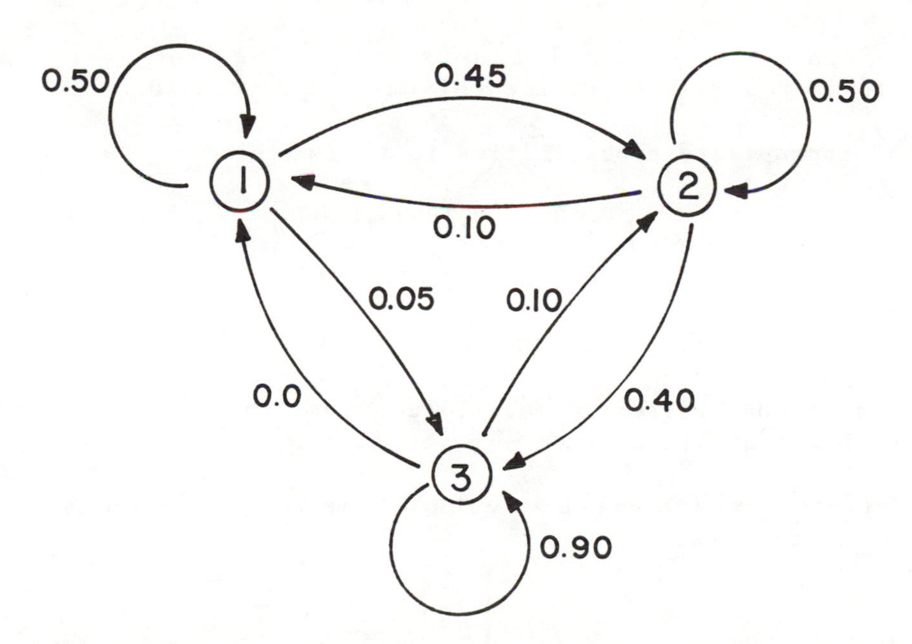

Figure 21.2. Transition diagram of simple secondary succession.

The hypothetical data presented here refer to the probability that a given small plot of ground will undergo a transition from one state to another during the next time interval. For example, the probability that a small plot of grass will change over and become dominated by aspen or related species is 0.45. The probability of it converting directly to hardwood species is only 0.05. Thus, the plot of ground is about 10 times as likely to convert to fast growing softwood as to hardwood. The probability of the plot remaining as grass is 0.5. Note that all possible occurrences have been considered, and that the sum of the probabilities is 1.

You could use the Monte Carlo technique to explore some of the transitions which might occur on a single plot of ground. This would be accomplished by assuming an initial state, such as grass (State 1), and generating a random number to compare with the probabilities of transition from this state. If a transition occurs generate a new random number in order to determine what happens next. Such a technique would result in a possible scenario for a given location. By keeping track of the amount of time spent in each state, you could get some idea of the average behavior of the system. Such a simulation would be similar to the queueing simulations discussed in Chapter 22.

A much more useful approach is to assign the probabilities as elements of a transition matrix, and to obtain information about future states by multiplying by a state vector. To use this technique on the above example, we first set up the transition matrix for the nine probabilities involved.

$$P = \begin{bmatrix} p_{11} & p_{12} & p_{13} \\ p_{21} & p_{22} & p_{23} \\ p_{31} & p_{32} & p_{33} \end{bmatrix}$$

where p_{21} refers to the probability of a transition from state 2 to state 1. See Chapter 16 if you wish to review this matrix convention.

Assigning the appropriate probabilities to this matrix gives

$$P = \begin{bmatrix} 0.50 & 0.10 & 0.00 \\ 0.45 & 0.50 & 0.10 \\ 0.05 & 0.40 & 0.90 \end{bmatrix}$$

Notice that the probabilities in each column must total 1, because they represent all possible transitions from a given state.

The number of plots which exist in each of the three states is assigned to a column vector as follows:

$$N = \begin{bmatrix} n_1 \\ n_2 \\ n_3 \end{bmatrix}$$

Alternatively, the column vector might represent the probability of being in each of the states at a given time. As such, the three probabilities must add up to 1.

The state of the system after one time interval is obtained by multiplying the transition matrix by the state vector. If the original vector contained the <u>number</u> of plots of each type at time t, then the product vector would contain the probable number of plots of each type at time t+1. If the original vector contained the <u>probability</u> of each state at time t, then the product vector would contain the probability of each state at time t+1.

Repeated multiplication of the transition matrix by the state vector ultimately gives the equilibrium condition for the system. In this case, it shows the proportion of each of the three classes of plants making up the climax community. Note that the probability of each state of the system at any time is purely a function of the previous state or state vector and the probability matrix. The history of the system previous to the last interval has no bearing on the state of the system during the next time interval. This behavior is a fundamental characteristic of all Markov processes.

<u>Exercise 21.6</u>**
Use the matrix multiplication procedure described above to model secondary succession. Use the hypothetical transition probabilities given. Start out with the probability of state 1 equal to 1 and the probability of the other two states equal to 0. Run the model for 20 time intervals to obtain the probabilities of each state for each time interval. The final probabilities should be very close to the equilibrium or climax condition.

<u>21.6 Genetic Drift</u>

Random fluctuations can occur in the frequencies of gene alleles when breeding populations are small. These fluctuations can result in fixation or homozygosity in one or more gene loci, where fixation is defined as the condition in which the recessive gene frequency, r, becomes either 1.00 or 0.00. Fluctuations are the result of the discrete random events involved in determining the genotypes of the individuals making up the parental gene pool, and the genotypes of the gametes contributed specifically by the heterozygous parents.

For a given generation, the process may be simulated by using uniform random numbers to determine the genotypes of each of the parents which will contribute genes to an offspring. After the genotypes of the parents for each new individual are established, a second random number is used to determine which of two possible alleles will be present in the gametes contributed by the heterozygous parent or parents.

The following assumptions are made about the population being modeled:

1) The model is limited to consideration of a single gene locus involving only two alleles.

2) There is no selection against either allele.

3) Generations are discrete in that the genotypes of a given generation are determined only by the genotypes of the previous generation.
4) The number of offspring per generation is constant.
5) The three genotypes in a given generation, aa, aA, and AA, have the respective frequencies S, T, and U.
6) The frequencies are the same for males and females.
7) Mating is random so that the probability of serving as a parent is the same as the frequency of potential parents in the population.

If we let the number of offspring per generation be designated by N, then 2N uniform random numbers must be examined to establish the genotypes of the 2N parents required. If a given random number, R_u, is less than S, it will be assumed to represent a homozygous recessive parent,and to contribute one recessive allele to the new offspring being considered. If R_u is greater than S+T then it is assumed to represent a homozygous dominant individual and to contribute a dominant allele to the new offspring. If R_u is greater than S, but less than S+T, then it is assumed to represent a heterozygous parent, and a second random number is used to determine whether the allele contributed is dominant or recessive. The program should be set up to examine random numbers in pairs, each pair being used to define the genotype of one new individual. The number of individuals of each genotype produced in this manner is summed in the same way as in the monohybrid cross simulation, Exercise 19.4. After the genotypes of all N offspring have been established, the totals for each genotype are used to calculate the frequencies, S, T, and U. These frequencies are then used to determine the parents for the following generation. The totals are also used to calculate the recessive gene frequency, r.

Exercise 21.7
Use the information provided in Section 21.6 to develop a Monte Carlo simulation of genetic drift. Write the program so that it is possible to vary generation size. Plot the frequency of the recessive allele, r, as a function of time to show the typical genetic drift pattern. It is also useful to write the program so that it may be repeated several times to determine the average number of time intervals required for fixation to occur for different generation sizes. Vary generation size from 10 to 100 and determine some average fixation times. Figure 21.3 shows some typical genetic drift simulations.

Exercise 21.8
In Chapter 10, we used deterministic models to describe the intensity of light on both an annual and a daily basis. Here we will explore some possible ways to provide a stochastic component to light intensity models.

Assume that storm fronts pass through a particular location about every 4 days. About 10 percent of the time they arrive after 3 days, and about 20 percent of the time 5 days elapse before the next front arrives. Assume 5 days is the maximum time between fronts. Make other reasonable assumptions as necessary, and develop a stochastic model to describe possible variations in light intensity. Your model should involve the following components based on random numbers:

a) the time between storm fronts,
b) the thickness of cloud cover and the effect on light intensity, and
c) the duration of a given storm front.

Use the stochastic terms which you develop in conjunction with the deterministic models from Chapter 10 to describe possible daily light intensities over a 30 day period.

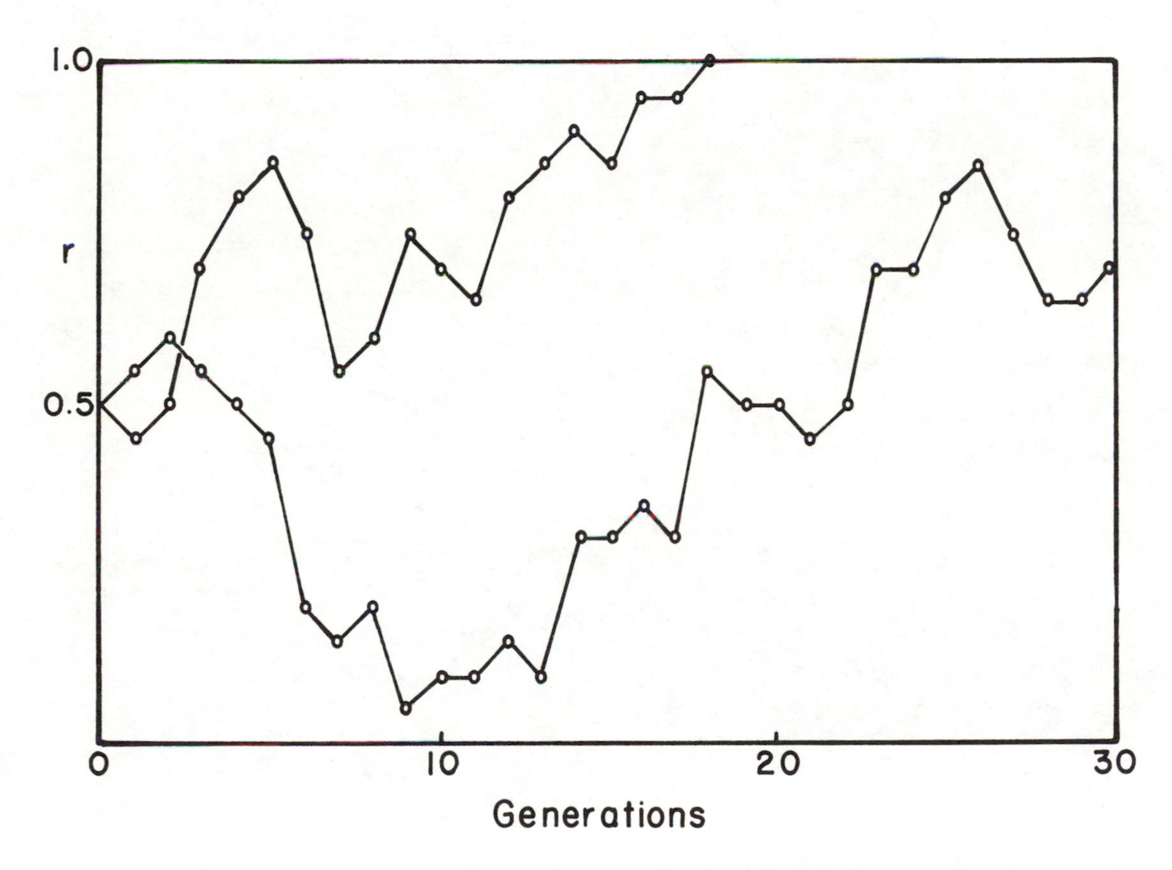

Figure 21.3. An example of simulated genetic drift.

Conclusion

This chapter has provided several examples of Markov processes which are of interest to biologists. In each process, the probability of being in a particular state at time $t+\Delta t$ depended only on the state occupied at time t, and the probabilities associated with the available states. This characteristic causes such systems to have unique behavior which sometimes runs counter to our expectations. Thus, it is useful to explore them in detail, by using the process of simulation. For additional examples, refer to Poole (1974).

Most of the simulations in this chapter employed the Monte Carlo technique to determine possible outcomes for various situations. However, you were also introduced to the very powerful transition matrix technique for calculating probable outcomes of transitions having fixed probabilities. Further applications of this technique will follow.

Some of the models discussed in the next chapter show the characteristics of Markov processes. However, the emphasis will be on examining them as examples of the queueing process, rather than Markov chains.

CHAPTER 22
QUEUEING SIMULATION IN BIOLOGY

A queue is defined as a line or file of people or vehicles waiting to be served in some manner. In a broader context, it may refer to any randomly arriving things, either animate or inanimate, which require a finite handling time by some processing facility. Thus, there does not need to be an actual line of things to constitute a queueing situation, only the potential for a line.

Because of the random nature of queueing systems, Monte Carlo simulation provides an important technique for the solution of queueing problems. Queueing simulation is especially important to the field of process design engineering because it provides one of the few means to explore alternative options to certain design problems. Classical examples of applications in this field include the simulation of such facilities as airports, supermarket checkout counters, toll booths, supply depots, and manufacturing assembly lines. The objective of each of these simulations is to achieve a design which optimizes the flow of people, vehicles or materials. Queueing simulation is so important for this purpose that several high level computer languages have been developed to handle the discrete event or process interaction type of simulation that these problems entail. These languages, which include SIMSCRIPT, GASP, and GPSS, all provide relatively simple programming techniques for individuals specializing in this type of simulation.

For the most part, a queueing simulation uses random numbers to determine the arrival and departure of things being processed. The computer keeps track of the status of individual items and moves them along according to certain algorithms. The results of a simulation are statistics describing the behavior of the system in terms of the number of interactions occurring in a given amount of simulated time, or the amount of time required to complete a given process or interaction.

James D. Spain, BASIC Microcomputer Models in Biology

ISBN 0-201-10678-7

After all this discussion about engineering and design, you are probably wondering how queueing simulations could possibly relate to biology. However, living organisms are basically processing systems, not unlike those designed by man. In addition, they often contain sub-systems and sub-sub-systems, which are themselves processing systems. The things being processed usually arrive randomly in time and sometimes arrive faster then they can be processed. Thus, they have the potential of saturating the system and forming a virtual queue.

Although queueing simulation is far from being the most active area of biological modeling, it has a high potential for developing models which can provide a whole new insight into certain types of biological systems.

22.1 The Toll Gate Design Problem

In order for you to have a better understanding of the nature of a typical queueing simulation, let us examine the classical toll gate design problem as described by Axelrod and Thomas (1977). Assume that we are simulating a toll gate with only three booths in operation. Lane 1 leads to an attended booth for drivers who do not have the correct change, while lanes 2 and 3 provide automatic booths for drivers who do have the correct change. There is a queue, or potential queue, associated with each booth.

Assume that 18 percent of the arriving customers do not have the correct change and must enter lane 1. The remainder do have the correct change and thus go into either lane 2 or lane 3, whichever is shorter. Also, assume that the cars arrive at the toll gate randomly. The probability of a car arriving during any one second is assumed to be 0.2, and only one car may arrive per second. The service time at gate 1 is assumed to be uniformly distributed between 10 and 30 seconds. Gates 2 and 3 have a constant service time of 5 seconds. The basic objective of this simulation is to determine the average and the maximum line length, the maximum wait for any customer, and the fraction of time each gate is being used.

The simulation is programmed to carry out the following operations in sequence:

1) Increment the time in seconds.

2) Compare a random number with the probability of arrival to determine if a car arrives at the toll gate.

3) If a car arrives, compare a second random number with the probability that the customer has change to determine whether the car enters queue 1, 2, or 3.

4) Decrement time on all cars being processed.

5) If the car being processed has finished, check to see if there are more cars in line for that booth. If so, move up the next car, and assign its processing time.

6) Add the existing queue length to the sum of the queue lengths for each booth. This will be used for subsequent calculation of the average queue length.

7) If the existing queue length exceeds the maximum queue length, make the existing queue length equal to the maximum for that queue.

8) Increment the booth utilization counter for each booth where a car is being processed.

9) Print out existing length of each queue.

10) Return to number 1.

At the end of the specified number of time intervals, the following summary data is calculated and printed:

1) the maximum queue length for each queue,

2) the average queue length for each queue,

3) the fraction of time each processing booth is utilized,

4) the average time spent waiting in each queue.

Exercise 22.1
Use the program provided in Figure 22.1 to simulate the utilization of the threc booth toll gate system. Vary the probability of arrival and the probability of having change in order to observe the behavior of the system under different conditions.

22.2 Queueing Simulation of Wolf-Moose Interaction

A simple queueing model may be used to describe some of the interactions taking place between a wolf pack and a moose herd in an isolated habitat, such as the Isle Royale National Park. This model views the moose as queueing up to be processed by the wolf pack. The wolf pack may be considered as a roving multicomponent processing facility.

The wolf pack is assumed to operate as a unit and to exist only in four states or modes of activity. The pack is either hunting for moose, testing a moose, eating a moose, or resting up after a kill. Presumably, social activities take place during all these states, but we are only concerned with wolf-moose interactions. For purposes of the model, the wolves are assumed to subsist totally on moose kills.

The meeting between the processors and an unwilling processee is assumed to be random. The probability of such a meeting is held constant, assuming that the total moose population does not change through time. What happens when the pack contacts a moose depends in large part on the age of the moose. Very young moose have little chance against the pack, as do very old moose, which often have an arthritic condition or other debilitating problems.

```
10  REM  THE CLASSICAL TOLL GATE QUEUEING SIMULATION
20  REM  PROGRAM BY J. SPAIN, MICHIGAN TECH. UNIV.
30  REM  BASED ON A PROBLEM BY HARVEY AXELROD
40  REM   STATE UNIVERSITY OF NEW YORK AT BUFFALO
50  REM  TIME IS EXPRESSED IN SECONDS
90 P = .3: REM  PROBABILITY OF A CAR ARRIVING IN NEXT SECOND
100  FOR T = 1 TO 500
110  IF  RND (1) > P THEN  GOTO 160: REM   NO CAR ARRIVED
115  REM  CAR ARRIVES...DO THEY HAVE CHANGE?
120  IF  RND (1) < .18 THEN  GOTO 150: REM   NO CHANGE
130  IF Q2 >  = Q3 THEN Q3 = Q3 + 1: GOTO 160: REM  ADD CAR TO QUEUE 3
140 Q2 = Q2 + 1: GOTO 160: REM  ADD CAR TO QUEUE 2
150 Q1 = Q1 + 1: REM  ADD CAR TO QUEUE 1
155  REM  DECREMENT PROCESSING TIME FOR CARS OCCUPYING BOOTH
160  IF T1 > 0 THEN T1 = T1 - 1
170  IF T2 > 0 THEN T2 = T2 - 1
180  IF T3 > 0 THEN T3 = T3 - 1
190  IF T1 = 0 THEN  GOSUB 230: REM   CUSTOMER IN BOOTH 1 FINISHED
200  IF T2 = 0 THEN  GOSUB 270: REM  CUSTOMER IN BOOTH 2 FINISHED
210  IF T3 = 0 THEN  GOSUB 300: REM  CUSTOMER IN BOOTH 3 FINISHED
220  GOTO 350: REM  BYPASS SUBROUTINES
230  IF Q1 = 0 THEN  RETURN : REM  NO CARS WAITING IN QUEUE 1
240 Q1 = Q1 - 1: REM  MOVE UP ONE CAR IN QUEUE 1
250 T1 =  INT ( RND (1) * 20 + 11): REM  TIME FOR NEXT CUSTOMER IN BOOTH

260  IF Q1 > M1 THEN M1 = Q1: REM  MAXIMUM LENGTH OF QUEUE 1
265  RETURN
270  IF Q2 = 0 THEN  RETURN : REM   NO CARS WAITING IN QUEUE 2
280 Q2 = Q2 - 1: REM  MOVE UP ONE CAR IN QUEUE 2
290 T2 = 5: RETURN : REM  ASSIGN PROCESSING TIME TO CAR IN BOOTH 2
300  IF Q3 = 0 THEN  RETURN : REM  NO ONE WAITING IN QUEUE 3
310 Q3 = Q3 - 1: REM  MOVE UP ONE CAR IN QUEUE 3
320 T3 = 5: RETURN : REM   ASSIGN PROCESSING TIME TO CAR IN BOOTH 3
350 L1 = L1 + Q1: REM   SUM OF LENGTHS FOR QUEUE 1
380 L2 = L2 + Q2: REM  SUM OF LENGTHS FOR QUEUE 2
390  IF Q2 > M2 THEN M2 = Q2: REM  MAXIMUM LENGHT FOR QUEUE 2
400 L3 = L3 + Q3: REM  SUM OF LENGHTS FOR QUEUE 3
410  IF Q3 > M3 THEN M3 = Q3: REM  MAXIMUM LENGHT FOR QUEUE 3
430  REM  TOTAL UTILIZATION TIME FOR EACH BOOTH
450  IF T1 > 0 THEN U1 = U1 + 1
460  IF T2 > 0 THEN U2 = U2 + 1
470  IF T3 > 0 THEN U3 = U3 + 1
590  PRINT Q1,Q2,Q3
600  NEXT T
603  PRINT : REM  SUMMARY STATISTICS
605  PRINT "MAXIMUM QUEUE LENGTHS"
610  PRINT "QUEUE 1 = ";M1;"  QUEUE 2 = ";M2;"  QUEUE 3 = ";M3
615  PRINT : PRINT "AVERAGE QUEUE LENGTHS"
620 A1 = L1 / 500:A2 = (L2 + L3) / 1000
622  PRINT "QUEUE 1 = ";A1;"    QUEUES 2 & 3 = ";A2
625  PRINT : PRINT "QUEUE UTILIZATION"
630  PRINT "QUEUE 1 = ";U1 / 500;"   QUEUES 2 AND 3 = ";(U2 + U3) / 1000
650  PRINT : PRINT "THE AVERAGE TIME SPENT IN THE QUEUE"
660  PRINT "QUEUE 1 = ";25 * A1;"    QUEUES 2 & 3 = ";5 * A2
670  PRINT : PRINT "THE MAXIMUM WAIT FOR ANY CUSTOMER"
680  PRINT "QUEUE 1 = ";M1 * 25;"   QUEUES 2 & 3 = ";M3 * 5
```

Figure 22.1. Computer program for the tollgate queueing simulation.

On the other hand, healthy young adult moose have a high probability of escaping or holding the pack at bay. Thus, the age class of the moose is important in determining the result of testing by the wolf pack. The probability of a given moose being in a given age class is assumed to be the same as the proportion of that age class in the population. When a kill does occur, the length of time required for processing by the pack varies as a function of the age class. The amount of food provided by a mature adult can be four times as great as that of a calf. The pack is assumed to rest for a constant period after each kill.

With these assumptions in mind, let us set up the general plan for a queueing simulation of this system. As with most queueing simulations the emphasis is on time. Each random occurrence is based on a fixed probability during a given time interval, and each deterministic component, such as resting and eating, is assumed to require a fixed number of time intervals.

Assume that the simulation starts with the wolf pack in the hunting mode. A random number is generated and compared with the probability of finding a moose. If the number is greater than the probability, it is assumed that no moose was found during that time interval. Time is then incremented, and the pack remains in the hunting mode. If a moose is found, then the pack moves to the testing mode in which a second random number is generated and compared with the proportions of each age class to determine which age class it represents. See Section 20.1 for discussion of this sampling technique. Once the age class of the moose is established by this procedure, a third random number is compared with the probability of a kill within that age class. If the random number is less than the probability, a kill is assumed to have occurred. If no kill occurs, the wolf pack reverts to the hunting mode and the moose presumably goes on its merry way. If a kill does occur, the pack goes into the eating mode and the age class is used to determine how much meat is available and how long the pack will remain in the eating mode. After each kill, the pack is assumed to remain one day in the resting mode and then return to the hunting mode.

The basic output of this model is a scenario describing the activities of a wolf pack at various times. Additional statistics of interest include the relative amount of time spent in each activity mode, the average amount of meat consumed by the pack per day, and to evaluate the extremes in pack nutrition, the average meat consumed per day during the most recent ten days. The primary objective of the model is simply to provide a means of gaining new insight into an important biological process.

Exercise 22.2

Use the general description provided in Section 22.2 to develop a queueing simulation for wolf-moose interaction. Assume that the population is divided into three age classes with the following characteristics:

Age class	I	II	III
Age range	0-1 years old	1-5 years old	6-20 years old
Average weight	200 lbs.	600 lbs.	800 lbs.
Fraction of population	0.32	0.44	0.24
Probability of being killed per test	0.62	0.02	0.34

Assume that time is incremented in tenths of days and that the probability of the wolves finding a moose during any time interval is 0.7. Assume that it takes 0.5 days to eat a moose of age class I, 1.5 days for age class II, and 2 days for age class III. Allow the model to run for 100 days and determine the average amount of meat consumed per day. Assuming that a healthy wolf pack requires about 200 pounds of meat per day, what strategy might the pack adopt in order to reach this goal?

22.3 Cattle Reproduction Simulation

Assume that a dairy farmer wants to maximize the efficiency of his dairy herd by reducing the number of bulls. Since he wants to avoid doing this in a way which would significantly affect the overall reproduction of the herd, he decides to have someone simulate the reproductive process using different numbers of cows per bull. The objective of the simulation is to determine the maximum number of cows which can be effectively served by one bull.

The following assumptions are made concerning the reproductive physiology of cows:

1) The temporal location of each cow in its estrous cycle is independent of all the other cows. That is, one cow has no effect on the time of ovulation of another.

2) The mean length of the estrous cycle is assumed to be 28 days with the following distribution:

   26% have a 28 day cycle
   68% fall in the range of 27 to 29 days
   95% fall in the range of 26 to 30 days
   99% fall in the range of 25 to 31 days
   100% fall in the range of 24 to 32 days

3) Cows may be successfully fertilized only on the day that they ovulate. If they are not successfully fertilized, they return to the beginning of their cycle.

4) The bull has a 90 percent probability of successfully fertilizing the first cow he mates with during a given day. However, because of reduced numbers of sperm, this probability drops to 30 percent on the second attempt at mating, and to 10 percent on the third. Thus, the probability is reduced to one third the previous value for each successive mating attempt during a given day.

Cattle reproduction is simulated by setting up an array equal in size to the number of cows. Then, a series of random numbers uniformly distributed between 1 and 28 is assigned to the array. This series represents the temporal location of each cow in its estrous cycle.

A second array is set up to retain the length of each cow's estrous cycle. A second series of uniform random numbers is generated to determine the cycle lengths using the probabilities in assumption number 2. Thus, if a given random number is less than 0.26, the subject cow is assumed to have a 28 day cycle, and the program assigns this number to the cycle length array. If this numbers is not less than 0.26, but it is less than 0.68, the cow is assumed to have a cycle of either 27 or 29 days. An additional random number is compared with 0.5 to decide between 27 and 29. This cycle length is assigned to the array, and so on. By this procedure, all the cows are assigned an estrous cycle length.

Once these two arrays are set up, the program can begin incrementing time by days. For any given day, the program searches through the array incrementing the time of the estrous cycle for each cow and comparing the new value against the length of the estrous cycle in the second array. If the time in the estrous cycle does not equal the cycle length the program simply moves on to the next cow in the array.

If the time in the estrous cycle does equal the length of estrous, then ovulation is assumed to have occurred, and a random number is generated and compared with the probability of mating success to determine whether mating has occurred. Whether it does or not, the probability of mating success is divided by 3 in case there is another attempt at mating during the same day. When a successful mating does occur, a pregnancy counter is incremented by one, and the cow's estrous time is assigned a value of -100 so that the cow will not come into the mating pool again during the course of the simulation. Thus, the simulation checks each cow each day and keeps track of the pregnancies which occur. The queueing aspect of this simulation occurs when by chance several cows ovulate on the same day and, as a result, may fail to become fertilized.

Exercise 22.3

Program the cattle queueing simulation described above. Run the simulation with an assumed number of cows and 100 days of simulated time. By repeating the simulation several times, determine the approximate number of cows a single bull can serve and still achieve a 90 percent mating success during the 100 day period.

22.4 Enzyme-Substrate Interaction

A single enzyme molecule may be thought of as a processing facility for substrate molecules. This point has been made in Chapter 2 and elsewhere in the book. Therefore, it is not surprising that queueing simulations have been applied to enzyme action. Bartholomay (1964) has considered the general catalytic queue process from a variety of viewpoints. Here, we will concern ourselves with the simple case of an enzyme interacting with a single species of substrate molecule to form the enzyme-substrate compound, and then, the product. The five transitions involved are diagrammed in Figure 22.2.

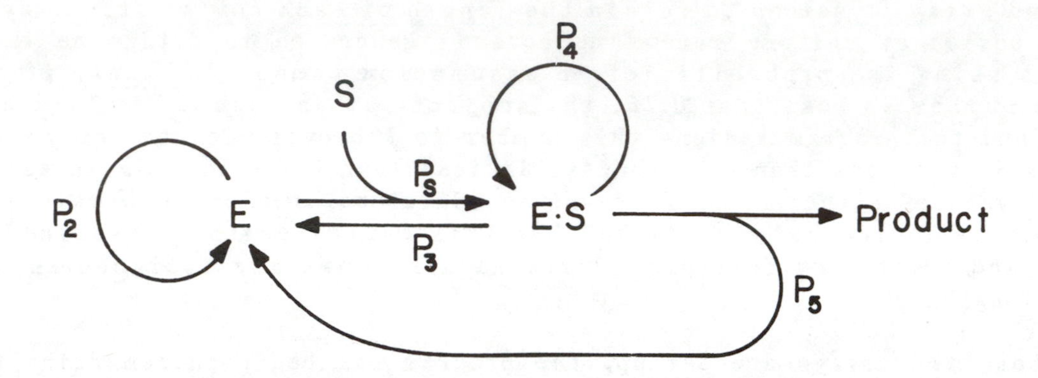

Figure 22.2. Probability transitions of the enzyme-substrate system.

The Monte Carlo simulation of this system is done by following a single enzyme molecule as it undergoes transitions between the two states, E, the free enzyme and, ES, the enzyme-substrate compound. The five probabilities involved in this process are as follows:

P_S is the probability of interacting with any one of S substrate molecules to form ES during one time interval.

P_2 is the probability of remaining as a free enzyme molecule during one time interval. $P_2 = 1 - P_S$

P_3 is the probability of the enzyme-substrate compound breaking down to form free substrate and enzyme during one time interval.

P_4 is the probability of the enzyme-substrate compound remaining in that form during one time interval.

P_5 is the probability of the enzyme breaking down to yield product and free enzyme during one time interval.

All these probabilities are constant, except P_S and P_2 which are a function of substrate concentration, here considered to be the number of substrate molecules available for interacting with the enzyme. This number is important because the response of enzyme to varying substrate concentration is the primary behavior we wish to simulate with this model.

If P_1 is the probability that a single enzyme molecule will react with a single substrate molecule, then $1-P_1$ is the probability that it will not react, during one time interval. The probability that the enzyme will not react given n opportunities is $(1-P_1)^n$. Thus, $1 - (1-P_1)^n$ is the probability that a single enzyme will react at least once given n opportunities, and the probability of an enzyme molecule reacting with any one of S substrate molecules is

$$P_S = 1 - (1-P_1)^S \quad (22.41)$$

As stated before,

$$P_2 = 1 - P_S \quad (22.42)$$

Thus, we have a means of varying the probability of enzyme-substrate interaction as a function of substrate concentration. In the simulation of this process, the substrate concentration is assumed to be constant as is the case when initial reaction velocity is measured experimentally.

Note that

$$(1-P_1)^S \approx e^{-SP_1} \quad (22.43)$$

Thus, the following equation may also be used to describe P_S.

$$P_S = 1 - e^{-SP_1} \quad (22.44)$$

The Monte Carlo simulation of enzyme substrate interaction employs random numbers to determine whether the enzyme makes a transition from one state to the other. The number of product molecules produced in a given number of time intervals would then provide a measure of enzyme activity for different numbers of substrate molecules. A flow chart for this simulation is provided in Figure 22.3. The objective of this simulation is to see if the probability assumptions made above can account for the saturating behavior of the simple Michaelis-Menten model.

Exercise 22.4

Write a program for the enzyme queueing simulation described in Section 22.4. Assume the following values for the probabilities involved:

$P_1 = 0.0002$, $P_3 = 0.01$, $P_4 = 0.95$, $t_{max} = 1000$, and $S_{max} = 500$.

Plot the number of product molecules produced in 1000 time intervals as a function of the number of substrate molecules.

Use the CURFIT program to see how closely the hyperbolic equation actually fits the simulation data. Use the statistics provided to interpret the goodness of fit. See if you are able to determine any relationship between the fitted values for K_m and V_{max}, and the probabilities assumed above.

Also keep track of the fraction of time that the enzyme spends in the form of the enzyme-substrate compound. This fraction should approach 1.0 as the substrate concentration is increased to higher and higher levels.

22.5 A Queueing Simulation of the Operon Model

A deterministic version of the Jacob and Monod gene repression model was described in Section 15.4. In that model, the rate of m-RNA transcription from a structural gene was assumed to be a function of the fraction of time the operon was in the unrepressed state. Since, in bacteria, there is only a single DNA molecule, and presumably only a single operon for a given

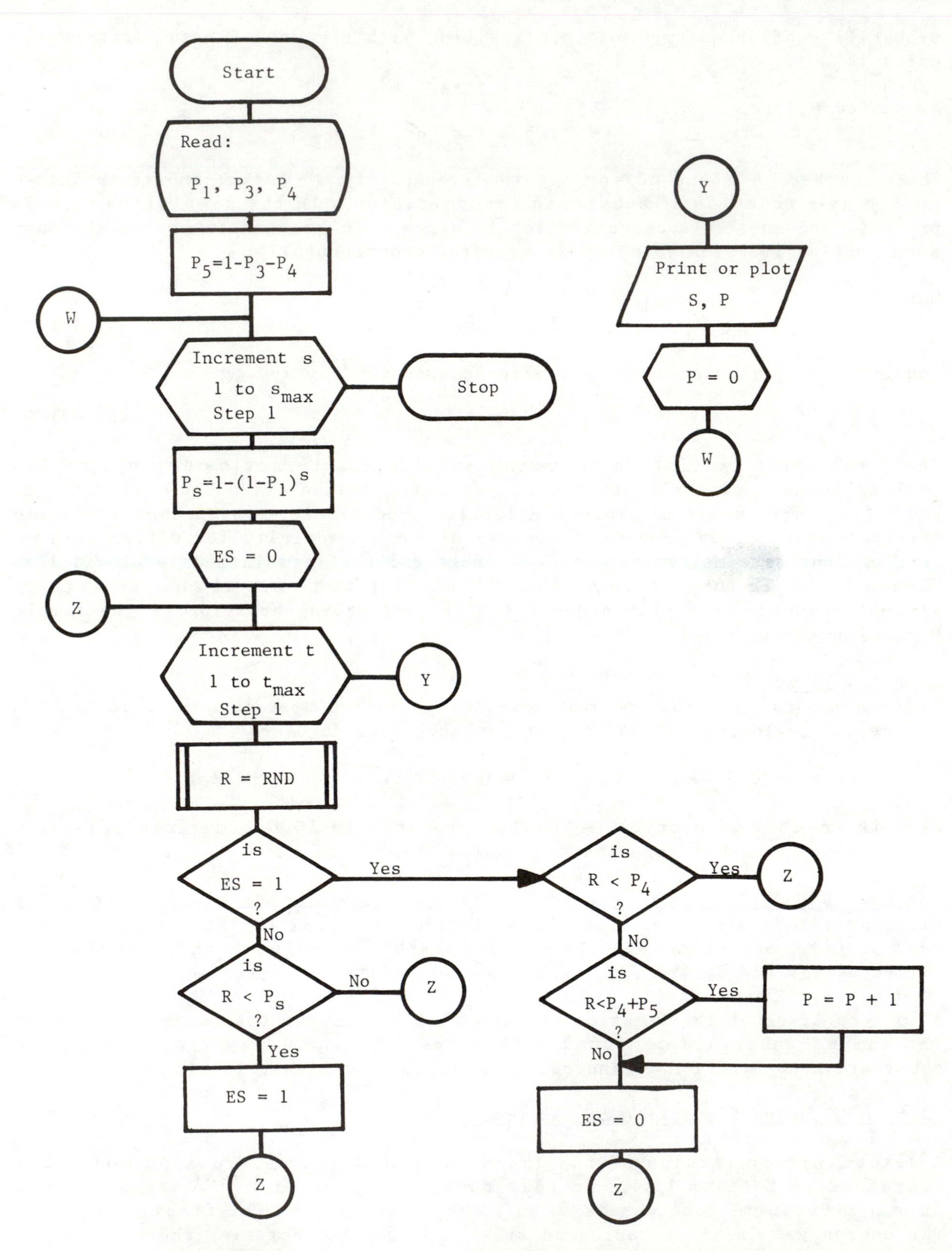

Figure 22.3. Flowchart of the enzyme queue simulation.

structural gene, m-RNA transcription is either fully on or fully off at any given time. Thus, the system provides an interesting example ofstatistical dynamics.

The operator controlling the operon may be considered to exist in either of two possible states. The transition probabilities are shown in Figure 22.4.

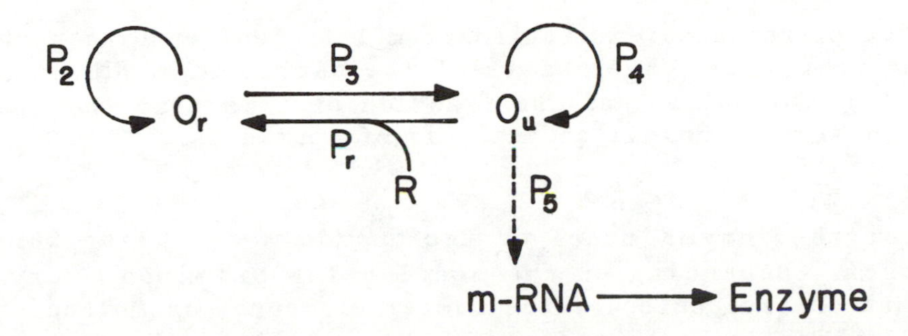

Figure 22.4. Transition probabilities for a gene repression system.

In this system, O_r is the operator in the repressed condition, and O_u is the operator in the unrepressed condition. The probability of converting from O_u to O_r is designated as P_r. This probability is a function of repressor concentration, according to the equation

$$P_r = 1 - (1-P_1)^R \tag{22.51}$$

where P_1 is the probability of interaction between O_u and a single repressor molecule, and P_r is the probability of interaction between O_u and at least one of R repressor molecules during one time interval. The probability of remaining in the repressed condition is P_2, the probability of reverting to the unrepressed condition is P_3, and the probability of remaining in the unrepressed condition is P_4.

Note that $$P_3 = 1 - P_2 \quad \text{and} \quad P_4 = 1 - P_r \tag{22.52}$$

The probability of an m-RNA molecule being transcribed during any time interval that the operon is in the unrepressed condition is P_5. This probability is constant, but results in the transcription of a single m-RNA molecule per time interval only when the operator is in the unrepressed state.

The m-RNA is assumed to be degraded at a rate described by the following decay expression:

$$Y_{t+\Delta t} = Y_t - k\cdot Y_t = Y_t(1-k) \tag{22.53}$$

where Y represents the number of m-RNA molecules. Enzyme synthesis is assumed to be a direct function of the number of m-RNA molecules available. An equation describing enzyme concentration is as follows:

$$Z_{t+\Delta t} = Z_t + eY_t - fZ_t \tag{22.54}$$

Further development of this model should be based upon the deterministic model described in Section 15.4.

Exercise 22.5
Program the operon model described above and implement it with the following probabilities:

$$P_1 = 0.001,\ P_2 = 0.99,\ P_3 = 0.01,\ \text{and}\ P_5 = 0.5$$

Vary the number of repressor molecules from 1 to 500, running each simulation for 1000 time intervals. Keep track of the number of m-RNA molecules, the number of enzyme molecules, and the fraction of time that the operator is in the unrepressed state. Submit an analysis of your data.

Exercise 22.6
Analyze either the enzyme model or the operon model using the transition matrix approach. Depending on the model being examined, vary either the number of substrate molecules or the number of repressor molecules and determine the equilibrium state by vector multiplication of the transition matrix. How do the results compare with the Monte Carlo simulation?

Conclusion

This chapter has provided several examples of models which, by the nature of the approach taken, fall into the category of queueing simulation. These models are distinguished from those in the other chapters in Part III by the fact that each involves the saturation of some type of facility by large numbers of animate or inanimate objects. If you examine the examples given, and consider the deterministic analogs of each, you will realize that there is a very close connection between the saturation or satiation models and queueing simulations. This should not be surprising as each involves similar underlying assumptions. Because of the applicability of saturation models to biological systems in general, it is likely that there are a large number of potential queueing simulations still waiting to be explored.

The objective of developing a queueing simulation is not much different from that of any other type of simulation. Each modeling approach provides a vehicle for exploring a given biological system from a new and different point of view. The new insight that results from development and implementation of computer models can lead to the ultimate understanding of the system being studied.

BIBLIOGRAPHY

Anderson, N. 1974. A mathematical model for the growth of giant kelp. Simulation. 22: 97-105.

Atkins, G. L. 1969. Multicomponent Models for Biological Systems. London: Methuen and Co., Ltd. 153 p.

Axelrod, H. and W. Thomas. 1977. Enriched understanding of simulation by usage and dissection of PASS IV. Proc. Conf. on Computers in Undergraduate Curricula, Michigan State University.

Bartlett, M. S. 1956. Deterministic and stochastic models of recurrent epidemics. Proc. Third Berkeley Symp. Math. Stat. Probab. 4: 81-109.

Bartholomay, A. F. 1964. The general catalytic queue process. In: Gurland, J. (ed.). Stochastic Models in Biology and Medicine. Madison: Univ. Wisconsin Press. 101-145.

Baum, R. F. 1967. Frequency of recessive traits in a population in which recessive homozygotes are not allowed to reproduce. In: R. M. Thrall, J.A. Mortimer, K. R. Rebman, R. F. Baum, (eds.) Some Mathematical Models in Biology, 40241-R-7; NIH-5-T01-GM01457-02, The University of Michigan.

Benyon, P. R. 1968. A review of numerical methods for digital simulation. Simulation. 11: 219-238.

Bittinger, M. L. 1976. Calculus, A Modeling Approach. Reading: Addison-Wesley Publishing Co. 430 p.

Boillot, M. H., Gleason, G. M., and Horn, L. W. 1975. Essentials of Flowcharting. Dubuque: W. C. Brown Company. 114 p.

Brylinsky, M. 1972. Steady-state sensitivity analysis of energy flow in a marine ecosystem. In: Patten, B. C., ed. Systems analysis and simulation in ecology, Vol. II. New York: Academic Press. 81-138.

Caperon, J. 1968. Population growth response of Isochrisis galbana to nitrate variation in limiting concentrations. Ecology. 49: 866-872.

Caperon, J., and Meyer, J. 1972. Nitrogen-limited growth of marine phytoplankton. Deep-Sea Research. 19: 601-632.

Chance, B. 1960. Analogue and digital representations of enzyme kinetics. J. Biol. Chem. 235: 2440-2443.

Chance, B., Garfinkel, D. Higgins, J. and Hess, B. 1960. Metabolic control mechanisms. V. A solution of equations for glycolysis and respiration in ascites tumor cells. J. Biol. Chem. 235: 2426-2439.

Chapin, N. 1971. Flowcharts. Princeton: Auerbach Publishers. 179 p.

Chen, C. W. 1970. Concepts and utilities of ecologic model. J. Sanitary Eng. Div., Proc. Am. Soc. Civil Eng., Vol. 96, No. SA5, Oct. 1970, 1085-1098.

Cleland, W. W. 1970. Steady-state kinetics. In: P. D. Boyer, (ed.), The Enzymes, 3rd ed., Vol 2. New York: Academic Press. 1-65 p.

Crow, J. F., and Kimura, M. 1970. An Introduction to Population Genetics Theory. New York: Harper and Row. 591 p.

Croxton, F. E., and Cowden, D. J. 1955. Applied General Statistics. New York: Prentice Hall, Inc.

Dabes, J. N., Finn, R. K., and Wilke, C. R. 1973. Equations of substrate-limited growth: a case for Blackman kinetics. Biotechnol. Bioeng. 15: 1159-1177

Dahlberg, B. L., and Guettinger, R. C. 1956. The white-tailed deer in Wisconsin. Wisc. Conserv. Dep. Tech. Wildl. Bull., No. 14, 282 p.

Dahlberg, G. 1948. Mathematical Methods for Population Genetics. New York: Wiley Interscience.

Daniel, C., and F. S. Wood. 1971. Fitting Equations to Data. New York: Wiley Interscience. 342 p.

Deevey, E. S. 1947. Life tables for natural populations of animals. Quarterly Rev. of Biology. 22: 283-314.

DeSapio, R. 1976. Calculus for the Life Sciences. San Francisco: W.H. Freeman and Company. 740 p.

DiToro D. M. 1980. Applicability of cellular equilibrium and Monod theory to phytoplankton growth kinetics. Ecological Modeling. 8: 201-218.

Droop, M. R. 1968. Vitamin B_{12} and marine ecology. IV. The kinetics, uptake and growth of Monochrysis lotherei. J. Mar. Biol. Assoc. U.K. 54: 825-855.

Droop, M. R. 1973. Some thoughts on nutrient limitation in algae. J. Phycol. 9: 264-272.

Droop, M. R. 1974. The nutrient status of algal cells in continuous culture. J. Mar Biol. Assoc. U. K. 54: 825-855.

Dugdale, R. C. 1967. Nutrient limitation in the sea: dynamics, identification and significance. Limnol. Oceanog. 12: 685-695.

Edwards, P. and Broadwell, B. 1974. Flowcharting and Basic. New York: Harcourt Brace Jovanovich, Inc. 214 p.

Elias, H. and Pauly, J. E. 1966. Human Microanatomy. Philadelphia: F. A. Davis Company. 380 p.

Ellgaard, E. G., Bloom, K. S., Malizia Jr., A. A., and Gunning, G. E. 1975. The locomotor activity of fish: an analogy to the kinetics of an opposed first-order reaction. Trans. Am. Fish. Soc. 104(4): 752-754.

Farner, D. S. 1945. Age groups and longevity in the American Robin. Wilson Bull. 57: 56-74. Also see Deevey (1947).

Fee, E. J. 1969. A numerical model for the estimation of photosynthetic production, integrated over time and depth, in natural waters. Limnol. Oceanogr. 14: 906-911.

Fee, E. J. 1973. A numerical model for determining integral primary production and its application to Lake Michigan. J. Fish Res. Board Can. 30: 1447-1468.

Feller, W. 1968. An Introduction to Probability and its Applications. Volume I. New York: John Wiley and Sons, Inc. 509 p.

Forrester, J. W. 1969. Urban Dynamics. Cambridge: M.I.T. Press. 285 p.

Gardner, E. J. 1972. Principles of Genetics. New York: John Wiley and Sons, Inc. 527 p.

Garfinkel, D. 1962. Digital computer simulation of ecological systems. Nature. 194: 856-857.

Garfinkel, D. and Sack, R. 1964. Digital computer simulation of an ecological system based on a modified mass action law. Ecology. 45: 502-507.

Garfinkel, D., Garfinkel, L., Pring, M., Green S. B., and Chance, B. 1970. Computer applications to biochemical kinetics. Ann. Rev. Biochem. 39: 473-498.

Gates, D. M. 1962. Energy Exchange in the Biosphere. New York: Harper and Row. 151 p.

Gause, G. F. and Witt, A. A. 1935. Behavior of mixed populations and the problem of natural selection. Am. Nat. 69: 596-609.

Gazis, D. C., Montroll, E. W., and Ryniker, J. E. 1973. Age-specific deterministic model of predator-prey populations: application to Isle Royale. I.B.M. J. Res. Develop. (Jan. '73) 47-53.

Gérardin, L. 1968. Bionics. New York: McGraw Hill Book Co.

Glass, B. 1963 The establishment of modern genetical theory as an example of the interaction of different models, techniques, and inferences. In: A. C. Crombie, (ed.), Scientific Change. New York: Basic Books, Inc.

Goel, N. S. and Richter-Dyn, N. 1974. Stochastic Models in Biology. New York: Academic Press. 269 p.

Gold, H. J. 1971. Mathematical Modeling of Biological Systems - An Introductory Guidebook. New York: John Wiley and Sons. Inc. 351 p.

Golley, F. B. 1965. Structure and function of an old-field broom sedge community. Ecol. Monograph. 35: 113-137.

Goodwin, B. 1963. Temporal Organization of Cells. New York: Academic Press.

Guyton, A. C. 1971. Basic Human Physiology. Philadelphia: W. B. Saunders. 721 p.

Hamil, W. H., Williams, R. R. and Mackay, C. 1966. Principles of Physical Chemistry. Englewood Cliffs: Prentice-Hall, Inc.

Hardin, G. 1966. Biology, its Principles and Implications. San Francisco: W. H. Freeman and Co. 771 p.

Harvey, H. W. 1963. The Chemistry and Fertility of Sea Waters. London: Cambridge Univ. Press. 240 p.

Hoerl, A. E. 1954. Fitting curves to data. In: J. H. Perry, (ed.), Chemical Business Handbook. New York: McGraw-Hill.

Holling, C. S. 1959. The components of predation as revealed by a study of small mammal predation of the European Pine Sawfly. Can. Entomol. 91: 293-320.

Holling, C. S. 1961. Principles of insect predation. Ann. Rev. Entomology. 6: 163-182.

Holyoak, G. W. 1969. Thermal Regulatory Responses of the Little Brown Bat, Myotis lucifigus, Following Seasonal Acclimatization and Temperature Acclimation. Michigan Technological Univ. M.S. Thesis. 63 p.

Horn, H. S. 1975. Forest succession. Sci. Amer. 232: 90-98.

Hougen, O. A. and Watson, K. M. 1943. Chemical Process Principles, Part One: Material and Energy Balances. New York: John Wiley and Sons. 436 p.

Humphrey, A. E. 1979. Fermentation process modeling: an overview. Ann. N.Y. Acad. Sci. 77: 17-33.

Hutchinson, G. E. 1957. A Treatise on Limnology, Vol. I Geography, Physics, and Chemistry. New York: John Wiley and Sons, Inc. 1015 p.

Hutchinson, G. E. 1978. An Introduction to Population Ecology. New Haven: Yale Univ. Press. 260 p.

Ivlev, V. S. 1961. Experimental Ecology of the Feeding of Fishes. New Haven: Yale Univ. Press. 302 p.

Jacob, F. and Monod, J. 1961. Genetic regulatory mechanisms in the synthesis of proteins. J. Mol. Biol. 3: 318.

Jeffers, J. N. R. 1978. An Introduction to Systems Analysis: With Ecological Applications. Baltimore: University Park Press. 198 p.

Jenkins, D. H. and Bartlett, I. H. 1959. Michigan Whitetails. Lansing: Michigan Dept. Cons., 40 p.

Jones, R. W. 1973. Principles of Biological Regulation: An Introduction to Feedback Systems. New York: Academic Press. 359 p.

Jones, R. and Hall, W. B. 1973. A simulation model for studying the population dynamics of some fish species. In: M. S. Bartlett, and R. W. Hiorns, (eds.) The Mathematical Theory of the Dynamics of Biological Populations. London: Academic Press. 35-59.

Justice, K. E. 1968. The Use of Computer Modeling in Biology Laboratory Instruction. Univ. of California, Irvine, Bulletin.

Kermack, W. O. and McKendrick, A. G. 1932. Contributions to the mathematical theory of epidemics. Proc. Roy. Soc. London, Series A. 138: 55-83.

Ketchum, B. H. 1939. The absorption of phosphate and nitrate by illuminated cultures of Nitzchia closterium. Am. J. Botany. 26: 399-407.

Kitchell, J. F., Koonce, J. F., O'Neill, R. V., Shugart, H. H., Magnuson, J. J. and Booth, R. S. 1974. Model of fish biomass dynamics. Trans. Amer. Fish. Soc. 103(4): 786-798.

Langmuir, I. 1916. Absorption of gases on solids. J. Am. Chem. Soc. 38: 2221.

Lehninger, A. L. 1975. Biochemistry. New York: Worth Publishers, Inc. 1104 p.

Leslie, P. H. 1945. The use of matrices in certain population mathematics. Biometrika. 33: 183-212.

Leslie, P. H. and Gower, J. C. 1960. Properties of a stochastic model for the predator-prey type of interaction between two species. Biometrika. 47: 219-301.

Lewontin, R. C. 1965. Selection for colonizing ability. In: Baker and Stebbins (eds.) The Genetics of Colonizing Species. New York: Academic Press. 77-91.

Lewontin, R. C. 1969. The meaning of stability. In: Diversity and Stability in Ecological Systems. Brookhaven Symposium in Biology, No. 22.

Lindeman, R. L. 1942. The trophic dynamic aspects of ecology. Ecology. 23: 399-418.

Li, C. C. 1976. First Course in Population Genetics. Pacific Grove: Boxwood Press. 631 p.

Lotka, A. J. 1956. Elements of Mathematical biology. New York: Dover Publ., Inc.

MacDuffee, C. C. 1933. The Theory of Matrices. Berlin: Verlag von Julius Springer.

Mahler, H. R. and Cordes, E. H. 1966. Biological Chemistry. New York: Harper and Row. 872 p.

Martin, F. F. 1968. Computer Modeling and Simulation. New York: John Wiley and Sons Inc. 331 p.

May R. M. 1976. Theoretical Ecology, Principles and Applications. Philadelphia: W. B. Saunders Co. 317 p.

Maynard Smith, J. 1968. Mathematical Ideas in Biology. Cambridge University Press. 152 p.

Maynard Smith, J. 1974. Models in Ecology. Cambridge University Press. 146 p.

McLay, C. 1970. A theory concerning the distance traveled by animals entering the drift of a stream. J. Fish. Res. Bd. Canada. 27: 359-370.

Meadows, D. H., Meadows, D. L., Randers, J. and Behrens, W. W. 1972. The Limits to Growth. New York: Universe Books. 205 p.

Messenger, P. S. and Flitters, N. E. 1958. Effect of constant temperature environments on the egg stage of three species of Hawaiian Fruitflies. Ann. Ent. Soc. Am. 51: 109-119.

Milhorn, H. T. 1966. The Application of Control Theory to Physiological Systems. Philadelphia: W. B. Saunders Co.

Milsum, J. H. 1966. Biological Control Systems Analysis. New York: McGraw-Hill. 466 p.

Monod, J. 1942. Recherches sur La Croissance Des Cultures Bacteriennes. Paris: Hermann. 210 p.

Monod, J., Wyman, J. and Changeux, J. P. 1965. On the nature of allosteric transitions: a plausible model. J. Mol. Biol. 12: 88.

Newsholme, E. A. and Start, C. 1976. Regulation in Metabolism. London: John Wiley and Sons.

Novick, A., and Szilard, L. 1950 Experiments with the chemostat on spontaneous mutations of bacteria. Proc. National Acad. Sci., U.S.A. 36: 708-719.

Odum, E. P. 1968. Energy flow in ecosystems: a historical review. Am. Zoologist. 8: 11-18.

Odum, E. P. 1971. Fundamentals of Ecology. 3rd ed. New York: W. B. Saunders Co. 420 p.

Odum, H. T. 1956. Primary production in flowing waters. Limnol. Oceanog. 1: 102-117.

Olinick, M. 1978. An Introduction to Mathematical Models in the Social and Life Sciences. Reading: Addison-Wesley Publishing Co. 466p.

O'Neill, R. V. 1969. Indirect estimation of energy fluxes in animal food webs. J. Theor. Biol. 22: 284-290.

O'Neill, R. V. 1968. Population energetics of a milipede, Narceus Americanus (Beavois). Ecology. 49: 803-809.

O'Neill, R. V., Goldstein, R. A., Shugart, H. H. and Makin, J. B. 1972. Terrestrial Ecosystem Energy Model. Eastern Deciduous Forest Biome Memo Report #72-19. 38 p.

Park, T. 1948. Experimental studies of interspecies competition Ecological Monographs. 18(2).

Park, T., Mertz, D. B., Grodzinski, W. and Prus, T. 1965. Cannibalistic predation in populations of flour beetles. Physiol. Zool. 38: 289-321.

Patten, B. C. 1971. A primer for ecological modeling and simulation with analog and digital computers. In: Patten, B. C., Systems Analysis and Simulation in Ecology, Vol. I. New York: Academic Press.

Patten, B. C. 1972. A simulation of the shortgrass prairie ecosystem. Simulation. 19: 177-186.

Patten, B. C., Egloff, D. A., Richardson, T. H., et al. 1975. Total ecosystem model for a cove in Lake Texoma. In: Patten, B. C., Systems Analysis and Simulation, Vol. III. 205-421.

Pauling, L. 1977. In: PBS NOVA. Linus Pauling: Crusading Scientist. 1 June 1977.

Pearl, R. 1927. The growth of populations. Quart. Rev. Biol. 2: 532-548.

Pearl, R. 1932. The influence of density of population upon the rate of reproduction in Drosophila. J. Exptl. Zool. 63: 57-85.

Pearl, R. and Miner, J. R. 1935. Experimental studies on the duration of life. XIV: Comparative mortality. Quart. Rev. Biol. 10: 60-79.

Peilou, E. C. 1977. Mathematical Ecology. New York: John Wiley and Sons. 385 p.

Poole, R. W. 1974. An Introduction to Quantitative Ecology. New York: McGraw-Hill Book Co. 532 p.

Powers, W. F. and Canale, R. P. 1975. Some applications of optimization techniques to water quality modeling and control. IEEE Trans. on Systems, Man, and Cybernetics. 5: 3.

Prosser, C. L. and Brown Jr., F. A. 1961. Comparative Animal Physiology. Philadelphia: W. B. Saunders Co. 688 p.

Ramkrishna, D., Frederickson, A. G. and Tuschiya, H. M. 1967. Dynamics of microbial propagation: Models considering inhibitors and variable cell composition. Biotechnol. Bioeng. 9: 129-170.

Randall , J. E. 1980. Microcomputers and Physiological Simulation. Reading: Addison-Wesley Publishing Co. Inc. 235 p.

Rand Corporation. 1955 A Million Random Digits with 100,000 Normal Deviates. Clencove, Ill.: Free Press. 200 p.

Rhee, G. Y. 1973. A continuous culture study of phosphate uptake, growth rate and polyphosphates in Scenedesmus spp. J.Phycol. 9: 495-506.

Rhee, G. Y. 1978. Effects of N:P atomic ratios and nitrate limitation on algal growth, cell composition, and nitrate uptake. Limnol. Ocenog. 23: 10-25.

Rhodes, D. G., Achs, J. J., Peterson, L. and Garfinkel, D. 1968. A Method of calculating time-course behavior of multi-enzyme systems from the enzymatic equations. Computers and Biomed. Res. 2: 45-50.

Richter, O. and Betz, A. 1976. Simulation of biochemical pathways and its application to biology and medicine. In: Levin, S. (ed.). Lecture Notes in Biomathematics. 11: 181-197.

Ricker, W. E. 1954. Stock and recruitment. J. Fish. Res. Bd. Canada. 11: 559-623.

Riggs, D. S. 1970. Control Theory and Physiological Feedback Mechanisms. Baltimore: Williams and Wilkins. 599 p.

Rosen, R. 1970. Dynamical System Theory in Biology. New York: John Wiley and Sons, Inc. 302 p.

Rozenzweig, M. L. and MacArthur, R. H. 1963. Graphical representation and stability conditions of predator-prey interactions. Am. Nat. 117: 209-223.

Ryther, J. H. 1956. Photosynthesis in the ocean as a function of light intensity. Limnol. Oceanog. 1: 61-70.

Ryther, J. H. 1959. Potential productivity of the sea. Science. 130: 602-608.

Savageau, M. A. 1976. Biochemical Systems Analysis. Reading: Addison-Wesley Publishing Co. 379 p.

Searle, S. R. 1966. Matrix Algebra for the Biological Sciences. New York: John Wiley and Sons, Inc. 296 p.

Schelske, C. L., Stoermer, E. F. and Feldt, L. E. 1971. Nutrients, phytoplankton productivity and species composition as influenced by upwelling in Lake Michigan. Proc. 14th Conf. Great Lakes Res. Internat. Assoc. Great Lakes Res.

Schuler, M. L., Leung, S. and Dick, C. C. 1979. A mathematical model for the growth of a single bacterial cell. Ann. N.Y. Acad. Sci. 77: 35-55.

Sias, F. R. and Coleman, T. G. 1971. Digital simulation of biological systems using conversational languages. Simulation. 16: 102-111.

Simon, W. 1972. Mathematical Techniques for Physiology and Medicine. New York: Academic Press. 267 p.

Smith, E. L. 1936. Photosynthesis in relation to light and carbon dioxide. Proc. Natl. Acad. Sci. U.S. 22: 504.

Spain, J. D. 1971. Desk top computer simulations. Simulation 16: 137-142.

Spain, J. D. 1973. The introduction of biological mensuration techniques through simulation. Proc. of Conf. on Computers in Undergraduate Curricula. 3: 80-84.

Spain, J. D. 1981. Teaching Simulation Techniques with Microcomputers. Proc. Nat. Ed. Computer Conf. 2: 225-227.

Spencer, D. D. 1975. A Guide to Basic Programming. Reading: Addison-Wesley Publishing Co. 242 p.

Spiegel, M. R. 1961. Theory and Problems of Statistics. New York: McGraw-Hill Book Co.

Steele, J. H. 1962. Environmental control of photosynthesis in the sea. Limnol. Oceanog. 7: 137-150.

Tanner, J. T. 1975. The stability and intrinsic growth rates of prey and predator populations. Ecology. 56: 855-867.

Usher, M. B. 1972. Developments in the Leslie matrix model. In: Jeffers, J. N. R., (ed.). Mathematical Models in Ecology. Oxford: Blackwell Scientific Publishing Co.

Vollenweider, R. A. 1965. Calculation models of photosynthesis-depth curves and some implications regarding day rate estimates in primary production measurements. Mem. Institute Ital. Idrobiol. 18(suppl.): 425-457.

Walter, C. F. 1974. Some dynamic properties of linear, hyperbolic and sigmoidal multi-enzyme systems with feedback control. J. Theor. Biol. 44: 219-240.

Walters, C. J. 1971. Systems ecology: the systems approach and mathematical models in ecology. In: Odum E. P., Fundamentals of Ecology. New York: W. B. Saunders Co.

Wangersky, P. J. and Cunningham, W. J. 1957. Time lag in prey-predator population models. Ecology. 38: 136-139.

Watson, J. D. 1968. The Double Helix. New York: Atheneum Press.

Watt, K. E. F. 1959. A mathematical model for the effect of densities of attacked and attacking species on the number attacked. Canadian Ent. 91: 129-144.

Watt K. E. F. (ed.) 1966. Systems Analysis in Ecology. New York: Academic Press, Inc.

Watt K. E. F. 1968. Ecology and Resource Management. New York: McGraw-Hill Book Co.

Watt, K. E. F. 1970. The systems point of view in pest management. In: Rabb, R. L. and F. E. Guthrie (ed.), Concepts of Pest Management. Raleigh: N. C. State Univ.

Webb, W. L., Schroeder, H. J. and Norris, L. A. 1975. Pesticide residue dynamics in a forest ecosystem: a compartment model. Simulation. 24: 161-169.

Wiebe, P. H. 1971. A computer model study on zooplankton patchiness and its effects on sampling error. Limnol. Oceanog. 16: 26-38.

Wills, C. 1970. Genetic load. Sci. Amer. 222: 98-107.

Wright, S. 1922. Coefficients of inbreeding and relationship. Am. Nat. 56: 330-338.

Yates, R. A. and Pardee, A. B. 1956. Control of pyrimidine biosynthesis in Escherichia coli by a feed-back mechanism. J. Biol. Chem. 221: 757-770.

Zar, J. H. 1974. Biostatistical Analysis. Englewood Cliffs: Prentice-Hall, Inc. 620 p.

APPENDIX 1

INTRODUCTION TO BASIC PROGRAMMING

What follows is a simplified description of BASIC programming language emphasizing those aspects which are particularly useful for modeling. It is assumed that most students who employ this book will also obtain a copy of a good general instructional manual dealing with the BASIC computer language, such as Spencer, _A Guide to BASIC Programming_, Addison-Wesley Publishing Company, Reading, Mass.

BASIC is an acronym for Beginner's All-purpose Symbolic Instruction Code. The language was developed at Dartmouth College by Professors Kemeny and Kurtz. The system was designed primarily for use on terminals in a time-sharing system. However, it has become the major resident language in most microcomputer systems. This section will emphasize the BASIC dialects employed by the Radio Shack TRS-80 and the Apple II Plus microcomputers. For full details on the language employed by any particular microcomputer, you should consult specific reference manuals.

A table of important command mode instructions for the Apple II and TRS-80 microcomputers is provided at the end of Appendix I.

BASIC Programs

A BASIC program consists of a series of lines of instruction each beginning with a line number followed by a command. Unless directed otherwise, the computer performs the commands in sequential order. The program commands must be stated precisely, because the computer will execute them exactly as they are written. If a given statement, for instance, statement 110, is not written in proper form, the computer will respond with a SYNTAX ERROR IN LINE 110 message.

Line Numbers

Each statement of a BASIC program is assigned a line number which must be an integer between 1 and 63999. These are usually assigned line number 10 for the first statement, 20 for the next, and so on, thus allowing you to add statements that have been forgotten. It is a good idea to start the actual program at line 100.

REM

The REMark statement provides a method for inserting comments into the program. The REM statement may be placed anywhere before the END statement in the program. The form of the statement is

```
040 REM ANY COMMENTS ABOUT THE PROGRAM
```

or:

```
160 Y = A * X + B : REM EQUATION FOR LINE
```

Typically it is used for naming the exercise, telling where it came from, and specifying some of the variables. Any characters may be used to follow REM. On the TRS-80, REM may be abbreviated by '.

READ and DATA

The READ statement is used to assign numerical values to variables or constants. The READ statement is designed to work in conjunction with the DATA statement. The form of the READ statement is as follows:

```
100 READ A, B, N1, N2, K1, K2, DT
```

The variables are listed, separated by commas. The numerical values for variables and constants are listed in the DATA statement in the same order as they were listed in READ. The general form of the DATA statement is

```
200 DATA 10, 32 1000, 3000, 2.5E5, 4.7E8, .1
```

There may be more data in the DATA statement (or more DATA statements) than constants or variables assigned in the READ statement, but there can never be less. Otherwise the computer responds with OUT OF DATA ERROR IN LINE 100. One may think of the data as being assigned to a data bank from which it is withdrawn in sequential order. For example, it does not matter whether the above data were entered as a single DATA statement or three DATA statements, it is the sequence of entry that is important. DATA statements are commonly placed together just before the END of the program, although they may be placed anywhere before END. Once data has been read, it cannot be used again unless the READ statement is followed by a RESTORE statement.

PRINT

The PRINT statement is used to print out the values of variables or explanatory terminology. The general form of the PRINT statement is:

```
110 PRINT "THE VALUE OF X IS ";X;"THE VALUE OF Y IS ";Y
```

A typical example for printing column headings is as follows:

```
100 PRINT "TIME","NO.ORGANISMS","DENSITY"
```

The number of characters and the number of print-columns in a line depend on the type of computer and printer being employed. The following table gives the values for some common equipment.

Equipment Type	Number of Characters	Number of Print Zones
Apple II, video display	40	3
Silentype printer	80	5
TRS-80, video display	64	4
Centronics printer	80	5

When using the standard print format, where zones are separated with commas, the size of the print zone is 15 characters, except for the last column which gets the residue. Thus, if you divide the number of characters by 15 you can find the number of print zones. For the Apple II video, 40/15 = 2.67. Hence, there are two full print zones and one 10 character short zone.

In the standard PRINT format, a comma causes the computer to skip to the next print zone, while a semicolon is interpreted as don't skip. The result is that the next item would be printed immediately after, with no space between. In using quote marks (") note that they must always be in pairs.

To cause the computer to print a space between lines simply provide a PRINT statement with nothing behind it, as follows:

```
120 PRINT
```

The statement required for printing the results of a repetitive calculation usually appears at the end of the calculation loop. For example the following statement would presumably follow the ones listed above after the number of organisms, N, and density, D, have been calculated for each time interval, T:

```
190 PRINT T, N, D
```

<u>LET</u>

Mathematical expressions are usually written as LET statements. The LET statement is used to assign a value to one or more variables. The computer executes the operation designated on the right side of the equals sign and assigns the result to the variable designated on the left side of the equals sign. This causes the value to be computed and used to replace the previous value for that variable. An example of a LET statement follows.

```
150 LET Y = 3*X^2/(2.55 + X) - 3E9
```

For most microcomputers, it is <u>not</u> necessary to include the word LET in this statement! The following symbols are used to denote each of the arithmetic operations in BASIC:

Addition +	Division /
Subtraction -	Exponentiation ^ or ↑
Multiplication *	Scientific notation, E (4×10^{-3} is 4E-3)

The example given in statement 150 has the meaning:

$$Y = \frac{3x^2}{(2.55 + x)} - 3 \times 10^9$$

Arithmetic expressions are evaluated according to the rules of precedence, in the following order:

1. items enclosed in parentheses
2. exponentiation
3. multiplication and division
4. addition and subtraction.

The computer scans from left to right, and evaluates all operations enclosed by parentheses. Next, it evaluates exponentiations, and so on. Since there cannot be too many parentheses, one way to prevent a violation of the rules of precedence is to use parentheses whenever there is a question. However, make sure parentheses are always used in pairs, otherwise the computer will respond with a SYNTAX ERROR!

Variables

Simple variables are designated either by a letter, a letter followed by a single digit, or by two letters. Examples would be A, B, P, Q, N1, N2, K1, K2, AB or PQ. Letters with double numbers like K11 or N21 are acceptable variables, but only the first digit is recognized by the computer. Certain double letters may not be used as variables since they have a special meaning to the computer and are called reserved words. These include AT, TO, IF, OR, ON, GR, and FN. Longer variable names may be used, such as TIME or DISTANCE. However, the computer only recognizes the first two letters as being unique. Thus, TIDE could not be distinguished from TIME by the computer.

Constants

Values for those constants which are not going to be changed during the use of a model could be written as part of the algebraic expression in which they are required. For example:

```
110 A = 3.1416*R^2
```

Another example would be:

```
110 Y = 3*X^2 + 5*X - 15
```

Those constants which one may wish to change during different runs of a simulation are best included in the algebraic expression as variables. As such, their value would be assigned by the READ and DATA statements. For example, the equation for numerical integration of population growth in a limited environment is

$$N_{t+\Delta t} = N_t + rN_t(1 - \frac{N_t}{K})$$

In a simulation based on this equation, one is frequently called upon to vary either the intrinsic growth rate, r, or the carrying capacity, K, or both. The variables in this equation are the population density, N, and time, T. Assuming time is incremented in units of 1, the value of N is replaced by the new value as follows:

$$N \longleftarrow N + rN(1 - \frac{N}{K})$$

This would be programmed

```
100 READ N, K, R
200 LET N = N + R*N*(1 - N/K)
300 DATA 20,500,.5
```

Remember that the computer evaluates the expression to the right of the equals sign, and assigns it to the variable on the left. Thus, the old value of N is replaced with the newly calculated value of N.

The program above would normally include many statements in addition to those listed.

<u>GO TO</u>

The GO TO statement is used to transfer control to another line of the program. This is sometimes called an unconditional jump statement. It is used to loop back so that a routine may be repeated, or to jump ahead if a segment of program is to be by-passed. A typical example of the use of the GO TO statement follows:

```
290 GO TO 110
```

Statement 110 might be a READ statement, in which case the computer would continue to read data from the DATA bank until the data were exhausted. The GO TO statement is often combined with the IF...THEN statement in branching programs.

<u>STOP</u>

The STOP statement terminates the execution of the program. It is often used to separate the subroutines from the main program.

<u>IF...THEN</u>

The IF...THEN statement is very important to the programmer as it allows operations to be executed only if certain conditions are met. The general form of the IF...THEN statement is as follows:

```
900 IF (expression) relation (expression) THEN (operation)
```

The expression may be an arithmetic expression, a variable, or a constant.

The relation between the expressions should be one of the following:

Relation	Meaning
<	is less than
<= or =<	is less than or equal to
>	is greater than
>= or =>	is greater than or equal to
=	is equal to
<> or ><	is not equal to

Some examples of simple IF...THEN statements follow:

```
200 IF N > 20 THEN GO TO 110
210 IF N < A*B THEN N = A*B
220 IF N < 30 THEN N = N*(1 - N/K) : R = 20 : GO TO 100
```

Note that in the last example all three operations are executed only if the condition N < 30 is met. If N => 30 then the computer moves on to the next statement and no operations to the right of THEN are executed.

The IF...THEN statement may also be combined with another IF...THEN statement in a logical manner as follows:

```
230 IF N > 20 OR N < 30 THEN GOTO 300
240 IF X = 5 AND Y > 10 THEN Z = 40 : GO TO 100
```

<u>FOR and NEXT</u>

The best way to cause a routine to be repeated a specified number of times is to employ the FOR and NEXT statements. The FOR statement specifies the number of times that a certain segment of the program is to be repeated. The NEXT statement tells the computer to loop back to the FOR statement and to check if the correct number of repeats has been attained. If it has not, the loop is repeated. Otherwise, control is passed to the statement following the NEXT statement. Time may be incremented or decremented in STEPs of varying size. If the STEPs are in units of 1, the STEP size need not be specified. Examples follow:

```
100 FOR T = 1 TO 30
110 etc...
190 NEXT T
```

These statements cause every instruction between 100 and 190 to be repeated 30 times.

STEP size other than 1 is specified as follows;

```
110 FOR N = 0 TO 10 STEP .5
```

To decrement,

```
115 FOR C = 100 TO 0 STEP -2
```

Every FOR statement must be paired with a NEXT statement. FOR and NEXT statements may be nested, but associated FOR and NEXT statements must either be completely inside, or completely outside, any other FOR...NEXT loops.

Nested FOR...NEXT loops are often used in programs where time needs to be changed by small, variable increments. In the following example, DT represents a small time increment, Δt, and IM is the number of increments in a major time interval.

```
100 DT = .1
110 IM = INT(1/DT + .1)
120 FOR T = 1 TO 100
130 FOR I = 1 TO IM
140 N = N + R*N*DT
150 NEXT I
160 PRINT T, N
170 NEXT T
180 END
```

Note that the inner loop, statements 130 to 150, allows you to vary the size of DT, and still obtain data output only for the major time intervals.

Statement 110 calculates the number of increments, IM, per major time interval, T, based on the size of the increment, DT.

<u>Numeric Functions</u>

These functions may be used in any algebraic expression and return a numeric value. Here X represents any variable or expression.

ABS(X) This function returns the absolute value of X.

ATN(X) This function returns the Arctangent of X, the angle in radians whose tangent is X.

COS(X) This function returns the cosine of X when X is expressed in radians.

EXP(X) This function returns e raised to the X power, antilog_e of X.

EXP(2.302585*X) This function gives the antilog of X if X is a Log_{10}.

INT(X) The greatest integer which is less than or equal to X is returned by this function.

LOG(X) This function returns log to the base e of X. To find the log to the base b, use the formula

$$\log_b(X) = \log_e(X)/\log_e(b)$$
$$\log_{10}(X) = \text{LOG}(X)/2.303$$

RND(1) This function returns a random number uniformly distributed between 0 and 1. Note that the random number will be a decimal fraction.

RND(X) This function produces random integers between 1 and X (TRS-80 only). On the Apple II use R = INT(RND(1)*X + 1).

RANDOM This statement normally precedes the RND(1) function in TRS-80 programs. The RANDOM instruction re-seeds the pseudo-random number routine so that a different string of numbers is produced each time.

SGN(X) This function returns 0 if X is 0, -1 if X is negative, and +1 if X is positive.

SIN(X) This function returns the sine of X when X is given in radians.

SQR(X) This function returns the square root of X.

TAN(X) This function returns the tangent of X when X is given in radians.

It is legitimate to include a mathematical expression involving X in place of X. For example, INT(2.5*X^3/(X-2)) would be calculated with no difficulty.

PRINT TAB

The PRINT TAB statement is used to provide a print format different from that described under PRINT. Using PRINT TAB, column headings and print zones may be placed any where you wish. The TAB system is analogous to a tab system on a typewriter in that it causes the computer to jump to a specified column position before printing. The general form used on most computers is

```
200 PRINT TAB (3);"X IS ";X;TAB(30);"Y IS ";Y
```

This statement would cause the computer to move over to position 3, and print out X IS, followed by value of X, and then move over to position 30, and print out Y IS, followed by the value of Y. The print positions are numbered from left to right.

HTAB

On the Apple II, the PRINT statement listed above could be accomplished by the following statement employing HTAB:

```
200 HTAB 3 : PRINT"X IS ";X; : HTAB 30 : PRINT"Y IS ";Y
```

PRINT USING (TRS-80 only)

Frequently, the standard four column print format which is provided by the PRINT statement does not provide enough columns for the amount of information which must be printed. It is very easy to set up an alternate format with the PRINT USING statement. This statement includes a form of the output and thus tells the computer the number of spaces, by using # symbols, and the position of the decimal. For example, the following would divide the 64 positions of the print line into six columns of 10, each containing 2 decimal places:

```
220 PRINT USING "#######.##":  A,B,C,D,E,F
```

In this example A,B,C,... are variables with less than seven digits to the left of the decimal point. This statement would have been preceded by a statement describing the positions of the column headings, using the PRINT TAB instruction

```
120 PRINT TAB(7);"A";TAB(17);"B";TAB(27);"C";TAB(37);"D";TAB(47);
    "E";TAB(57);"F"
```

GOSUB and RETURN

The GOSUB instruction is used to call up a subroutine. A subroutine is a sequence of instructions used repeatedly, or used at several points in the same program. Control is transferred to the statement number that is included in the GOSUB instruction. For example:

```
110 GOSUB 240
```

This statement would transfer control to statement 240 which is presumably the first statement number of the subroutine. The instructions of the subroutine would be followed in order until a RETURN instruction causes control to be transferred back to the statement following GOSUB, normally, statement 120.

GRAPH Subroutine

Because of the importance of graphical output in most of the exercises in this book, a GRAPH subroutine has been provided. See Appendix 6. The following description is based on the Apple II version of the GRAPH subroutine. For the TRS-80 version, refer to the user's guide which accompanies the TRS-80 disk.

Load the GRAPH subroutine into the computer just before you begin to enter your program. You may then build your program, taking into account that lines 0 - 21 and 3000 to 4100 are reserved for the GRAPH subroutine. It is recommended that you use lines 30 - 90 for REM statements identifying the exercise number, exercise name, your name, and the date. Thus, a listing of the program would have a proper heading for submission to your instructor.

The actual program should start at line 100. Typically, it would be written so as to plot data as it is calculated although it is possible to store the data in an array and plot it later. Thus, the axes for the graph would be drawn before the main calculation loop is begun. To draw axes, first call for HGR to place the computer in the high resolution graphics mode. Then specify the names of the variables on the x and y axes, for example, X$ = "TIME" : Y$ = "MONEY". Then specify the maximum units on the x-axis, such as, XM = 100, and the maximum units on the y-axis, such as, YM = 900. Note that the maximum number of digits allowed for XM is 3 and the maximum for YM is 4. If you wish to specify minimum values for the x or y axis, they would be defined as XN and YN. If these minimums are not specified, they will be zero by default. Finally, call for the axes subroutine by GOSUB 3000. The complete set of instructions required to draw and label axes may be placed on a single line as follows:

```
130 X$ = "TIME" : Y$ = "MONEY" : XM = 100 : YM = 900: GOSUB 3000
```

The actual plotting process is executed by specifying the values of X and Y and then calling for the subroutine 4000. The statement is normally placed just before the NEXT T at the end of the time loop. The following example assumes that money, M, is a function of time, T.

```
150 FOR T = 1 TO 100
160 M = 3.95*T
170 X = T : Y = M : GOSUB 4000
180 NEXT T
```

If you want to plot lines rather than points, the following 'flag' statement should be placed just before the time loop starts:

```
140 ZF = 1 : ZG = 0
```

If you want points marked with circles or crosses, set flag ZO = 1 to indicate you want circles and ZP = 1 to indicate you want crosses.

The graph subroutine may also be used to write messages on HGR. Simply specify the message as follows:

```
L$ = "MESSAGE"
```

Then specify the x and y coordinates for the beginning of the message as X0 and Y0, and call for the subroutine by GOSUB 3500. The complete statement would be

```
135 L$ = "TOTAL AMOUNT" : X0 = 50 : Y0 = 100 : GOSUB 3500
```

Note that coordinate positions on HGR go from 0 to 279 on the x-axis and 0 to 159 on the y-axis. The origin is the top left corner.

Arrays and Subscripted Variables

Subscripted variables are designated by one or two characters followed by one or more integers or letters enclosed in parentheses. The following are examples of subscripted variables:

A(1), A(2), A(3), ... are all variables in a list or array called A.
B(1), B(2), B(3), ... are all variables in a list or array called B.

On the other hand, C(1,1), C(1,2), C(1,3), C(2,1), ... are elements of a two dimensional array called C. Use of these subscripted variables does not preclude the use of regular scalar variables having the same letters. A, B, CC and A1, B3, C0 may all be used in the same program with the above subscripted variables.

BASIC automatically saves space for arrays which involve no more than 10 elements. However, for larger arrays, one must first reserve storage by the use of a dimension or DIM statement. The DIM statement is placed at the beginning of the program, as re-dimensioning will result in an error. The following dimension statement would reserve storage for 3 lists of varying length designated A, B, and I; and two 6 by 6 matrices, designated C and D:

```
100 DIM A(25),B(100),I(1000),C(6,6),D(6,6)
```

As illustrated, lists may be of almost any length. In dimensioning matrices, the number of rows are specified first, and the number of columns second. The largest matrix that can be accommodated by microcomputers is about 50 by 50.

One of the more valuable applications for a list of subscripted variables is a result of the ability to define one variable by another variable. Let us consider list "A" with subscript "I", and use it to obtain a new list, "B". We could read data into the A-list sequentially by varying I from 1 to 26

```
100 DIM A(26)
110 LET I = 1 to 26
120 READ A(I)
130 NEXT I
140 DATA 16,12,7,8,...,19,23
```

Lists such as this are conveniently employed to perform certain repetitive operations. For example, we could obtain the mean between each successive pair of values in the above list and record these in a new list called "B".

```
150 PRINT "I", "A(I)", "MEAN"
160 DIM B(25)
170 FOR I = 1 TO 25
180 LET B(I) = (A(I) + A(I + 1))/2
190 NEXT I
```

With the subscripted variable, it is also possible to store data and print it out at the end of all calculations. For example, we could print out the previous data as follows:

```
200 FOR I = 1 TO 25
210 PRINT I, A(I), B(I)
220 NEXT I
```

Subscripted variables may also be used to sort numbers in order to produce a histogram. Assume we have 1000 integers between, 1 and 25, which have been listed as data or produced during a simulation. These may be defined as I, and counted in the following way:

```
100 PRINT "NUMBER", "FREQUENCY"
110 DIM A(25)
120 FOR N = 1 TO 1000
130 READ I
140 LET A(I) = A(I) + 1
150 NEXT N
160 FOR I = 1 TO 25
170 PRINT I, A(I)
180 NEXT I
190 DATA 17,21,3,...,6,22
200 END
```

Sorting or counting routines of this type are very valuable for Monte Carlo simulations. These simulations normally must be run a great many times to give meaningful results, and thus usually require some data analysis.

Matrix Subroutines

The following section provides some standard matrix subroutines.

Inputting Data to a Matrix

```
100 DIM A(R,C), B(R,C), C(R,C), Y(C), X(C)
110 FOR I = 1 TO R : REM ROWS
120 FOR J = 1 TO C : REM COLUMNS
130 READ A(I,J)
140 NEXT J
150 NEXT I
160 DATA 6,1,3,8,...
```

Printing a Matrix

```
200 FOR I = 1 TO R
210 FOR J = 1 TO C
220 PRINT A(I,J);" ";
230 NEXT J : PRINT " "
240 NEXT I
```

Scalar Multiplication by a Constant, K

```
300 FOR I = 1 TO R
310 FOR J = 1 TO C
320 B(I,J) = A(I,J)*K
330 NEXT J
340 NEXT I
```

Post-multiplication of a Matrix by a Vector, X(C)

```
400 FOR I = 1 TO R
410 Y(I) = 0
420 FOR J = 1 TO C
430 Y(I) = Y(I) + A(I,J)*X(J)
440 NEXT J
450 NEXT I
```

When multiplying a matrix by a vector, R must equal C.

Important Command Mode Instructions for Apple II and TRS-80

The following table lists some of the important commands which are used when interacting with the computer. Analogous commands are provided for both the Apple II Plus and Radio Shack TRS-80 microcomputers.

(See relevant reference manuals for details.)

Apple II Plus	Command Operation Performed	TRS-80
<RETURN>	Causes any keyboard entry to be transferred to memory.	<ENTER>
<RESET> or <CTRL>C	Stop program execution and places computer in Command mode.	<BREAK>
HOME	Clears text display, and returns cursor to top left position.	<CLEAR>
LOAD NAME	Loads program called NAME from diskette to memory.	LOAD "NAME"
SAVE NAME	Saves program called NAME from memory to diskette.	SAVE "NAME"
RUN NAME	Loads and runs program called NAME.	RUN "NAME"
RUN	Clears all variables and causes program to be executed.	RUN
RUN 100	Starts RUN at line 100.	RUN 100
LIST	Displays complete program from the beginning.	LIST
LIST 10-500	Displays lines 10 through 500 only.	LIST 10-500
LIST 300-	Displays program starting at line 300	LIST 300-
DEL 10, 200	Deletes lines 10 through 200	DELETE 10-200
310	Deletes line 310 only...(gone forever).	310
CONT	Continues program execution following STOP, <BREAK> on the TRS-80, or <CTRL>C on the Apple II Plus.	CONT
TRACE	Causes line numbers for each statement to be displayed as program is executed.	TRON
NO TRACE	Turns off TRACE.	TROFF
<ESC> I, J, K, M	Allows you to edit a program statement. Refer to manuals for full details on editing program statements.	EDIT

APPENDIX 2

CURFIT: A PROGRAM FOR FITTING THEORETICAL EQUATIONS TO DATA

```
1  IF  PEEK (104) = 64 THEN  GOTO 5
2  PRINT "PROGRAM LOADING ERROR.......                RE-ENTER 'RUN CURFIT' AN
    D PRESS <RETURN>"
3  PRINT  CHR$ (4);"EXEC MODE II": REM  RESET PROGRAM START POINTERS AT 16
    384
4  STOP
5  HOME : REM  LOAD CHARACTERS FOR HGR
10  POKE 232,0: POKE 233,08: SCALE= 1: HCOLOR= 3: ROT= 0
20 D$ =  CHR$ (4): REM  CONTROL-D CHAR
30  PRINT D$;"BLOAD SMALL CHARACTERS,A$0800": REM

40  REM  TITLE
50 L$ = "CURFIT":X0 = 90:Y0 = 80: SCALE= 2:SC = 2: HCOLOR= 2: HGR : GOSUB
    3500
60  FOR I = 1 TO 800: NEXT I
70  DIM X(100),Y(100),R(100): REM

110  REM  EXPANDED TITLE
120  TEXT : HOME
125  PRINT : PRINT : PRINT : PRINT
130  PRINT : PRINT "LINEAR LEAST SQUARES CURVE FITTING": PRINT : PRINT
140  PRINT "DEVELOPED FOR THE TEXT:                  'BASIC MICROCOMPUTER M
     ODELS IN BIOLOGY'   BY JAMES SPAIN                           ADDISON-W
     ESLEY PUBLISHING COMPANY"
145  VTAB 22: PRINT "FOR FURTHER INFORMATION ..PRESS RETURN": INPUT " ";Q$

146  REM

147  HOME : REM  INTRODUCTION
150  PRINT : PRINT "THE PROGRAM PERFORMS THE FOLLOWING TASKSAFTER A SET OF
      'X,Y' DATA HAS BEEN       ENTERED:"
155  PRINT
160  PRINT "1. THE DATA ARE PLOTTED FOR EXAMINATION"
170  PRINT "2. THE USER SELECTS ONE OF TEN STANDARD     EQUATION TYPES."
180  PRINT "3. THE DATA IS TRANSFORMED AND FITTED BY    LINEAR LEAST SQUARE
     S."
190  PRINT "4. THE COEFFICIENTS FOR THE FITTED MODEL    ARE PROVIDED."
200  PRINT "5. X, Y , MODEL Y, AND RESIDUALS ARE        CALCULATED AND LIST
     ED."
210  PRINT "6. SEVERAL STATISTICS ARE PROVIDED."
220  PRINT "7. THE RESIDUALS MAY THEN BE PLOTTED."
```

```
230  PRINT "8. THE CURVE MAY BE PLOTTED ALONG WITH     ORIGINAL DATA."
240  PRINT "9. SUMMARY DATA AND STATISTICS FOR THE     BEST FITTING MODEL
     MAY BE PRINTED OUT   FOR A PERMANENT RECORD."
250  PRINT : INPUT "TO ENTER DATA.....PRESS <RETURN>";Q$
253  REM

255  HOME : PRINT : REM  DATA ENTRY OPTIONS
257  PRINT "YOU MAY ENTER DATA IN EITHER OF THE     FOLLOWING WAYS:
                     1. FROM THE KEYBOARD                          2. FRO
     M A PREVIOUSLY SAVED FILE."
260  PRINT : PRINT : INPUT "WHICH MODE OF DATA ENTRY ( 1 OR 2 )     ? ";A$

261 A$ =  LEFT$ (A$,1)
263  IF A$ = "1" OR A$ = "K" THEN  GOTO 280: REM

264  REM  DATA ENTRY FROM DISK FILE
265  INPUT "NAME OF TEXT FILE TO BE USED? ";N$
266 D$ =  CHR$ (4)
267  PRINT D$;"OPEN";N$: PRINT D$;"READ";N$: INPUT M: FOR I = 1 TO M: INPUT
     X(I),Y(I): NEXT : PRINT D$;"CLOSE";N$
270  GOTO 385: REM

275  REM  DATA ENTRY FROM KEYBOARD
280  PRINT : INPUT "ENTER THE NUMBER OF DATA POINTS TO BE   FITTED  ";M
295  HOME
300  PRINT "NOW ENTER THE DATA POINTS, FIRST X,     THEN Y."
310  PRINT : PRINT "TO CORRECT AN ERROR AFTER DATA ENTRY............TYPE
     (E).
320  PRINT : PRINT "ZERO VALUES ARE NOT ALLOWED IN CERTAIN  TRANSFROMATION
     S........ DELETE THE DATA POINT."
330  PRINT
340  FOR I = 1 TO M
350  PRINT "ENTER X,Y DATA POINT NUMBER ";I;: INPUT " ";X$,Y$
355  IF  LEFT$ (X$,1) <  > "E" OR  LEFT$ (Y$,1) <  > "E" THEN  GOTO 365
360 I = I - 1: PRINT " ": GOTO 350
365 Y(I) =  VAL (Y$)
370  LET X(I) =  VAL (X$)
380  IF X(I) =  < 0 OR Y(I) =  < 0 THEN  PRINT "PROGRAM WILL NOT TAKE VALU
     ES EQUAL TO OR LESS THAN ZERO !!": GOTO 350
382  NEXT I
385  PRINT : INPUT "DO YOU NEED TO CORRECT ANY OF THE DATA  ? ";A$
386  IF  LEFT$ (A$,1) = "Y" THEN  GOSUB 6000: REM

387  REM  FINDING THE MAXIMUM VALUES OF X AND Y
388  FOR I = 1 TO M
390  IF XM < X(I) THEN XM = X(I)
400  IF YM < Y(I) THEN YM = Y(I)
405  NEXT I
415  HOME
420  HGR :X$ = "X VALUE":Y$ = "Y VAL.": GOSUB 3000
470  FOR I = 1 TO M
473 X = X(I):Y = Y(I):Z0 = 1: GOSUB 4000
474  NEXT I
475  HOME : VTAB 22: REM

480  INPUT "TO CHOSE AN EQUATION FOR FITTING.......PRESS <RETURN>";Q$
485  TEXT : HOME
490  PRINT "THE FOLLOWING EQUATIONS MAY BE FITTED BY THIS PROGRAM"
500  PRINT
```

```
510  PRINT "1. STRAIGHT LINE      Y = A*X + B "
515  PRINT
520  PRINT "2. EXPONENTIAL        Y = A*EXP(N*X)"
525  PRINT
530  PRINT "3. POWER FUNCTION     Y = A*X©N "
535  PRINT
540  PRINT "4. HYPERBOLIC         Y = A*X/(B + X)"
545  PRINT
550  PRINT "5. EXP. SATURATION    Y = A*(1 - EXP(N*X))"
560  PRINT "6. SIGMOID            Y = A/(1 + B*X©N)"
565  PRINT
570  PRINT "7. EXP. SIGMOID    Y = A/(1 + B*EXP(N*X))"
580  PRINT "8. MODIFIED INVERSE  Y = A/(B + X)"
585  PRINT
590  PRINT "9. MODIFIED POWER FN.Y = A*X©N + B"
600  PRINT
605  PRINT "10.MAXIMA FUNCTION    Y = A*X*EXP(N*X)"
607  PRINT
610  INPUT "ENTER EQUATION NUMBER TO BE FITTED..";Q
620  LET LP = 0: REM  LINE PRINTER OFF
630  IF Q > 10 THEN  PRINT "ERROR....THIS IS NOT ONE OF THE EQUATION NUMBE
     RS !!": GOTO 610
640  LET X1 = 0:X2 = 0:X3 = 0:Y1 = 0:Y2 = 0:B = 0:A = 0:N = 1
643  HOME : REM

645  PRINT : INPUT "IF YOU WANT TO OFF-SET THE VALUES OF X  THEN ENTER OFF
     -SET VALUE................OTHERWISE PRESS <RETURN>";Q$
646  IF Q$ = "" THEN OS = 0: GOTO 650
647 OS =  VAL (Q$)
650  IF Q < 5 OR Q > 7 THEN 660
652  INPUT "ENTER ESTIMATED VALUE OF A ";A
655  IF A =  < YM THEN  PRINT "ESTIMATE OF A IS TOO LOW...TRY AGAIN !": GOTO
     650
660  IF Q = 9 THEN  INPUT "ENTER              ESTIMATED VALUE OF B ";B: REM


663  PRINT : PRINT "NOW TRANSFORMING DATA AND FITTING TO THELINE BY LEAST
     SQUARES. PLEASE BE PATIENT"
667  IF LP = 1 THEN  PR# 1: PRINT : PRINT : PRINT "STATISTICS FOR THE MODE
     L WITH THE          HIGHEST VALUE FOR 'F' :"
670  FOR D = 1 TO M
680  REM  TRANSFORMATION OF X AND Y BEGIN***
690 X = X(D) - OS:Y = Y(D)
700  IF Q = 2 THEN Y =  LOG (Y(D))
710  IF Q = 3 THEN Y =  LOG (Y(D))
720  IF Q = 4 THEN Y = X / Y(D)
730  IF Q = 5 THEN Y =  LOG (A - Y(D))
740  IF Q = 6 THEN Y =  LOG (A / Y(D) - 1)
750  IF Q = 7 THEN Y =  LOG (A / Y(D) - 1)
760  IF Q = 8 THEN Y = 1 / Y(D)
770  IF Q = 9 AND Y(D) - B = 0 THEN  PRINT "YOUR ESTIMATE OF B MUST BE OUT
     SIDE THE  RANGE OF Y VALUES !!! ": GOTO 660
780  IF Q = 9 THEN Y =  LOG ( ABS (Y(D) - B))
785  IF Q = 10 THEN Y =  LOG (Y(D) / X)
790  IF Q = 3 THEN X =  LOG (X)
800  IF Q = 6 THEN X =  LOG (X)
810  IF Q = 9 THEN X =  LOG (X)
815  REM
```

```
820  REM  CALCULATING THE SUMS OF SQUARES************
830  LET X1 = X1 + X
840  LET X2 = X2 + X © 2
850  LET X3 = X3 + X * Y
860  LET Y1 = Y1 + Y
870  LET Y2 = Y2 + Y © 2
880  NEXT D
885  REM

890  REM  LEAST SQUARES INTERCEPT (I) AND SLOPE (S)
900  LET I = (Y1 * X2 - X1 * X3) / (M * X2 - X1 © 2)
910  LET S = (M * X3 - X1 * Y1) / (M * X2 - X1 © 2)
920  LET R2 = ((X3 - X1 * Y1 / M) © 2) / (X2 - (X1 © 2) / M) / (Y2 - (Y1 ©
     2) / M)
930  HOME : PRINT "THE FOLLOWING ARE STATISTICS FOR THE    TRANSFORMED DAT
     A:"
940  PRINT : PRINT "THE BEST FITTING STRAIGHT LINE FOLLOWS"
950  PRINT " Y  =  "; INT (I * 1000) / 1000;"  +  "; INT (S * 1000) / 1000
     ;" * X"
960  PRINT : PRINT "COEFFICIENT OF DETERMINATION (R©2) IS:  "; INT (R2 * 1
     0000) / 10000
965  REM

970  REM  DEFINING A, B, AND N BY SLOPE AND INTERCEPT ******
990  IF Q = 1 THEN B = I:A = S
1000  IF Q = 4 THEN A = 1 / S:B = I * A
1010  IF Q = 3 THEN A =  EXP (I):N = S
1020  IF Q = 2 THEN A =  EXP (I):N = S
1030  IF Q = 5 THEN A =  EXP (I):N = S
1040  IF Q = 7 THEN B =  EXP (I):N = S
1050  IF Q = 6 THEN B =  EXP (I):N = S
1060  IF Q = 9 THEN A =  SGN (Y(D - 1) - B) *  EXP (I):N = S
1070  IF Q = 8 THEN A = 1 / S:B = I * A
1080  IF Q = 10 THEN A =  EXP (I):N = S
1090  PRINT
1100  PRINT "THE VALUE OF A IS "; INT (A * 10000) / 10000
1110  IF Q = 2 OR Q = 3 OR Q = 5 THEN  GOTO 1140
1120  PRINT "THE FITTED VALUE OF B IS "; INT (B * 10000) / 10000
1130  IF Q = 1 OR Q = 4 OR Q = 8 THEN  GOTO 1150
1140  PRINT "THE FITTED VALUE OF N IS "; INT (N * 10000) / 10000
1150  PRINT : IF Q = 1 THEN  PRINT "    Y = A*X + B"
1160  IF Q = 2 THEN  PRINT "    Y = A * EXP(N*X)"
1170  IF Q = 3 THEN  PRINT "    Y = A * X©N"
1180  IF Q = 4 THEN  PRINT "    Y = (A * X)/(B + X)"
1190  IF Q = 5 THEN  PRINT "    Y = A * (1 + EXP(N*X))"
1200  IF Q = 6 THEN  PRINT "    Y = A / (1 + B*X©N)"
1210  IF Q = 7 THEN  PRINT "    Y = A / (1 + B * EXP(N*X))"
1220  IF Q = 8 THEN  PRINT "    Y = A / (B + X)"
1230  IF Q = 9 THEN  PRINT "    Y = A * X©N  +  B"
1235  IF Q = 10 THEN  PRINT "    Y = A * X * EXP(N * X )"
1240  IF LP = 1 THEN  PRINT : PRINT : GOTO 1430
1390  PRINT : REM

1410  PRINT "TO COMPARE ORIGINAL DATA WITH MODEL"
1420  INPUT "       PRESS.....<RETURN>";Q$
1425  HOME
1430  PRINT "COMPARISON OF MODEL DATA WITH ORIGINAL"
1440  PRINT "X       Y DATA     PREDICTED Y   RESIDUALS"
1450  LET Y0 = 0:Y2 = 0:Y3 = 0:Y4 = 0:Y6 = 0: FOR D = 1 TO M
```

```
1460  REM  CALCULATING Y FROM MODEL EQN.******
1465 X = X(D) - OS
1470  IF Q = 1 THEN Y = A * X + B
1480  IF Q = 2 THEN Y = A *  EXP (N * X)
1490  IF Q = 3 THEN Y = A * X © N
1500  IF Q = 4 THEN Y = A * X / (B + X)
1510  IF Q = 5 THEN Y = A * (1 -  EXP (N * X))
1520  IF Q = 6 THEN Y = A / (1 + B * X © N)
1530  IF Q = 7 THEN Y = A / (1 + B *  EXP (N * X))
1540  IF Q = 9 THEN Y = A * X © N + B
1550  IF Q = 8 THEN Y = A / (B + X)
1555  IF Q = 10 THEN Y = A * X *  EXP (N * X)
1560  LET R(D) = Y(D) - Y
1570  PRINT X(D);: HTAB 8: PRINT Y(D);: HTAB 18: PRINT  INT (1000 * Y) / 1
     000;: HTAB 32: PRINT  INT (1000 * (Y(D) - Y)) / 1000
1580  REM

1590  REM  CALCULATING SUMS OF Y , MODEL Y , SQUARES , ETC..****
1600  LET Y0 = Y0 + Y: LET Y2 = Y2 + Y © 2: LET Y3 = Y3 + Y(D) © 2
1610  LET Y4 = Y4 + Y(D): LET Y6 = Y6 + (Y - Y(D)) © 2
1620  NEXT D
1630  IF LP = 1 THEN  PRINT : PRINT : GOTO 1660
1640  PRINT : INPUT "FOR ADDITIONAL STATISTICS PRESS..RETURN";Q$
1650  REM

1660  PRINT "SUMMARY OF STATISTICS ON FITTED MODEL": PRINT
1670  PRINT "THE RESIDUAL SUM OF THE SQUARES IS "; INT (Y6 * 100) / 100: PRINT

1690 P = 2
1700  PRINT "THE VARIANCE FOR ";M - P;" DEGREES OF FREEDOM IS" INT (Y6 / (
     M - P) * 1000) / 1000
1710  PRINT : PRINT "THE STANDARD ERROR OF THE ESTIMATE IS   "; INT ( SQR
     (Y6 / (M - P)) * 1000) / 1000
1720  LET SE(Q) =  SQR (Y6 / (M - P))
1750  LET F(Q) = (Y2 - (2 * Y4 / M) * Y0 + (Y4 © 2) / M) / (Y6 / (M - P))
1760  PRINT : PRINT "THE VALUE FOR F IS "; INT (F(Q))
1770  IF LP = 1 THEN  PR# 0: GOTO 1930
1790  PRINT
1810  PRINT : PRINT : PRINT : REM

1820  INPUT "DO YOU WANT A PLOT OF THE RESIDUALS ? ";Q$
1830  IF  LEFT$ (Q$,1) <  > "Y" THEN  GOTO 1905
1840  HGR : HPLOT 23,0 TO 23,159: HPLOT 23,80 TO 279,80
1845 L$ = "RESIDUAL":Z0 = 1:X0 = 10:Y0 = 75: GOSUB 3500
1860  FOR D = 1 TO M
1863 X = 20 + X(D) * 256 / XM:Y = 80 - R(D) * 300 / YM
1865  IF Y > 0 AND Y < 159 THEN  DRAW 43 AT X,Y
1870  NEXT D
1905  HOME : VTAB 22: REM

1910  INPUT "DO YOU WANT TO COMPARE DATA WITH THE    MODEL REGRESSION PLOT
     ? ";Q$
1920  IF  LEFT$ (Q$,1) <  > "Y" THEN  GOTO 2220
1930  HGR :X$ = "X VALUES":Y$ = "Y VAL.": GOSUB 3000
1935 ZF = 1:ZG = 0
1960  REM  CALCULATIONS FOR THE MODEL REGRESSION PLOT ***
1970  FOR D = 0 TO 100:X = D * XM / 100
1980  IF Q = 1 THEN Y = A * X + B
1990  IF Q = 2 THEN Y = A *  EXP (N * X)
2000  IF Q = 3 THEN Y = A * X © N
```

```
2010  IF Q = 4 THEN Y = A * X / (B + X)
2020  IF Q = 5 THEN Y = A * (1 -  EXP (N * X))
2030  IF Q = 6 THEN Y = A / (1 + B * X © N)
2040  IF Q = 7 THEN Y = A / (1 + B *  EXP (N * X))
2045  IF Q = 8 AND D = 0 THEN  GOTO 2100
2050  IF Q = 8 THEN Y = A / (B + X)
2060  IF Q = 9 THEN Y = A * X © N + B
2070  IF Q = 10 THEN Y = A * X *  EXP (N * X)
2080 X = X + OS
2090  GOSUB 4000
2100  NEXT D
2160  FOR D = 1 TO M
2170 X = X(D):Y = Y(D):ZO = 1: GOSUB 4000
2180  NEXT D
2183  IF LP = 0 THEN  GOTO 2190
2184  PRINT  CHR$ (4);"EXEC HARDCOPY": TEXT
2185  PRINT "THE COMPUTER WILL NOW MAKE HARD-COPY OF THE GRAPH FOR THE BES
     T FITTING MODEL... TO CONTINUE WITH THE PROGRAM..          ENTER 'GOT
     O 2580' AND PRESS <RETURN>.": STOP
2187  IF LP = 1 THEN  STOP
2190  HOME : VTAB 22
2200  INPUT "TO CONTINUE....PRESS <RETURN>";Q$
2220  TEXT : HOME : PRINT "DO YOU WANT TO FIT ANOTHER EQUATION ? ": INPUT
     Q$
2230  IF  LEFT$ (Q$,1) = "Y" THEN  GOTO 485
2235  REM

2240  PRINT "SUMMARY OF STATISTICS ON CURVE-FITTING"
2250  PRINT "EQUATION      STD.ERROR     F VALUE"
2260  PRINT
2270  PRINT "STR. LINE ", INT (SE(1) * 100) / 100, INT (F(1))
2280  PRINT "EXP. GROW.", INT (SE(2) * 100) / 100, INT (F(2))
2290  PRINT "PWR.FUNCT.", INT (SE(3) * 100) / 100, INT (F(3))
2300  PRINT "HYPERBOLIC", INT (SE(4) * 100) / 100, INT (F(4))
2310  PRINT "EXP.SATURAT", INT (SE(5) * 100) / 100, INT (F(5))
2320  PRINT "SIGMOID", INT (SE(6) * 100) / 100, INT (F(6))
2330  PRINT "EXP.SIGMOID", INT (SE(7) * 100) / 100, INT (F(7))
2340  PRINT "MOD.INVERSE", INT (SE(8) * 100) / 100, INT (F(8))
2350  PRINT "MOD. PWR. FN.", INT (SE(9) * 100) / 100, INT (F(9))
2353  PRINT "MAXIMA FN.", INT (SE(10) * 100) / 100, INT (F(10))
2355  IF PR = 1 THEN  GOTO 2390
2360  PRINT : INPUT "WOULD YOU LIKE A PRINT-OUT OF SUMMARY ? ";Q$
2370  IF  LEFT$ (Q$,1) = "N" THEN  GOTO 2505
2380  PR# 1
2385 PR = 1: GOTO 2240
2390  REM
2500  PR# 0
2505  PRINT : REM

2510  INPUT "DO YOU WANT A PRINT-OUT OF THE COMPLETE STATISTICS ON THE BES
     T FITTING MODEL      EQUATION ? ";Q$
2520  IF  LEFT$ (Q$,1) = "N" THEN  HOME : GOTO 2580
2530  LET FM = 0
2540  FOR C = 1 TO 9
2550  IF F(C) > FM THEN FM = F(C):Q = C
2560  NEXT C
2570 LP = 1: GOTO 640
```

```
2580  PRINT : INPUT "WHAT DO YOU WANT TO DO NOW ?                    1. FIT MOR
    E EQUATIONS TO THIS DATA.     2. ENTER NEW DATA FOR FITTING.
     3. SAVE DATA ON A DISK FILE.              4. EXIT PROGRAM.
                  WHICH ? ";QS
2590  IF QS = 1 THEN  GOTO 510
2600  IF QS = 2 THEN  RUN 255
2605  IF QS = 3 THEN  GOTO 7000
2610  PRINT "HOPE CURFIT HAS BEEN HELPFUL.....PROGRAM TERMINATED": STOP
2620  REM

3000  REM  AXES AND UNITS FOR GRAPHS
3010  REM    SUBROUTINE DEVELOPED  BY J. SPAIN,MICHIGAN TECH UNIV.,AND B.J
    . WINKEL,ROSE-HULMAN INST.OF TECHNOLOGY
3050  HCOLOR= 3: SCALE= 1:SC = 1: ROT= 0
3055  REM    LIST OF RESERVED VARIABLES:X,X0,XM,X$,Y,Y0,YM,Y$,Z,ZF,ZG,ZP,Z
    0,I,L$,SC
3060  HPLOT 23,0 TO 23,149
3070  HPLOT 25,149 TO 279,149
3075  FOR I = 9 TO 116 STEP 35
3080  HPLOT 26,I TO 23,I: NEXT I
3085  FOR I = 55 TO 279 STEP 32
3090  HPLOT I,146 TO I,149: NEXT I
3095  REM  WRITE VARIABLE NAME ON X-AXIS
3100 Z0 = 0:L$ = X$:X0 = 175:Y0 = 150
3110  GOSUB 3500
3120  REM  WRITE VARIABLE NAME ON Y-AXIS
3130 Z0 = 1:L$ = Y$:X0 = 10:Y0 = 60
3140  GOSUB 3500
3150  REM  WRITE UNITS ON THE X-AXIS
3155 Y0 = 150:Z0 = 0
3160 L$ =  LEFT$ ( STR$ (XM),3):X0 = 260: GOSUB 3500
3165 L$ =  LEFT$ ( STR$ (XM / 2),3):X0 = 146: GOSUB 3500
3170 L$ =  LEFT$ ( STR$ (XM / 4),3):X0 = 82: GOSUB 3500
3172  REM  WRITE UNITS ON Y-AXIS
3175 X0 = 1
3180 L$ =  LEFT$ ( STR$ (YM),3):Y0 = 6: GOSUB 3500
3185 L$ =  LEFT$ ( STR$ (YM / 2),3):Y0 = 76: GOSUB 3500
3190 L$ =  LEFT$ ( STR$ (YM / 4),3):Y0 = 111: GOSUB 3500
3195  DRAW 48 AT 15,150
3200  RETURN : REM

3500  REM  ALPHANUMERIC CHARACTERS FOR HGR
3520  FOR Z = 1 TO  LEN (L$):Z3 =  ASC ( MID$ (L$,Z,1))
3530  IF Z0 <  > 0 THEN  GOTO 3570
3540  IF Z3 > 64 THEN  DRAW Z3 - 64 AT X0 + (Z - 1) * 7 * SC,Y0: GOTO 3600

3550  DRAW Z3 AT X0 + (Z - 1) * 7 * SC,Y0: GOTO 3600
3570  ROT= 48: IF Z3 > 64 THEN  DRAW Z3 - 64 AT X0,Y0 - (Z - 1) * 7: GOTO
    3600
3580  DRAW Z3 AT X0,Y0 - (Z - 1) * 7
3600  ROT= 0: NEXT Z
3610  RETURN : REM

4000  REM  PLOTTING X AND Y VALUES
4010 X0 = 23 + X * 253 / XM
4020 Y0 = 149 - Y * 140 / YM
4030  IF X0 > 279 OR X0 < 0 OR Y0 > 149 OR Y0 < 0 THEN  GOTO 4050
4033  IF ZP = 1 THEN  DRAW 43 AT X0 - 2,Y0 - 3:ZP = 0: GOTO 4050
4034  IF Z0 = 1 THEN  DRAW 79 AT X0 - 2,Y0 - 4:Z0 = 0: GOTO 4050
4035  IF ZG = 1 THEN  HPLOT  TO X0,Y0: GOTO 4050
4040  HPLOT X0,Y0: IF ZF = 1 THEN ZG = 1
4050  RETURN : REM
```

```
5990  REM  CORRECTION SUBROUTINE
6000  INPUT " WHICH DATA POINT NUMBER TO CHANGE? ";AN
6005  IF AN > M THEN  PRINT "";: PRINT "THAT NUMBER DOESN'T EXIST IN YOUR
    FILE, TRY AGAIN.": GOTO 6000
6010  PRINT "ENTER NEW X AND Y VALUES FOR DATA POINT NUMBER ";AN;: INPUT X
    (AN),Y(AN)
6020  INPUT "MORE CORRECTIONS? (Y/N) ";AN$
6030  IF  LEFT$ (AN$,1) = "Y" THEN 6000
6040  RETURN : REM
6990  REM  WRITING DATA ON DISK FILE
7000  PRINT : INPUT "ENTER A NAME THAT CAN BE USED TO DEFINE THE NEW DISK
    FILE.";N$
7010 D$ =  CHR$ (4)
7015  PRINT D$;"MON C,I,O"
7020  PRINT D$;"OPEN "N$
7025  PRINT D$;"DELETE"N$
7027  PRINT D$;"OPEN"N$
7030  PRINT D$;"WRITE "N$
7040  PRINT M
7050  FOR I = 1 TO M
7060  PRINT X(I);",";Y(I)
7070  NEXT
7080  PRINT D$;"CLOSE "N$
7090  GOTO 2580
7108  END
```

SUMMARY OF STATISTICS ON CURVE-FITTING

EQUATION	STD.ERROR	F VALUE
STR. LINE	.76	487
EXP. GROW.	2.29	88
PWR.FUNCT.	.52	954
HYPERBOLIC	0	0
EXP.SATURAT	0	0
SIGMOID	0	0
EXP.SIGMOID	0	0
MOD.INVERSE	0	0
MOD. PWR. FN.	.46	1231
MAXIMA FN.	0	0

STATISTICS FOR THE MODEL WITH THE
HIGHEST VALUE FOR 'F' :
THE FOLLOWING ARE STATISTICS FOR THE
TRANSFORMED DATA:

THE BEST FITTING STRAIGHT LINE FOLLOWS
Y = -.157 + 1.297 * X

COEFFICIENT OF DETERMINATION (R©2) IS:
.9959

THE VALUE OF A IS .8551
THE FITTED VALUE OF B IS .1
THE FITTED VALUE OF N IS 1.2979

Y = A * X©N + B

COMPARISON OF MODEL DATA WITH ORIGINAL

X	Y DATA	PREDICTED Y	RESIDUALS
1	1	.955	.044
2	2	2.202	-.203
3	4	3.659	.34
4	5	5.27	-.271
5	7	7.006	-7E-03
6	9	8.85	.149
7	10	10.789	-.79
8	13	12.812	.187
9	15	14.911	.088
10	18	17.082	.917

SUMMARY OF STATISTICS ON FITTED MODEL

THE RESIDUAL SUM OF THE SQUARES IS 1.76

THE VARIANCE FOR 8 DEGREES OF FREEDOM IS
.22

THE STANDARD ERROR OF THE ESTIMATE IS
.469

THE VALUE FOR F IS 1231

APPENDIX 3

POLYFIT: A PROGRAM FOR FITTING POLYNOMIAL EQUATIONS TO DATA

```
1  IF  PEEK (104) = 64 THEN  GOTO 5
2  TEXT : HOME : PRINT "PROGRAM LOADING ERROR..................RE-ENTER '
    RUN POLYFIT'...PRESS <RETURN>"
3  PRINT  CHR$ (4);"EXEC MODE II": REM  RESET PROGRAM START POINTERS AT 16
    384
5  TEXT : HOME : REM

55  REM  LOAD CHARACTERS FOR HGR
60  POKE 232,0: POKE 233,08
70 D$ = "": REM  CONTROL-D CHAR
80  PRINT D$;"BLOAD SMALL CHARACTERS,A$0800"
90  SCALE= 2:SC = 2: ROT= 0:X0 = 90:Y0 = 50:L$ = "POLYFIT": HCOLOR= 2: HGR
     : GOSUB 3500
93  FOR I = 1 TO 1000: NEXT I
95  SCALE= 1:SC = 1: TEXT
97  DIM X(100),Y(100): REM

100  HOME : PRINT : PRINT : PRINT "         POLY-FIT"
110  PRINT : PRINT "A PROGRAM FOR FITTING VARIOUS POLYNOMIALEQUATIONS TO R
     EAL DATA"
120  PRINT : PRINT "DEVELOPED FOR THE TEXT:                   BASIC MICROCOM
     PUTER MODELS IN BIOLOGY   BY JAMES D. SPAIN."
125  PRINT : PRINT "COPYRIGHT (C) 1981                        ADDISON-WESLEY
      PUBLISHING COMPANY,INC.  READING, MASSACHUSETTS"
127  VTAB 22: INPUT "FOR INTRODUCTION...PRESS <RETURN>";Q$: HOME
130  PRINT : PRINT "THE PROGRAM ALLOWS THE USER TO ENTER A  NUMBER OF X,Y
     DATA POINTS WHICH MAY BE  FITTED BY LEAST SQUARES TO DIFFERENT    POL
     YNOMIALS."
140  PRINT : PRINT "THE COEFFICIENTS ARE PROVIDED ALONG WITHSTATISTICS NEC
     ESSARY TO JUDGE HOW WELL  THE EQUATION FITS THE DATA."
150  DIM A(6,12),S(11),Z(12),C(10)
160  VTAB 22: INPUT "PRESS RETURN TO CONTINUE";A$
163  HOME : PRINT "YOU MAY ENTER DATA IN EITHER OF THE     FOLLOWING WAYS:
                            1. FROM THE KEYBOARD.                2
     . FROM A PREVIOUSLY SAVED DISK FILE."
165  PRINT : PRINT : INPUT "WHICH MODE OF DATA ENTRY (1 OR 2) ? ";A$:A$ =
      LEFT$ (A$,1)
166  IF A$ = "1" OR A$ = "K" THEN  GOTO 190: REM

167  REM  DATA ENTRY FROM DISK FILE
168  IF A$ < > "2" THEN  PRINT "PLEASE TYPE EITHER (1) OR (2)": INPUT A$:
```

```
     GOTO 166
170  PRINT : INPUT "     NAME OF TEXT FILE TO BE USED? ";N$
172 D$ =  CHR$ (4)
174  PRINT D$;"OPEN";N$: PRINT D$;"READ";N$: INPUT N: FOR I = 1 TO N: INPUT
     X(I),Y(I): NEXT : PRINT D$;"CLOSE";N$
176 XM = 0:YM = 0: FOR T = 1 TO N: IF X(T) > XM THEN XM = X(T)
178  IF Y(T) > YM THEN YM = Y(T)
179  NEXT
180  GOTO 292: REM

185  REM  DATA ENTRY FROM KEYBOARD
190  PRINT
200  INPUT "HOW MANY DATA POINTS ARE YOU GOING TO   ENTER (AT LEAST 7 PLEA
     SE) ? ";N
220  HOME : PRINT "IF YOU MAKE AN ERROR TYPE IN 'E' AND    PRESS <RETURN>"

223  PRINT
225  PRINT "BECAUSE OF THE EXPONENTIATION REQUIRED  IN THIS PROGRAM, ALL D
     ATA SHOULD BE    NORMALIZED TO NUMBERS BETWEEN -100 AND  +100.  PART
     ICULARLY IF YOU ARE FITTING  HIGHER POLYNOMIALS."
226  PRINT : PRINT "NOTE...THERE IS NO LIMITATION AGAINST   ZERO OR NEGATI
     VE NUMBERS IN THIS PROGRAM"
227  PRINT
230  FOR C = 1 TO N: REM   ENTERING N DATA POINTS
240  PRINT "ENTER (X,Y) DATA POINT NUMBER "C" ";: INPUT " ";X$,Y$
242  IF  LEFT$ (X$,1) = "E" OR  LEFT$ (Y$,1) = "E" THEN C = C - 1: PRINT "
      ": GOTO 240
250 X =  VAL (X$)
260 Y =  VAL (Y$)
270  REM  IF X OR Y IS TO BE TRANSFORMED, PUT TRANSFORM EQUATIONS HERE
280  LET X(C) = X:Y(C) = Y
290  NEXT C: REM

292  PRINT : INPUT "DO YOU NEED TO CORRECT ANY OF THE DATA ?";Q$
294  IF  LEFT$ (Q$,1) = "Y" THEN  GOSUB 6000
300  HOME : PRINT "CALCULATING THE SUMS OF SQUARES FOR DATA...PLEASE STAND
      BY"
305  FOR I = 1 TO 10:Z(I) = 0:S(I) = 0: NEXT I
310  FOR C = 1 TO N
313  IF Y(C) > YM THEN YM = Y(C)
315  IF X(C) > XM THEN XM = X(C)
320  LET Z(1) = Z(1) + X(C)
330  LET Z(2) = Z(2) + X(C) © 2: REM  © SYMBOLIZES EXPONENTIATION
340  LET Z(3) = Z(3) + X(C) © 3
350  LET Z(4) = Z(4) + X(C) © 4
360  LET Z(5) = Z(5) + X(C) © 5
370  LET Z(7) = Z(7) + X(C) © 7
380  LET Z(6) = Z(6) + X(C) © 6
390  LET Z(8) = Z(8) + X(C) © 8
400  LET Z(9) = Z(9) + X(C) © 9
410  LET Z(10) = Z(10) + X(C) © 10
420  LET S(1) = S(1) + Y(C)
430  LET S(2) = S(2) + X(C) * Y(C)
440  LET S(3) = S(3) + Y(C) * X(C) © 2
450  LET S(4) = S(4) + Y(C) * X(C) © 3
460  LET S(5) = S(5) + Y(C) * X(C) © 4
470  LET S(6) = S(6) + Y(C) * X(C) © 5
480  LET S(9) = S(9) + Y(C) © 2
490  NEXT C: REM
```

```
500  TEXT : HOME : PRINT "THE FOLLOWING POLYNOMIAL EQUATIONS MAY  BE FITTE
     D TO THE DATA:"
502  PRINT
503  PRINT "1. STRAIGHT LINE   Y = A0 + A1*X"
504  PRINT
505  PRINT "2. QUADRATIC      Y = A0 + A1*X + A2*X©2"
507  PRINT
510  PRINT "3. CUBIC Y = A0 + A1*X + A2*X©2 + A3*X©3"
515  PRINT : PRINT "4. QUARTIC Y = A0 + A1*X + A2*X©2
       + A3*X©3 + A4*X©4"
517  PRINT
520  PRINT "5. QUINTIC Y = A0 + A1*X + A2*X©2                          + A3*X©
     3 + A4*X©4 +A5*X©5"
530  PRINT : INPUT "ENTER THE NUMBER OF THE EQUATION TO BE  FITTED. ";R
540  IF R + 2 > N OR R > 5 THEN  PRINT "SORRY...THE ";R;" ORDER POLYNOMIAL
      CAN NOT BE FITTED TO ....... ";N;" DATA POINTS ! ": GOTO 530: REM

550  REM  SETTING MATRIX TO ZERO
560  FOR I = 1 TO 6
570  FOR J = 1 TO 12
580  LET A(I,J) = 0
590  NEXT J
600  NEXT I: REM

610  REM  LOADING THE MATRIX WITH THE SUMS OF SQUARES
620  LET A(1,1) = N:A(1,2) = Z(1):A(2,1) = Z(1):A(2,2) = Z(2)
630  IF R = 1 THEN  GOTO 730
640  LET A(3,1) = Z(2):A(1,3) = Z(2):A(2,3) = Z(3):A(3,2) = Z(3):A(3,3) =
     Z(4)
650  IF R = 2 THEN  GOTO 730
660  LET A(4,1) = Z(3):A(1,4) = Z(3):A(2,4) = Z(4):A(4,2) = Z(4):A(3,4) =
     Z(5):A(4,3) = Z(5):A(4,4) = Z(6)
670  IF R = 3 THEN  GOTO 730
680  LET A(5,1) = Z(4):A(1,5) = Z(4):A(2,5) = Z(5):A(5,2) = Z(5):A(3,5) =
     Z(6):A(5,3) = Z(6):A(4,5) = Z(7):A(5,4) = Z(7):A(5,5) = Z(8)
690  IF R = 4 THEN  GOTO 730
700  LET A(6,1) = Z(5):A(1,6) = Z(5):A(2,6) = Z(6):A(6,2) = Z(6):A(3,6) =
     Z(7):A(6,3) = Z(7):A(4,6) = Z(8):A(6,4) = Z(8)
710  LET A(5,6) = Z(9):A(6,5) = Z(9):A(6,6) = Z(10)
715  REM

720  REM  SETTING UP THE IDENTITY MATRIX NEEDED FOR INVERSION
730  LET M = R + 1
740  FOR I = 1 TO M
750  LET A(I,I + M) = 1
760  NEXT I
770  IF PF = 0 THEN  PRINT "CURVE FITTING CALCULATIONS IN PROGRESS."
775  REM

780  REM  INVERSION OF MATRIX...A
790  FOR T = 1 TO M
800  FOR B = 1 TO M
810  LET F(B) = A(B,T)
820  NEXT B
830  FOR J = T TO 2 * M
840  LET A(T,J) = A(T,J) / F(T)
850  NEXT J
860  IF T = M THEN  GOTO 930
870  LET C = T + 1
```

```
880  FOR I = C TO M
890  FOR J = 1 TO 2 * M
900  LET A(I,J) = A(I,J) - A(T,J) * F(I)
910  NEXT J
920  NEXT I
930  NEXT T
940  FOR I = M TO 2 STEP  - 1
950  FOR B = M TO I STEP  - 1
960  LET F(B) = A(I - 1,B)
970  NEXT B
980  FOR B = M TO I STEP  - 1
990  FOR J = I TO 2 * M
1000  LET A(I - 1,J) = A(I - 1,J) - A(B,J) * F(B)
1010  NEXT J
1020  NEXT B
1030  NEXT I
1040  FOR I = 1 TO M
1050  FOR J = M + 1 TO 2 * M
1060  LET K = J - M
1070  LET A(I,K) = A(I,J)
1080  NEXT J
1090  NEXT I
1095  REM

1100  REM  MULTIPLICATION OF MAT C = A * S
1110  FOR I = 1 TO 6
1120  LET C(I) = 0
1130  NEXT I
1140  FOR I = 1 TO M
1150  FOR J = 1 TO M
1160  LET C(I) = C(I) + A(I,J) * S(J)
1170  NEXT J
1180  NEXT I
1182  IF PF = 0 THEN  GOTO 1187: REM

1183  PRINT : PRINT "THE BEST FITTING POLYNOMIAL EQUATION IS:"
1184  PRINT "    Y = A0 + A1*X ";: IF R > 1 THEN  PRINT "+ A2*X©2 ";: IF R
 > 2 THEN  PRINT "+ A3*X©3 ";: IF R > 3 THEN  PRINT "+ A4*X©4 ";: IF
R > 4 THEN  PRINT "+ A5*X©5";
1185  PRINT " ": GOTO 1200
1187  HOME : REM

1190  PRINT "FOLLOWING ARE THE COEFFICIENTS OF THE   POLYNOMIAL OF ORDER "
;R
1195  PRINT
1200  PRINT "A0 = ";C(1):A0(R) =  INT (C(1) * 10000) / 10000
1205  PRINT
1210  PRINT "A1 = ";C(2):A1(R) =  INT (C(2) * 10000) / 10000
1215  PRINT
1220  IF R > 1.5 THEN  PRINT "A2 = ";C(3):A2(R) =  INT (C(3) * 10000) / 10
000
1225  PRINT
1230  IF R > 2.5 THEN  PRINT "A3 = ";C(4):A3(R) =  INT (C(4) * 10000) / 10
000
1235  PRINT
1240  IF R > 3.5 THEN  PRINT "A4 = ";C(5):A4(R) =  INT (C(5) * 10000) / 10
000
1245  PRINT
1250  IF R > 4.5 THEN  PRINT "A5 = ";C(6):A5(R) =  INT (C(6) * 10000) / 10
000
```

```
1255  PRINT : REM   © SYMBOLIZES EXPONENTIATION
1260  LET R2 = C(1) * S(1) + C(2) * S(2) + C(3) * S(3) + C(4) * S(4) + C(5
     ) * S(5) + C(6) * S(6) - (S(1) © 2) / N
1270  LET R2 = R2 / (S(9) - (S(1) © 2) / N)
1280  LET CR =  SQR (R2)
1290  PRINT "THE COEFFICIENT OF DETERM.,R©2 IS:"; INT (R2 * 1000) / 1000
1295  PRINT
1300  PRINT "THE CORRELLATION COEFFICIENT  IS:"; INT (CR * 1000) / 1000
1310  LET RC(R) =  INT (CR * 1000) / 1000
1312  IF PF = 1 THEN  GOTO 1330
1315  PRINT : REM

1320  INPUT "TO COMPARE REAL DATA WITH MODEL DATA ....PRESS <RETURN>";Q$: HOME

1330  PRINT : PRINT "REAL X    REAL Y    MODEL Y    RESIDUAL": PRINT
1340  LET D2 = 0
1350  FOR C = 1 TO N
1360  LET Y1 = C(1) + C(2) * X(C) + C(3) * X(C) © 2 + C(4) * X(C) © 3 + C(
     5) * X(C) © 4 + C(6) * X(C) © 5
1370  LET D = Y(C) - Y1
1380  LET D2 = D2 + D © 2
1390  PRINT X(C);: HTAB 8: PRINT Y(C);: HTAB 18: PRINT  INT (Y1 * 1000) /
     1000;: HTAB 32: PRINT  INT (D * 1000) / 1000
1400  NEXT C
1405  PRINT : PRINT
1407  IF PF = 1 THEN  GOTO 1420
1410  INPUT "FOR ADDITIONAL STATISTICS..PRESS RETURN";Q$
1415  HOME : REM

1420  PRINT "THE RESIDUAL SUM OF SQUARES IS:"; INT (D2 * 1000) / 1000
1430  LET RM(R) =  INT (D2 * 1000 / (N - M)) / 1000
1435  PRINT
1440  PRINT : PRINT "THE RESIDUAL MEAN SQUARE IS:";RM(R)
1450  PRINT : PRINT "THE STANDARD ERROR OF THE REGRESSION    ESTIMATE:"; INT
     ( SQR (RM(R)) * 10000) / 10000
1455  PRINT
1460  LET F = R2 * (S(9) - (S(1) © 2) / N) / (D2 / (N - M))
1470  PRINT : PRINT "THE VALUE OF F IS:"; INT (F * 100) / 100
1475  IF PF = 1 THEN  GOTO 2500
1480  LET VF(R) =  INT (F * 100) / 100: REM

1483  PRINT : INPUT "DO YOU WANT TO SEE THE MODEL GRAPHED IN RELATION TO T
     HE DATA POINTS ? ";Q$
1485  IF  LEFT$ (Q$,1) = "Y" THEN  GOTO 2500
1490  INPUT "DO YOU WANT TO FIT ADDITIONAL EQUATIONS (YES OR NO) ";Q$
1500  IF  LEFT$ (Q$,1) = "Y" THEN  GOTO 500
1545  TEXT : HOME : REM

1550  PRINT "SUMMARY OF CURVE FITTING RESULTS"
1560  PRINT : PRINT "EQUATION                  F VALUE"
1570  PRINT : PRINT "1. STR. LINE             "; INT (VF(1))
1575  PRINT : PRINT "2. QUADRATIC             "; INT (VF(2))
1580  PRINT : PRINT "3. CUBIC                 "; INT (VF(3))
1585  PRINT : PRINT "4. QUARTIC               "; INT (VF(4))
1590  PRINT : PRINT "5. QUINTIC               "; INT (VF(5))
1600  PRINT : PRINT : INPUT "WOULD YOU LIKE HARD-COPY OF THE SUMMARY ? ";Q
     $
1610  IF  LEFT$ (Q$,1) < > "Y" THEN  GOTO 1715
```

```
1620  PR# 1: REM   TURN PRINTER ON ...ADDITIONAL INSTRUCTIONS MAY BE REQUI
     RED DEPENDING ON PRINTER
1630  PRINT "SUMMARY OF RESULTS OF FITTING POLYNOMIAL EQUATIONS TO A SET O
     F DATA:"
1640  PRINT : PRINT "    LINE","QUADRATIC","CUBIC","QUARTIC","QUINTIC"
1645  PRINT : PRINT "A0   ";A0(1),A0(2),A0(3),A0(4),A0(5)
1650  PRINT "A1   ";A1(1),A1(2),A1(3),A1(4),A1(5)
1655  PRINT "A2   ",A2(2),A2(3),A2(4),A2(5)
1660  PRINT "A3   ","   ",A3(3),A3(4),A3(5)
1665  PRINT "A4   ","   ","   ",A4(4),A4(5)
1670  PRINT "A5   ","   ","   ","   ",A5(5)
1675  PRINT
1680  PRINT "R   ";RC(1),RC(2),RC(3),RC(4),RC(5)
1685  PRINT
1690  PRINT "F   ";VF(1),VF(2),VF(3),VF(4),VF(5)
1695  PRINT
1700  PRINT "RMS ";RM(1),RM(2),RM(3),RM(4),RM(5)
1710  PR# 0
1715  TEXT : HOME : INPUT "DO YOU WANT HARD-COPY OF THE STATISTICS ON THE
     BEST FITTING MODEL ? ";Q$
1720  IF  LEFT$ (Q$,1) = "Y" THEN  GOTO 1730
1722  PRINT : INPUT "DO YOU WANT TO FIT ANOTHER EQUATION TO  DATA ? ";A$
1723  IF  LEFT$ (A$,1) = "Y" THEN  HOME : GOTO 500: REM
1725  PRINT : INPUT " WOULD YOU LIKE TO SAVE YOUR DATA IN A  TEXT FILE? ";
     Q$
1726  IF  LEFT$ (Q$,1) = "Y" THEN  GOTO 7000
1728  PRINT : INPUT "DO YOU WANT TO FIT ANOTHER SET OF DATA ? ";Q$: IF  LEFT$
     (Q$,1) = "Y" THEN  TEXT : HOME : GOTO 200
1729  STOP
1730  FOR C = 1 TO 5
1733  IF VF(C) > VF THEN R = C:VF = VF(C)
1735  NEXT C
1740 PF = 1: PR# 1: GOTO 550: REM

2499  REM  GRAPH OF DATA ALONG WITH MODEL CURVE
2500 X$ = "X VALUE":Y$ = "Y VALUE": HCOLOR= 3: HGR : GOSUB 3000
2505 ZF = 1
2510  FOR I = 0 TO 260 STEP 2
2520 X = I * XM / 260
2530 Y = C(1) + C(2) * X + C(3) * X © 2 + C(4) * X © 3 + C(5) * X © 4 + C(
     6) * X © 5
2540  GOSUB 4000
2550  NEXT I
2555 ZG = 0
2560  FOR C = 1 TO N
2570 Z0 = 1:X = X(C):Y = Y(C): GOSUB 4000
2580  NEXT C
2585  IF PF = 1 THEN  PR# 0: VTAB 21: PRINT "TYPE....EXEC HARD COPY....PRE
     SS <RETURN>": PRINT : PRINT "AFTER HARD COPY IS OBTAINED............
     TYPE ....GOTO 1490  ...PRESS <RETURN>": TEXT : STOP
2590  VTAB 22
2600  GOTO 1490: REM

3000  REM  AXES AND UNITS FOR GRAPHS
3010  REM   SUBROUTINE DEVELOPED BY J. SPAIN,MICHIGAN TECH UNIV.,AND B.J.
     WINKEL,ALBION COLLEGE
3050  HCOLOR= 3: SCALE= 1:SC = 1: ROT= 0
3055  REM    LIST OF RESERVED VARIABLES:X,X0,XM,X$,Y,Y0,YM,Y$,Z,ZF,ZG,ZP,Z
     O,I,L$,SC
3060  HPLOT 23,0 TO 23,149
```

```
3070  HPLOT 25,149 TO 279,149
3075  FOR I = 9 TO 116 STEP 35
3080  HPLOT 26,I TO 23,I: NEXT I
3085  FOR I = 55 TO 279 STEP 32
3090  HPLOT I,146 TO I,149: NEXT I
3095  REM  WRITE VARIABLE NAME ON X-AXIS
3100 Z0 = 0:L$ = X$:X0 = 175:Y0 = 150
3110  GOSUB 3500
3120  REM  WRITE VARIABLE NAME ON Y-AXIS
3130 Z0 = 1:L$ = Y$:X0 = 10:Y0 = 60
3140  GOSUB 3500
3150  REM  WRITE UNITS ON THE X-AXIS
3155 Y0 = 150:Z0 = 0
3160 L$ =  LEFT$ ( STR$ (XM),3):X0 = 260: GOSUB 3500
3165 L$ =  LEFT$ ( STR$ (XM / 2),3):X0 = 146: GOSUB 3500
3170 L$ =  LEFT$ ( STR$ (XM / 4),3):X0 = 82: GOSUB 3500
3172  REM  WRITE UNITS ON Y-AXIS
3175 X0 = 1
3180 L$ =  LEFT$ ( STR$ (YM),3):Y0 = 6: GOSUB 3500
3185 L$ =  LEFT$ ( STR$ (YM / 2),3):Y0 = 76: GOSUB 3500
3190 L$ =  LEFT$ ( STR$ (YM / 4),3):Y0 = 111: GOSUB 3500
3195  DRAW 48 AT 15,150
3200  RETURN
3500  REM  ALPHANUMERIC CHARACTERS FOR HGR
3520  FOR Z = 1 TO  LEN (L$):Z3 =  ASC ( MID$ (L$,Z,1))
3530  IF Z0 <  > 0 THEN  GOTO 3570
3540  IF Z3 > 64 THEN  DRAW Z3 - 64 AT X0 + (Z - 1) * 7 * SC,Y0: GOTO 3600

3550  DRAW Z3 AT X0 + (Z - 1) * 7 * SC,Y0: GOTO 3600
3570  ROT= 48: IF Z3 > 64 THEN  DRAW Z3 - 64 AT X0,Y0 - (Z - 1) * 7: GOTO
     3600
3580  DRAW Z3 AT X0,Y0 - (Z - 1) * 7
3600  ROT= 0: NEXT Z
3610  RETURN : REM

4000  REM  PLOTTING X AND Y VALUES
4010 X0 = 23 + X * 253 / XM
4020 Y0 = 149 - Y * 140 / YM
4030  IF X0 > 279 OR X0 < 0 OR Y0 > 149 OR Y0 < 0 THEN  GOTO 4050
4033  IF ZP = 1 THEN  DRAW 43 AT X0 - 2,Y0 - 3:ZP = 0: GOTO 4050
4034  IF ZO = 1 THEN  DRAW 79 AT X0 - 2,Y0 - 4:ZO = 0: GOTO 4050
4035  IF ZG = 1 THEN  HPLOT  TO X0,Y0: GOTO 4050
4040  HPLOT X0,Y0: IF ZF = 1 THEN ZG = 1
4050  RETURN : REM

5990  REM  CORRECTION SUBROUTINE
6000  PRINT : INPUT "      WHICH DATA POINT NUMBER DO YOU WANTTO CHANGE? ";
     AN
6005  IF (AN > N) OR (AN < 0) THEN  PRINT "";: PRINT "THAT NUMBER DOESN'T
     EXIST IN YOUR FILE. TRY AGAIN.": GOTO 6000
6007  PRINT "CURRENT VALUES FOR DATA POINT NUMBER    "AN" ARE "X(AN)", "Y(
     AN)"."
6010  PRINT "ENTER NEW X AND Y VALUES FOR DATA POINT NUMBER ";AN;: INPUT X
     (AN),Y(AN)
6020  INPUT "MORE CORRECTIONS? (Y/N) ";AN$
6030  IF  LEFT$ (AN$,1) = "Y" THEN 6000
6040  RETURN : REM
```

```
6990  REM  WRITING DATA TO A NEW DISK FILE
7000  PRINT : INPUT "ENTER NAME OF NEW FILE  ";N$
7010 D$ =  CHR$ (4)
7015  PRINT D$;"MON C,I,O"
7020  PRINT D$;"OPEN "N$
7025  PRINT D$;"DELETE";N$
7027  PRINT D$;"OPEN";N$
7030  PRINT D$;"WRITE "N$
7040  PRINT N
7050  FOR I = 1 TO N
7060  PRINT X(I);",";Y(I)
7070  NEXT
7080  PRINT D$;"CLOSE "N$
7090  GOTO 1728
7100  END
```

```
THE BEST FITTING POLYNOMIAL EQUATION IS:
    Y = A0 + A1*X + A2*X©2 + A3*X©3 + A4*X©4 + A5*X©5
A0 = 2.82996452

A1 = -5.18179631

A2 = 4.39001048

A3 = -1.11890078

A4 = .12018507

A5 = -4.48117033E-03

THE COEFFICIENT OF DETERM.,R©2 IS:1

THE CORRELLATION COEFFICIENT  IS:1

REAL X    REAL Y    MODEL Y    RESIDUAL

1       1        1.034        -.035
2       3        2.854        .145
3       5        5.23         -.231
4       7        6.911        .088
5       8        7.92         .079
6       9        9.011        -.012
7       11       11.134       -.135
8       15       14.898       .101
9       20       20.031       -.032

THE RESIDUAL SUM OF SQUARES IS:.118


THE RESIDUAL MEAN SQUARE IS:.039

THE STANDARD ERROR OF THE REGRESSION    ESTIMATE:.1974


THE VALUE OF F IS:7103.66
```

APPENDIX 4

KEY TO CURVE TYPES

The following key is similar to those employed in the field of taxonomy to identify organisms based upon various morphological characteristics. At each stage you are provided with two alternatives. Pick the one which best describes your curve. It will then tell you to either move on to the next statement number, or try a particular equation number based on the CURFIT program.

(1a) A linear relationship exists between x and y.............Try Equation 1
(1b) A non-linear relationship exists between x and y..............Go to (2)
(2a) y is a simple monotonic function of x (No maximum or minimum)..Go to (3)
(2b) The curve has a maximum or a minimum or both.................Go to (11)
(3a) y decreases as x increases....................................Go to (4)
(3b) y increases as x increases....................................Go to (9)
(4a) y approaches zero as x approaches infinity....................Go to (5)
(4b) y approaches some nonzero value as x approaches infinity.......Go to (8)
(5a) y approaches infinity as x = 0...............................Equation 8
(5b) Curve intercepts the y axis as x = 0..........................Go to (6)
(6a) Curve has an inflection point resulting in a sigmoid shape....Equation 6
(6b) Curve has no inflection.......................................Go to (7)
(7a) Curve decreases in an exponential fashion, i.e. y = 0.5 at a given value x, y = 0.25 at 2x, etc...............Equation 2
(7b) Curve does not decrease exponentially........................Equation 8
(8a) y approaches a value B, as x approaches infinity.........Equation 8 or 9
(8b) y approaches negative infinity as x approaches infinity.......Equation 9
(9a) y approaches some limiting value, A, as x approaches infinity.Go to (10)
(9b) y approaches infinity as x approaches infinity......Equations 2, 3, or 9
(10a) Curve has an inflection point causing it to have a sigmoid shape...............Equations 6 or 7
(10b) Curve has no inflection.............................Equations 4 or 5
(11a) The curve has a single maximum or a single minimum............Go to (13)
(11b) The curve has both maxima and minima..........................Go to (12)
(12a) The curve shows periodic oscillations.............................. See discussion of sine curves..............................Chapter 10
(12b) The curve has a single maximum or a single minimum...........See POLYFIT or try compound equations, Section 3.5.
(13) Try compound equations...............................See Section 3.5

APPENDIX 5
POISSON DISTRIBUTION PROGRAM

```
100  REM  POISSON DISTRIBUTION
110  REM  P IS PROB. OF OCCURANCE IN A UNIT OF SPACE
120  REM  M IS THE MEAN NUMBER OF OCCURANCES IN A UNIT SPACE
130  REM  T IS THE TOTAL NUMBER OF OCCURANCES
140  REM  N IS THE NUMBER OF SPACES
150 N = 100:T = 500
160 M = T / N
170  PRINT "X     ";"  P(X) ","          N - P(X)"
180 P = N /  EXP (M)
190 P1 = P
200  PRINT X;: HTAB (5): PRINT P;: HTAB (25): PRINT N - P1
210  FOR X = 1 TO 4 * M
220 P = P * M / X
230 P1 = P1 + P
240  PRINT X;: HTAB (5): PRINT P;: HTAB (25): PRINT N - P1
250  NEXT X
```

```
]RUN
X       P(X)              N - P(X)
0   .6737947            99.3262053
1   3.3689735           95.9572318
2   8.42243375          87.5347981
3   14.0373896          73.4974085
4   17.546737           55.9506715
5   17.546737           38.4039345
6   14.6222808          23.7816537
7   10.4444863          13.3371674
8   6.52780394          6.80936346
9   3.62655774          3.18280572
10  1.81327887          1.36952683
11  .824217669          .545309156
12  .343424029          .201885134
13  .132086165          .0697989762
14  .0471736303         .0226253569
15  .0157245434         6.90081716E-03
16  4.91391983E-03      1.98689103E-03
17  1.44527054E-03      5.41627407E-04
18  4.01464038E-04      1.40160322E-04
19  1.05648431E-04      3.45110893E-05
20  2.64121078E-05      8.10623169E-06
```

APPENDIX 6

GRAPH: A SUBROUTINE FOR PLOTTING DATA

```
30  REM  READING HIGH RESOLUTION CHARACTER SET
40  POKE 232,0: POKE 233,64: SCALE= 1: HCOLOR= 3: ROT= 0
50  PRINT  CHR$ (4)"BLOAD SMALL CHARACTERS,A$4000"
60  REM

100  REM  EXAMPLE PROGRAM
110  REM  SINE CURVE VS. TIME
120 X$ = "TIME":Y$ = "SINE":YM = 1:YN =  - 1:XM = 1: GOSUB 3000
125 ZF = 1:ZG = 0: REM  FLAGS FOR LINE PLOT
130  FOR T = 0 TO 1 STEP .01
140 S =  SIN (6.28 * T)
150 X = T:Y = S: GOSUB 4000
160  NEXT T
170  STOP
2990  REM

3000  REM    AXES AND UNITS FOR GRAPHS REV. SEPT.,1981
3010  REM    SUBROUTINE DEVELOPED BY J. SPAIN,MICHIGAN TECH UNIV.,AND B.J.
      WINKEL,ROSE-HULMAN INST. TECH.
3020  REM  X$=VARIABLE PLOTTED ON X AXIS
3030  REM  Y$=VARIABLE PLOTTED ON Y AXIS
3040  REM  YM =MAXIMUM UNITS ON THE Y AXIS
3042  REM  YN = MINIMUM UNITS ON THE Y AXIS
3045  REM XM = MAXIMUM UN ITS ON THE X AXIS
3047  REM   XN = MINIMUM UNITS ON X-AXIS
3050  HCOLOR= 3: SCALE= 1:SC = 1: ROT= 0
3055  REM       LIST OF RESERVED VARIABLES:X,XO,XM,XN,X$,Y,YO,YM,YN,Y$,Z,ZF
     ,ZG,ZP,ZO,I,L$,SC
3060  HGR : HPLOT 23,0 TO 23,149
3070  HPLOT 25,149 TO 279,149
3075  FOR I = 9 TO 116 STEP 35
3080  HPLOT 26,I TO 23,I: NEXT I
3085  FOR I = 54 TO 275 STEP 31
3090  HPLOT I,146 TO I,149: NEXT I
3093  REM

3095  REM  WRITE VARIABLE NAME ON X-AXIS
3100 ZO = 0:L$ = X$:XO = 175:YO = 150
3110  GOSUB 3500
3120  REM  WRITE VARIABLE NAME ON Y-AXIS
3130 ZO = 1:L$ = Y$:XO = 10:YO = 68
3140  GOSUB 3500
3145  REM
```

```
3150  REM  WRITE UNITS ON THE X-AXIS
3155 Y0 = 150:Z0 = 0
3160 L$ =  LEFT$ ( STR$ (XM),3):X0 = 260: GOSUB 3500
3165 L$ =  LEFT$ ( STR$ ((XM - XN) / 2 + XN),3):X0 = 146: GOSUB 3500
3170 L$ =  LEFT$ ( STR$ ((XM - XN) / 4 + XN),3):X0 = 82: GOSUB 3500
3171 L$ =  LEFT$ ( STR$ (XN),2):X0 = 23: GOSUB 3500
3172  REM  WRITE UNITS ON Y-AXIS
3175 X0 = 1
3180 L$ =  LEFT$ ( STR$ (YM),3):Y0 = 6: GOSUB 3500
3185 L$ =  LEFT$ ( STR$ ((YM - YN) / 2 + YN),3):Y0 = 76: GOSUB 3500
3190 L$ =  LEFT$ ( STR$ ((YM - YN) / 4 + YN),3):Y0 = 111: GOSUB 3500
3192 L$ =  LEFT$ ( STR$ (YN),3):Y0 = 147: GOSUB 3500
3200  RETURN
3205  REM

3500  REM  ALPHANUMERIC CHARACTERS FOR HGR
3510  REM  THE FOLLOWING MUST BE DEFINED
3511  REM  BEFORE ENTERING THE SUBROUTINE
3512  REM  L$ = "CHARACTER STRING"
3513  REM  Y0 = THE INITIAL Y POSITION
3514  REM  X0 = THE INITIAL X POSITION
3515  REM  SET Z0 = 0 IF PRINTING HORIZONTAL
3516  REM  SET Z0 = 1 IF PRINTING VERTICALLY
3520  FOR Z = 1 TO  LEN (L$):Z3 =  ASC ( MID$ (L$,Z,1))
3525  IF X0 > 275 OR X0 < 0 OR Y0 > 153 OR Y0 < 0 THEN  GOTO 3600
3530  IF Z0 <  > 0 THEN  GOTO 3565
3540  IF Z3 > 64 THEN  DRAW Z3 - 64 AT X0 + (Z - 1) * 7 * SC,Y0: GOTO 3600

3550  DRAW Z3 AT X0 + (Z - 1) * 7 * SC,Y0: GOTO 3600
3565  IF Y0 - (Z - 1) * 7 < 7 THEN  GOTO 3600
3570  ROT= 48: IF Z3 > 64 THEN  DRAW Z3 - 64 AT X0,Y0 - (Z - 1) * 7: GOTO
     3600
3580  DRAW Z3 AT X0,Y0 - (Z - 1) * 7
3600  ROT= 0: NEXT Z
3610  RETURN
3615  REM

4000  REM  PLOTTING X AND Y VALUES
4005  REM   FOR LINE PLOT MAKE ZF=1:ZG=0:AT BEGINNING OF PLOT
4006  REM   TO PLOT (+) SET ZP = 1 BEFORE EACH POINT
4007  REM   TO PLOT (O) SET Z0 = 1 BEFORE EACH POINT
4010 X0 = 23 + (X - XN) * 252 / (XM - XN)
4020 Y0 = 149 - (Y - YN) * 140 / (YM - YN)
4030  IF X0 > 276 OR X0 < 0 OR Y0 > 149 OR Y0 < 4 THEN ZG = 0: GOTO 4050
4033  IF ZP = 1 THEN  DRAW 43 AT X0 - 2,Y0 - 3:ZP = 0: GOTO 4050
4034  IF Z0 = 1 THEN  DRAW 79 AT X0 - 2,Y0 - 4:Z0 = 0: GOTO 4050
4035  IF ZG = 1 THEN  HPLOT  TO X0,Y0: GOTO 4050
4040  HPLOT X0,Y0: IF ZF = 1 THEN ZG = 1
4050  RETURN
4055  END
```

APPENDIX 7

TABLE OF CHI SQUARE

Column headings indicate probability, P, of deviations greater than those occurring between observed and expected data.

D.F. \ P	0.99	0.90	0.75	0.50	0.25	0.10	0.01
1	-	0.02	0.10	0.46	1.32	2.71	6.64
2	0.02	0.21	0.58	1.39	2.77	4.61	9.21
3	0.12	0.58	1.21	2.37	4.11	6.25	11.35
4	0.30	1.06	1.92	3.36	5.39	7.78	13.28
5	0.55	1.61	2.68	4.35	6.63	9.24	15.09
6	0.68	2.20	3.46	5.35	6.63	9.24	15.09
7	1.24	2.83	4.26	6.35	9.04	12.02	18.48
8	1.65	3.49	5.07	7.34	10.22	13.36	20.09
9	2.09	4.17	5.90	8.34	11.39	14.68	21.67
10	2.56	4.87	6.74	9.34	12.55	15.99	23.21
15	5.23	8.55	11.04	14.34	18.25	22.31	30.58
20	8.26	12.44	15.45	19.34	23.83	28.41	37.57
30	14.95	20.60	24.48	29.34	34.80	40.26	50.89
40	22.16	29.05	33.66	39.34	45.62	51.81	63.69
50	29.71	37.69	42.94	49.34	56.33	63.17	76.15
99	69.23	81.45	89.18	98.34	108.09	117.41	134.64
200	156.43	174.84	186.17	199.33	213.10	226.02	249.45
300	245.97	269.07	283.14	299.33	316.14	331.79	359.91
500	429.39	459.93	478.32	499.33	520.95	540.93	576.49

Adapted from Table of Chi Square appearing in Handbook of Statistical Tables by D. B. Owen, Addison-Wesley, 1962, page 50. Reprinted by permission of U.S. Atomic Energy Commission.

INDEX

A

Absorption
   of gas 31
   of light 140
Activation energy 151
Active transport 193,195,196
Adenosine triphosphate 231,238
Age class 107,224,225
Allosteric enzymes 39,237,238
Anaerobic glycolysis 228,231,238
Analog computer 4
Analytical models 16,28,43,46
Annual cycle
   of light 135,164
   of temperature 146,168
Aortic valve 199
Aquatic systems
   absorption of light 140
   diversity 265
   food chain 162,167
   productivity 143,241,248
   reflection of light 138
   temperature variation 148,158,168
Arrays 217
Arrhenious model 151
Assimilation 33,102,243
Autocatalytic reaction 87

B

Bacteria 175,214,241,295
Balanced polymorphism 127
BASIC 10,113,220,309
Batch culture 175,244
Beer's law 140,143
Bertalanffy fish model 23
Bimolecular reactions 24,81
Biochemical kinetics 80,227,241
Biomass 160,173,242
Birth and death process 277,278
Bisubstrate reactions 236,246
Block diagram 162,204,223
Blood flow 196
Blood space 196
Body temperature 150,209,211
Brownian motion 257

C

Cannibalism 120
Caperon model 242
Capture coefficient 33
Carbon dioxide 206
Carrying capacity 92,96,222
Cattle reproduction 292
Cause and effect 13
Cell growth 241
Chance-Cleland model 86,88,228,232
Chemical equilibrium 83
Chemical kinetics 24,80
Chemostat 176,244,246
Chen model 178,246
Chi-square test 253,260,347
Cleland model 86,88,179,236
Climax community 283
Cloud cover and light 137
Cohort 108,118
Coin tossing 255,256,275

Column vector 218
Community sampling 264
Compartment model 89,162,187,217
Competition 97,180,183,222
Compound equations 55
Conceptual models 2
Concerted enzyme model 39,237,238
Conductance 198
Contagious 118
Continuous culture 176,244,246
Control mechanism 39,204,207,209, 211,237,238,246
Cooling process 20,209,211
Correlation coefficient 47,58
Counterintuitive behavior 10,227
Countercurrent diffusion 191
CURFIT 46,245,323
Curve fitting 43,93,200,245

D

Decay 20,258
Decomposers 162
Deer management 114,225,226
Denaturation 152
Density of water 61,158
Dependent variable 12
Design engineering 288
Deterministic models 77,251
Diatoms 178
Dice throw 255
Difference equation 71
Differential equation 16,70
Diffusion 22,89,187,189,191,196, 220,257
Digital computer 1,4
Dihybrid cross 260,261
Dilution model 22
Discrete event process 250,263,287
Disease 118,280
Distribution 253,254,269,274
Diurnal variation
  light 137,144
  temperature 147
Diversity 265
Drift in stream 23
Droop model 244
Droop plot 245

E

Ecology 91,241,264
Ecosystem 160,222
Empirical models 14,43,56,200,272
Endocrines 207
Energy flow 160,220
English Channel 169
Environment 135,146,284
Enzymes
  kinetics 28,86,227,293
  pH 35
  temperature 152,156
  synthesis 214,295
Enzyme-substrate complex 28,36,86,152, 232,293
Epidemic 118,280
Equilibrium 27,83
Eugenics 126
Euler integration 71
Evolution 127
Exponential decay 20,258
Exponential distribution 274
Exponential growth 16,91,277,278
Extracellular space 196

F

Fecundity 60
Feedback control 203,204,214,237, 238,246
Feeding coefficient 165
Finite difference 71
First order kinetics 25
Fish models 23,116
Flowcharts 63
  rules 63
  symbols 67
  templates 64
Flow of fluids 197,199
Fluid flow 197,199
Fluid spaces 196
Flux equations 163,230
Food chain 160,220
Forcing function 164,220
Frequency distribution 254,264,269,274
Fresnel's law 139
Function 12,16

G

Gause competition model 97,222
Gaussian distribution 270

Gene frequencies 125,259,283
Gene repression 214
Genetic drift 283
Genetic equilibrium 41
Genetics 41,124,259,260,283
Glycolysis 228,238
Graphing 12,345
Growth curve 16,92
Growth rates 16,92,222,241

H

Half-life of decay 258
Half-saturation constant 30,173,184, 243
Hardy-Weinberg model 41,124
Heart valve 199
Heat balance 148,210,211
Heat transfer 20,148,150,209
Henderson-Hasselbalch model 35
Herbivore 162
Hierarchy 9
Holling-Tanner model 33,104
Homogeneous populations 91
Hormones 207
Hunter-deer model 114
Hyperbolic expressions 28,30,33,104, 114,173
Hypothesis 5

I

Inbreeding 132
Independent variable 12
Infective 118,280
Inhibited growth model 24,91,246
Inhibition
  allosteric 39,237,238
  microbial growth 184
  nutrient uptake 246
  photosynthesis 141,142
Input-output 13
Integration
  numerical 70
  rules 17
Internal nutrient 242
Interspecific action
  competition 97,180,183,222
  parasitism 39,222
  predation 59,102,222

Instability 93
Iodine and thyroid 200
Isoclines 99,103
Iterative process 75
Ivlev model 105

J

Jacob-Monod model 214,295
Jones model 204

K

Kermack-McKendrick model 118

L

Lags in feedback 93,237,238
Langmuir absorption model 31
Law of mass action 80,93,97,102,106 228
Least squares 44,52
Leslie-Gower model 105
Leslie matrix 224,225
Liebig's law 178,247
Life table 109
Light intensity 135,284
Limiting factors 60,61,241
Linear algebra 217
Linear diffusion 89,220
Linear regression 44,52
Logistic growth 24,91,278
Lotka-Volterra
  competition model 97,222
  predation model 102
Luxury consumption 242

M

Management 8,114,116,225
Markov chain 275,281,283,286
Mark-recapture method 266,267
Mass action approach 228,232
Mass action law 80,93,97,102,106,228
Material transport 160,187,191,197,242
Mathematical model 3
Matrix algebra 217,281
Matrix multiplication 219,220,282,283

Mean square 47,58
Mediated transport 195,242
Membrane 193
Metabolic heat 150,209,211
Metabolic rate 59,60
Michaelis-Menten model 28,174,234, 242,293
Microbial growth 173,241
Microcomputer 1,10,309,321
Models 172
Monod model 173,241
Monohybrid cross 258,284
Monte Carlo models 250,252,255
Moose-queue model 289
Mortality coefficient 165,277
Multicomponent models 8,77
Multienzyme systems 227,248
Multistage nutrient limitation 241,248
Mutational load 129

N

Natality rates 107
Negative feedback 204,237,238,295
Newton's dot notation 163
Newton's law of cooling 20,148,211
Non-linear models 166,221,223,227
Non-metabolizable substances 168
Numerical integration 71
Nutrients
- bound to biomass 174
- growth limitation 173,241
- multiple 178,246
- uptake 173,246

O

Objectives 11,14,78,251
Old field community 170
O'Neill predation model 104
O'Neill temperature model 153
Operon theory 214,295
Optimum temperature 153,155
Oscillations 93,206,216,237,238
Osmotic pressure 189

P

Parasitism 39
Parsimony principle 123,241
Patten 70,161,164,166,172
Pearl 60,91,108
Personal computer 1,10
Peterson estimate 266
pH and enzymes 35
Phase space diagram 99,103
Phenotypic ratio 260
Phosphofructokinase 238
Photosynthesis 60,61,141,180
Physiological control 187,203
Physiology 187,203
Phytoplankton 140,143,178,183,245, 246,265
Poisson distribution 254,269,270,343
POLYFIT 56,333
Polynomial regression 56
Population
- estimate 266
- genetics 41,124
- growth 16,93,113,277,278
- sampling 264
- trajectory 100

Positive feedback 206,213
Predator-prey 59
- Holling-Tanner 33,104
- Leslie-Gower 105
- Lotka-Volterra 102
- O'Neill model 104
- Watt-Ivlev 105

Pressures and flows 190,198,199
Probabilistic modeling 250,252,263,275
Probability 253,282,291
Probability distribution
- normal 270
- Poisson 270,343

Productivity 60,61,143
Protein synthesis 214
Pseudorandom numbers 252

Q

Q_{10} approximation 152
Quantitative inheritance 131
Quantitative research 7
Queueing 287,298

R

Radioactive decay 258

Random deviates 272
Random number generators 252
Random process 250,252,260,275
Random walk 257,275
Randomness test 252,262
Rate equations 16,70
Rate law approach 234,236
Rates
    effect of temperature 151
    growth 16,92,105,241
    reaction 24,80
Reaction kinetics 24,80
Reaction pathway 227,232,234,237,241
Recruitment 116
Reflection of light 138
Regression
    linear 44,52
    polynomial 56
Replacement principle 13
Repression of protein synthesis 214, 295
Reproduction 292
Research, role of modeling 5
Respiration 165,175,206,243
Ribose nucleic acid 214
Richer model 116
Ryther-Steele model 61,141
Runge-Kutta method 73

S

Sample size 265
Sampling 263
Satiation model 33,104,173
Saturation fraction 36,40
Saturation models 28,30,33,104,114, 242,287,293,298
Schnabel estimate 266
Schumacher-Eschmeyer estimate 266
Second order kinetics 25,166
Segregation 259,260
Selection 124
Sex-linked recessives 130
Sex-differentiated survival 114
Shannon-Weaver index 265
Silver Springs model 161,164,218,220, 222
Simulation 1,4,5,12,77,250
Sine curves 136,146,164
Single substrate reactions 232,234
Smith model 141
Sodium space 197
Spatial distribution 269
Species abundance 264
Stability 93
Standard deviation 272,274,279
State variable 91,162
Statistics 47,58
Steady-state 28,163,176,178
Stefan-Boltzmann law 149
Stochastic models 250,252,263,275
Stream drift 23
Structured growth model 248
Substitution principle 13
Substrate 28,36,232,234,236
Succession 222,281
Summation principle 77
Survival patterns 108
Survivorship 108
Susceptible 118,280
Sweating 209

T

Table-lookup method 270,273
Technological control 204
Temperature 20,146,156,209,211
Template 64
Terrestrial systems 146
Test of randomness 252
Theoretical models 14,46,50,52
Thermoregulation 150,209,211
Thyroid 200,207
Thyroxine 59,200,207
Time increment 72
Time lags 93,216,237
Tollgate 288
Toxic agents 184
Transfer coefficients 164,187,218,294, 297
Transformation 44,47,50,52
Transition diagram 281,294,297
Transition matrix 281,282,298
Transport 193,195
Trophic level 162,220
Two stage integration 81
Two stage nutrient uptake 242

U

United States population 113
Unpredictability and randomness 251
Uniform random numbers 252,272,284

V

Validation 7,14,252,272
Variance 278
Vectors 218,282
Ventilation rate 206
Verhulst-Pearl model 91
Verification 14
Vollenweider model 141
Von Bertalanffy model 23

W

Watt-Ivlev model 105
Weak acid ionization 35
Weather 284
Wright inbreeding model 132
Wolf-moose interaction 289

Y

Yates-Pardee model 237